Environmental Quality and Safety
Vol. II

Georg Thieme Publishers, Stuttgart
Academic Press, Inc., New York, N.Y.

Environmental Quality and Safety

Global Aspects of Chemistry, Toxicology and Technology as Applied to the Environment

Vol. II

64 Figures
92 Tables

Contributors:

H. Aebi, Bern, Switzerland
P. C. Blokker, The Hague, Netherlands
F. Coulston, Albany, USA
Velmar W. Davis, Annandale, Va., USA
H. Egan, London, England
R. Franck, Berlin, Germany
M. Goto, Bonn, Germany
N. Gruener, Jerusalem, Israel
H. Hurtig, Ottawa, Canada
N. A. Illiff, London, England
T. Jaakkola, Helsinki, Finland
E. Kenaga, Midland, Mich., USA
W. Klein, St. Augustin, Germany
W. Kreuzer, München, Germany
U. Lohmar, Bonn, Germany
Frank C. Lu, Geneva, Switzerland
L. Mendia, Napoli, Italy
J. K. Miettienen, Helsinki, Finland
R. Mollenhauer, Bonn - Bad Godesberg
  Germany

W. B. Murphy, Tolworth, Surrey, England
S. N. H. Naqvi, Marburg/Lahn, Germany
W. Pohlit, Frankfurt a. M., Germany
S. Rashid, Marburg/Lahn, Germany
C. Resnick, Jaffa, Israel
L. J. Revallier, Geelen, Netherlands
R. Roll, Berlin, Germany
R. A. Roohi, Marburg/Lahn, Germany
W. Salzer, Wuppertal, Germany
H. I. Shuval, Jerusalem, Israel
S. F. Singer, Washington D.C., USA
H. Takahashi, Helsinki, Finland
T. Teeuwen, Amersfoort, Netherlands
W. Tombergs, Bonn, Germany
R. Truhaut, Paris, France
E. E. Turtle, Tolworth, Surrey, England
J. J. Went, Arnhem, Netherlands
K. Yanagisawa, Tokyo, Japan
I. Ziegler, München, Germany

1973

Georg Thieme Publishers, Stuttgart
Academic Press, Inc., New York, N.Y.

© Georg Thieme Verlag, Stuttgart 1973 – Printed in Germany by Druckerei Lauk KG, Altensteig.
For Georg Thieme Verlag ISBN 3 13 498001 0
For Academic Press ISBN 0–12 227002–9
Library of Congress Catalog Card Number: 70–145669

Contents

VI   Contents

Foreword

Recently, a group of distinguished academic scientists, public officials, and industrial scientists, met in Munich on May 28, 1971, and they established the International Academy of Environmental Safety (IAES). The Academy was formed to provide a forum for discussion of the various aspects of the evaluation of safety of environmental chemicals, drugs, physical agents, pesticides, and food additives. The need for this kind of evaluation is apparent, when one considers the present dilemma confronting regulatory bodies of the world regarding DDT, 2,4,5-T, and Pb, to mention just a few examples. IAES will sponsor important symposia on timely topics, such as "The Scientific Basis for the Establishment of Air Quality Standards", to be held in Paris this October, 1972. In addition, the members of the Academy will sponsor specific papers and speeches of interest to people concerned with the environment.

In this issue of EQS, there appear 3 such statements, by Vice President Agnew of the USA, by Professor H. Aebi of Switzerland, and by W. B. Murphy of the USA. Each paper deals with a specific problem of Environmental Quality and defines clearly and respectively, the role of youth in combatting the antics of older people who wish to prevent progress by stopping the olderly process of changes in our environment (V. P. Agnew), the role of man in his adaption to the technical age (H. Aebi), and the role played by the scientist in providing safe and nutritive food for the world (W. B. Murphy).

Each of these statements are fully endorsed by the IAES and it is hoped that each reader will benefit from them, regardless of his general viewpoint as to how best to protect the environment.

F. COULSTON

Introduction

Address by the Vice President of the United States of America to the Boy Scouts Annual Dawn Patrol Breakfast

Spiro T. Agnew

New York, 3 February 1972

Like most Americans, I have always had a high regard for the Boy Scouts. I admire the Scouting movement because it builds boys into men — into the kind of men who have made this Nation what it is, the kind of men we now need — perhaps more than ever before in our history. And I applaud the Boy Scouts themselves because they are dedicated to making America an even better place in which to live.

Back when "Ecology" was just a tough word on a grammar school spelling exam, the Boy Scouts of America were practicing ecology. From the moment William Boyce introduced the Scouts to these shores in 1910, the accent has been on the outdoors — on the environment — on understanding nature and learning to get along in it and with it.

You might even say that the Boy Scouts anticipated by more than half a century President Nixon's environmental program of "making our peace with nature". My own earliest recollection of the Boy Scouts is of a couple of youngsters with a wagon collecting old newspapers and magazines, which were then to be recycled. So there's nothing new about recycling. The Boy Scouts have been at it for decades.

And they are still at it, as part of their *Project SOAR* — Save Our American Resources. Project SOAR was inaugurated last year, and I am told that in a single day, on "Scouting Keep America Beautiful Day", Boy Scouts across the Nation collected over one million tons of litter from 200,000 miles of streams and highways and from 400,000 acres of parks and other public places.

And a subsequent check showed, that once litter was removed from a specific area, the area tended to remain relatively clean.

I am delighted to hear that the Boy Scouts of America will continue Project SOAR in 1972. On behalf of President Nixon and the *Environmental Protection Agency*, I want to commend you and to wish you every success.

Since the environment obviously represents an area of mutual interest for the Boy Scouts and the Administration, I would like to address my remarks this morning to this very vital issue and its role in public affairs.

As you may know, ladies and gentlemen, this Administration has been pictured in some quarters as being less than enthusiastic in its attitude toward environmental cleanup. Just last week, for example, the New York Times ran an article in which one of its Washington correspondents suggested that President Nixon was being tight-fisted with the money Congress had appropriated for environmental improvement.

Anyone who read the article got the impression of a huge pile of Federal funds sitting around collecting dust, with Congress urging the President to use the money for anti-pollution purposes and the President saying, no thanks, he'd rather not. The main thrust of the piece was an implication that when it came to ecology, President Nixon — in the words of Sam Goldwyn — would like to be included out.

Well, I know only one antidote for this kind of deceptive innuendo and that's a strong dose of the facts. So here *are* the facts, and you can draw your own conclusion about President Nixon's posture on the environment.

In the four fiscal year budgets submitted by the Nixon Administration, Federal outlays for major environmental programs have totaled four billion, four hundred and eighty-one million dollars. That sum, ladies and gentlemen, is considerably more than three times as much as the total environmental outlay in eight years of the Kennedy and Johnson Administrations combined. Let me repeat those figures. Our budgets had three times as much muscle in four years as theirs had in eight.

And what about the media's contention — through veiled hints and open accusations — that the Nixon Administration sides with Big Business in thwarting environmental reforms? Consider these facts:

In the entire eight years of the Kennedy and Johnson Administrations, the Federal government filed 298 criminal actions against private industry for having violated anti-pollution provisions of the *Refuse Act,* which has been in effect since 1899. Two hundred and ninety-eight criminal actions over eight years. This prosecution total was exceeded in just the first *two* years of the Nixon Administration. In other words, the government is currently moving against industrial polluters at a rate more than four times greater than that of the preceding two administrations.

Furthermore, neither of the other administrations ever filed a *civil* suit under the Refuse Act. But since 1970, the Nixon Administration has filed 92 such suits.

Twenty-nine of them have been completed, and the government has won the verdict in every single case.

Now this is hardly the record of an Administration that is indifferent to the environment. It is the record of an Administration carrying out the promises of a President who is deeply concerned — a President who said in his Inaugural Address three years ago that, in protecting our environment, "we will and must press urgently forward".

In fulfilling this commitment, President Nixon in 1970 created the *United States Environmental Protection Agency,* with broad powers to investigate causes of pollution and to develop, institute and enforce the necessary remedies. In its first year, EPA did more to improve the environment of this Nation than has any other agency in any other Administration, without exception.

Since 1970, EPA has initiated 188 enforcement actions to protect water quality in rivers, lakes, and streams. It has initiated a permit system to control dumping of industrial wastes into the Nation's waterways — a system which already can inventory more than 90 percent of all industrial pollutant discharges.

At the same time, it has set nationwide standards of air quality. Implementation of these standards by all 50 states is now underway, with technical assistance from EPA.

Some $ 2 billion — 80 percent of EPA's first-year budget — was allocated to cities for the construction of modern sewage treatment plants. And to enhance prospects for a cleaner tomorrow, EPA is creating a system of *National Environmental Research Centers.*

The President has also established a new *National Oceanic and Atmospheric Administration* and has proposed an entire new Department of National Resources.

An indication of the progress already made can be seen in the latest annual report of the *Council on Environmental Quality* — a three member board of environmental experts formed two years ago to advise the President on ecology matters. Their report, issued last August, states in part:

"Federal progress in environmental protection is evident in several broad areas. Organization for action is better. Water quality enforcement is stronger. Standards in areas such as air quality radiation have been tightened. Pesticides are regulated more stringently. The Nation's financial commitment to environmental improvement has been substantially increased."

I think it's revealing to go back over that brief summary and analyze the comparative phrases. "Organization is better." "Enforcement is stronger." "Standards have been tightened." Then ask yourself, better than what? Stronger than what? Tighter than what? The obvious answer is that conditions are better, stronger, and tighter than they were under previous administrations.

Here, then, is eloquent testimony to just where this Administration stands relative to those that preceded it on the question of environmental improvement.

President Nixon is amply aware of the tremendous role in environmental reform that has been played by the Nation's youth, including the remarkable work of the Boy Scouts. To encourage even greater efforts, he inaugurated in October of 1971 the *President's Environmental Merit Awards Program*. Over 2,000 high schools — public and private — across the country have already responded and are involved in the program. The stated objective is "to provide a vehicle through which the Federal government may recognize and reward constructive, respon-

sible environmental services performed by American high school youth".

As these environmental awards are made by local judges and advisory panels, I have no doubt that among the recipients will be a large number of Boy Scouts.

The President's determination to clean up the environment and keep it clean has been reflected in every one of his State of the Union messages. He has devoted a substantial part of each message to his general environmental aims and to specific recommendations. His latest message, delivered two weeks ago today, carries on the tradition.

In it, the President cites steps already taken by this Administration to "make our peace with nature". He also notes that, of the 21 major legislative proposals on the environment submitted by him to Congress last winter, 18 are still awaiting final Congressional action.

They include measures to regulate pesticides and toxic substances, to control noise pollution, to restrict dumping in the oceans, in coastal waters and in the Great Lakes, to create an effective policy for the use and development of land, to regulate the siting of power plants, to control strip mining and to help achieve many other important environmental goals.

The President told the Congress:

"The need for action in these areas is urgent. The forces which threaten our environment will not wait while we procrastinate. Nor can we afford to rest on last year's agenda in the environmental field. For as our understanding of these problems increases, so must our range of responses."

As a further demonstration of the President's high-priority interest in environmental matters, he plans to send a special message to Congress on that subject within the next few days. It will point toward new horizons in mankind's relationship with his native planet, and I hope that all of you will have a chance to read it. It many suggest additional areas for work by the Scouts.

It most certainly will provide a sharp focus on how all of us can help in making this a cleaner and healthier Nation in a cleaner and healthier world.

And despite the mournful cries of the doomsayers, as faithfully reported in the press and over the airwaves day after day after day, we are succeeding in our efforts. This isn't to say that the problem is solved — or even nearly solved. We still have a tremendous amount of work to do before we can truthfully say that we are out of the woods — or, to be less figurative and more accurate, out of the smog.

But we have made a beginning. A solid and substantial beginning. In that 1971 report of the *President's Council on Environmental Quality*, it was noted that in the last year for which figures were available, 1970, pollution from transporation and solid waste disposal sources declined slightly. It is true that it was only a slight decline. But there was a decline.

In addition, ambient levels of air pollution have dropped in urban areas. In a number of major cities, including New York, there have been decreases in pollution from suspended particles and sulfur dioxide, while carbon monoxide incidence is beginning to decline in the wake of more stringent controls over automobile emissions.

Meanwhile, regulations of pesticide usage have resulted in a lowering of dietary intake of *DDT* and related substances of potential harm to humans and animals. Human intake is now at acceptable levels as set by the *World Health Organization*. I don't believe accurate measurement has been yet made with regard to animals.

And the Council also noted that, due to improved waste treatment and increased treatment capacity, the pollution factors of organic wastes discharged into our waterways have remained roughly constant in recent years, despite the continuing growth in our population.

At the same time, the dumping of wastes into U.S. coastal waters has decreased sharply as a result of new and anticipated anti-pollution regulations. A recent study showed that such dumping off the U.S. Pacific Coast had dropped from one million tons in 1968 to 24,000 tons in 1971. A similar decrease is predicted for the rest of the Nation.

I cite these figures not to imply that we have solved our pollution problems, but to show that in our efforts to solve them, we are at least beginning to make progress. Encouraging progress. Progress, I might add, that seems to refute the grim predictions of those highly publicized pessimists who see ecological doomsday around every corner.

Unlike the prophets of inevitable catastrophe — and I am afraid every generation has had them — the scientists working in the Nixon Administration are convinced that there are feasible solutions to our major environmental problems. The President, of course, has given the highest priority to the attainment of these solutions.

But I should point out that a high priority does not mean fanatical obsession. It does not mean an abdication of common sense. It does not mean surrender to the voices of panic and irrationality.

There are, unfortunately, some prominent Americans who believe that we must achieve *zero pollution* overnight, in every aspect of the environment, in every area of the country. They demand the most extreme measures imaginable, even if such a program means economic chaos and the pigeonholing of every other worthwhile project on the government agenda.

Well, I for one do not concur. I agree that an all-out effort must be made immediately to reduce pollution levels to a point where they no longer constitute a threat to health. But beyond that point, environmental programs can and should be more selective because they are competing for our limited societal resources with all our other domestic and international priorities.

I also deplore the activities of some environmentalists who believe that sabotage and horseplay are the answers to what ails us ecologically. I am thinking particularly of an article I read in last Wednesday's Washington Post about an organization of 25 teachers, students and other amateur ecologists known as the Miami-Eco-Commandos. These self-appointed vigilantes claim they are trying to save the country through what they describe as Ecotage. They define "eco-tage" as "non-violent sabotage which can be executed without injury to life systems".

Here are some of the things they have done, according to the Washington Post writer, who wrote as though he were captivated by the group and its shenanigans.

During the past few months, in a paroxysm of brilliancy, they have cemented shut the sewage outlet of a bottling company.

They have hurled packets of yellow dye into the machinery of six sewage treatment plants.

They have tossed 700 bottles into the seas off Miami.

They have suggested publicly that everyone send bundles of garbage to persons they suspect — *suspect,* mind you — of being polluters.

Now aren't these dandy contributions to our country's environmental cleanup campaign? Dye in the sewage plants. Bottles in the ocean. Garbage in the mails. Just what we all need.

Contrast these actions, if you will, with the very positive accomplishments of Governor Rockefeller's Administration in reducing air and water pollution in New York State. Substantial improvements have already been recorded, and additional gains are in prospect. To help achieve these recorded, and additional gains are in prospect. To help achieve these gains, *New York State's Commissioner of Environmental Conservation* several weeks ago released plans for more direct action in meeting strict new Federal air pollution standards for motor vehicle exhaust emissions.

And compare the exploits of the Miami Eco-Commandos with the constructive — and productive — efforts of the Boy Scouts of America and other responsible citizens. As an American vitally interested in the environment, I wouldn't trade you one level-headed, serious-minded, service-oriented Boy Scout for all of the publicity-seeking environmental dilettantes the news media can dig up between now and Hallowe'en.

Our Nation owes a great debt to its Boy Scouts, and this would be an ideal time for part of that debt to be repaid. Your national organization has launched its Boy Power '76 national drive to enlist into Scouting an additional million and a half American boys between the ages of 8 and 16 by 1976, America's bicentennial year. That would give Scouting a membership of one-third of all the boys in the nation between those ages.

And national headquarters has set as its goal a representative balance in membership along age, ethic and economic lines.

To achieve these commendable aims, the Boy Scouts seek to raise $ 65 million to go to local councils on a matching fund basis.

Here, in New York City, your council has accepted the very rugged challenge of raising its membership to 96,800 by the end of 1972 and to 135,000 by 1976. You are in need of public support on a local basis.

If Scouting did nothing more than sponsor the SOAR cleanup programs, you would merit that support, and I would encourage it.

But Scouting is more than environmental reform.

Scouting is the building of character into generations of future Americans.

Scouting is teaching boys that there is nothing wrong with honor and patriotism and decency and fair play ... and that there is nothing right with deceit and cowardice and the shirking of one's duty.

Scouting is a reaffirmation of the qualities that established this great Nation and helped make it the strongest and richest and most charitable Nation on earth.

Scouting, in brief, is a synthesis of much of what is good and right about America. Today, there are some people in this Nation who are prone to weep and wail and wring their hands whenever a crisis arises. And there are others, like those ecology saboteurs, who think disruption is the answer to everything.

As long as there are Boy Scouts of America, our country can rest assured that a substantial number of its citizens will march to a different drummer — to the upbeat of confidence and cooperation and constructive self-help.

There is an old Chinese proverb that says: "It is better to light one candle than to curse the darkness." I have listened long and painfully to the laments of the darkness-cursers. It is a privilege and a pleasure to be here today among the candle-lighters.

Man in the Technological Age and His Environment*

H. Aebi

University of Berne, Switzerland

Address given as Rector of the University of Berne

Summary. The enormous success achieved by science and technology has placed mankind in a position where it can transform the environment at will, and to a degree and at a rate which would probably have been inconceivable at the beginning of this century. It appears that in this process man's power and will to effect far-reaching changes in physical nature have increased more rapidly than his understanding for this world he lives in.

Zusammenfassung. Die großen Fortschritte, die durch die Wissenschaft und Technologie erreicht wurden, haben den Menschen in eine Lage gebracht, durch die er die Umwelt nach Belieben verändern kann, und zwar in einem Ausmaß, das zu Beginn des Jahrhunderts noch als unvorstellbar galt.
Es scheint, daß durch diese Entwicklung die Macht und der Wille des Menschen, weitreichende Veränderungen der stofflichen Umwelt zu bewirken, schneller gewachsen sind als das Verständnis für die Welt, in der er lebt.

Is the Environment Threatened?

There is a great deal of discussion nowadays about environmental problems. Some seek to dismiss the discussion as an intellectual speculation, as an hysterical outcry calculated to prevent people in the modern industrial civilization from peacefully enjoying the great accomplishments of technical progress. Others believe that man, as a consequence of this careless and shortsighted behavior, has already inflicted irreparable damage on his natural environment, and that for this reason the continuance of the human race is endangered. The spokesmen of both extreme positions agree that man, thanks to his inventive genius, has come a long way; for one group, this is a realization giving a feeling of pride and satisfaction, and for the other group, any statement regarding human progress is met with mordant irony[1].
Which of these two extremes has the best chance of corresponding most closely to the facts? Those who believe in the responsibility and obligation of science towards human society are deeply concerned when they consider what is happening — all public authorities are. The Council of Europe declared 1970 as European Conservation Year[2], and the Swiss Confederation and Cantons have also acted accordingly. The plenary meeting of the World Health Organization decided, on May 22, 1970, to declare war on the pollution of soil, water and air, on noise and on the contamination of foodstuffs with artificial products and other factors affecting the health of mankind. This campaign aims at the setting

up of a worldwide system of supervision[3]. It is to be hoped that these fine intentions will be followed by concrete action. Switzerland can pride itself on taking action and by having passed a law regulating the use of deleterious substances and by proceeding with constitutional measures to protect man and his natural environment. The report of the Federal Council of May 6, 1970 proposes the amendment of the Swiss Federal Constitution by an Article 24 septies[4]. This forms a point of departure and a basis for a series of urgent measures for combating air pollution, noise and other disturbing factors. At the same time, the law on water pollution is subjected to revision. The numerous motions made in the Cantonal and Federal chambers as well as the general tenor of the press permit us to consider the future of our country with moderate optimism.

The Present Situation

Whatever attitude one adopts regarding the deterioration of the natural environment, one has no choice but to come to grips with the problems which are increasingly pressing[5, 6, 7]. The most vital demands man has on his environment can be summarized in the following way, well known to anybody engaged in first aid and rescue work: Without oxygen, man can manage to live for about three minutes, without water for about three days, and without food for about three weeks. What do we find if these three factors, crucial for life, are studied more closely?

Air

Although the atmosphere surrounding our planet contains enormous quantities of gas, the composition of the air has undergone profound transformations in the course of time. The primitive atmosphere, about three billion years ago, was made up of different substances. The main ingredients were not nitrogen and oxygen, but carbon dioxide, methane, hydrogen, and ammonia. Due to the process of photosynthesis in green plants during millions of years, the atmosphere has become enriched with oxygen. Today, the atmosphere changes its composition as a result of industrialization. With the exception of nuclear fission, all power-producing processes, human labor, steam engine or gas and diesel engines, always entail the consumption of oxygen and the production of carbon dioxide. A medium-sized private car after 1,000 km of driving has consumed approximately the same amount of oxygen as a man engaged in moderate physical activity throughout an entire year. It is obvious that the power needs of our high-level consumption society geared to production and comfort are increasing in an alarming fashion, and that the equilibrium between oxygen production by plants and consumption by machines, human beings, and animals, which has existed for millennia, is shifted towards disturbance in an increasing degree.

There are approximately 100 million motor vehicles currently on the road in the USA, which consume about twice as much oxygen as the vegetation of the entire North American continent is capable of producing. On the other side, the carbon dioxide content of the atmosphere has gone up somewhat more than 10% since the turn of the century, and will presumably go up to 25% by the year 2000. This increase is small and no reason for alarm. The scientist, however, is confronted with the difficult task of evaluating these signs which may have long-range fateful consequences[8].

Other products which are discharged into the atmosphere are causing much greater worry, although considerably smaller quantities are involved. The pollution of the air above densely populated regions by products of incomplete combustion, such as soot, carbon monoxide, sulphur dioxide, polycyclic aromatic hydrocarbons, aldehydes, nitrous oxides, is familiar enough to the layman — usually from personal experience. Two examples may serve to illustrate the problems bound up with the discharge of such substances into the atmosphere:

The habits of our industrial society are characterized by waste and the consumption of luxuries. It is regarded as a sort of progress when more and more goods are consumed, to be used once and then thrown away. The unfriendly formula, the "throw-away society", has been coined to describe us. It is not surprising, then, that in Switzerland an ever growing amount of plastics are sold: in 1968, the figure was 180,000 tons of plastic material, of which 30,000 tons of polyvinylchloride. No one would dispute the fact that the vast progress achieved in the field of polymers has brought advantages and comforts. The problem is: how to get rid of this blessing. In burning these plastics — and this is probably the usual method of disposal — one kg of polyvinyl-chloride produces around 500 gm of hydrochloric acid, i.e. one half of the weight. The approximately 10,000 tons of hydrochloric acid which annually get into the atmosphere in our country are actually negligible compared to the ten times greater quantities of sulphur dioxide or sulphuric acid, combustion products of sulphur-containing heating oil. Human beings are not the only ones to be affected, but also their corrosion-sensitive buildings — and artistic monuments. Even those who would not like to do without the great advantages and comforts offered by a car must know that the present-day degree of motorization has entailed an alarming pollution of the air

and the soil with lead. Tetraethyl-lead, a valued anti-knock factor in use for nearly 50 years, is mainly responsible for the fact that the lead content of the air is steadily increasing. Even those who are inclined to belittle this fact by answering that, even in the air of big cities, only minute quantities of around 10 micrograms per cubic meter accumulate, must realize that most of these substances get into the soil by precipitation. While hydrochloric acid and sulphuric acid become harmless in the soil, lead as an element is persistent and actively accumulated by plants. Measurements made in the vicinity of highways with heavy traffic have revealed that quantities of 200 to 500 mg of lead per kg of plant material are not rare. In contrast to iron, cobalt or molybdenum, lead is not a vital trace element, but is relatively toxic for all organisms. The problem of chronic lead poisoning in man is known to the doctors, but hard to get at.

This state of affairs has induced public authorities and manufacturers in the country of the automobile to eliminate this toxic source[8]. New construction designs are now in preparation, and may bring the end of the lead-petrol era to that country and, we can only hope, also to us in Europe. No doubt every reasonable person will gladly accept a slight decrease in efficiency and increased cost in order to have a "clean car".

In contrast to the big industrial nations with their vast urban agglomerations and production centers we, in Switzerland, are still comparatively well off. So far, we have had no direct experience of smog (smoke plus fog), which spreads out over whole regions in the form of a mantle of haze or veil of dust. The Swiss Federal Commission for Atmospheric Hygiene, in its most recent findings dated 1970, made favorable forecasts, at least for the near future[9]. But we must be careful against any feeling of false security! The contamination of the atmosphere, especially with long-lasting chemical substances, does not stop at national frontiers or at the borders of continents. The industrial nations of Europe, therefore, have every reason in their deliberations, in the Parliament of Strasbourg, to grant priority to the coordination and intensification of the campaign against air pollution. Unfortunately, the way is long from well-meant decisions to the realization of effective corrective measures! It would, however, be an illusion to believe that with the present rate of economic development it will be possible to avoid any contamination of the atmosphere at all. For the sake of fairness, it may be pointed out that endeavors are under way among the large industrial companies to reduce air pollution to a bearable minimum. Although the air we breathe in Swiss towns is not ideally pure, the limits are only in exceptional cases exceeded [10]. It is important, however, to devote close attention to further developments.

Water

Examination of the general situation of our water resources leads to a much less favorable picture. Switzerland, with its numerous rivers and lakes, has to cope seriously with water pollution [11]. The situation is threatening. Reports on bathing prohibitions, poisoning of fish and oil alarms, increase.

Just consider how fundamentally conditions have changed, for instance, in the town where I grew up: At the beginning of this century, the yield of the Rhine fishery was so various and abundant that an unwritten law forbade the serving of salmon more than three times a week to the staff. At present, the catch of high-quality fish on the lower Rhine has declined to a few percent of the yield of a few decades ago. If a salmon is taken from the Rhine in Basel today, this is an event of local importance. The international commission installed to combat the pollution of the Rhine, finds itself confronted with a difficult task, especially if it does not want to restrict itself merely to carrying out measurements and issuing recommendations. The Rhine is a particularly serious case, but things are not much better in the beautiful lakes of Switzerland.

What has happened? Standing water is capable, to a limited extent, of self-purification, but this capacity has been hopelessly overtaxed by all kinds of waste that is poured increasingly into rivers and lakes. The far-reaching paralysis of the biological regeneration process is due, among other things, to the increasing introduction of organic materials, of minerals (especially phosphates and nitrates) as well as of toxic substances. There is no agreement to which extent the increase in minerals can be made responsible for the increase in the growth of algae, the intensification of decaying processes and the shortage of oxygen. On the other hand, there is no disagreement that toxic substances in industrial waste water, such as cyanide, phenol, sulphur, chromium, and copper compounds as well as pollution by petroleum products are having a fateful effect on the state of our water resources as biotope and as drinking water reserve.

Particularly disadvantageous is the fact that the micro-organism to be found in phyto- or zoo-plankton, owing to the lack of suitable enzymes, are for the most part unable to metabolize hydrocarbons, i.e. CH_2 chains. The result is that all water resources, including the oceans, are more or less exposed to pollution by oil. For the same reason, the detergents used in household and industry, which are largely persistent owing to their chemical structure, should be replaced by substances which can be easily broken down by micro-organisms. There are, however, a number of species which, owing to their well-developed enzyme outfit, are capable of breaking down hydrocarbons and even converting them into protein. This places into the hands of those scientists who have declared war on hunger throughout the world a new and efficient weapon. Already at the present time, thousands of tons of high-grade protein are being annually produced by microbe- and yeast-culture grown on petroleum-water mixtures. We hope that the introduction of such agents will make it possible to improve the biological self-purification of our water resources.

Such hopeful prospects, however, must not keep us from fighting water pollution with the traditional methods. The intermediate balance for this year of water resources protection in Switzerland yields the following picture: There are now about 330 central purification plants for 462 townships with 4 million inhabitants in operation; 69 additional ones are under construction, and 95 projects are ready for construction. This means that about two thirds of the population are connected up with such plants. If, in this way, it has been possible to close the gap somewhat, we have to be grateful to those farseeing men in the government who took on themselves the unpopular task of pointing out inadequacies. These initial advances do not solve the problem in a distant future, and will most probably bring a still higher degree of industrialization, and then we want to have sufficient clean water. Mankind has to realize the fact that an ever decreasing minority will enjoy the use of absolutely pure spring or ground water. As a substitute, technology can offer us water supplies from lakes and rivers. This can be done with the aid of purification processes (seepage and filtration plants yielding artificial ground water) and subsequent sterilization (e.g. by means of ozone or chlorine); thus, water can be produced, that is of acceptable quality and, above all, in sufficient quantity. Only this way will it be possible to satisfy the increased demands of modern industrial society (500 l per person per day) [5].

Food

Considering the food situation, the fact stands out that food consumption patterns, like the ways of living in general, have radically changed in the course of the last hundred years. Engines of all kinds are increasingly relieving us of any kind of physical work; thus we require fewer calories. The main worry is over-nourishment [12]. Urbanization and the extension of the food markets have increased the distance between producer and consumer. The uncertain political situation induces us, in following the Biblical example, to store up food reserves for the lean years. Thus, the factor of food-stability is becoming more and more important. Refined food products, i.e. those which are processed technically, such as white flour, oil, sugar, etc., are being consumed in proportionately greater amounts than natural foods. Increased stability, increased nutritional value, improvement in appearance and flavor (= food appeal) as well as the commercial fabrication of ready-to-serve meals are probably the main reasons why the inhabitants of industrial countries now consume increasing amounts of food-additives such as antioxydants, stabilizing agents, emulgators, thickening agents, artificial coloring, artificial flavoring, synthetic sweetening agents, organic acids, and the like. The average American, for instance, consumes 1.5 kg of such food-additives per capita per year [13]. No precise figures are available for Switzerland, but it may be assumed that the amounts consumed are less than in America. Food-additives are for some people cosmetics of the palate, for others a necessary evil or simply food ingredients in the technological age.

Foreign Residues

The situation is quite different with the residues that appear in food. These are for the most part biologically highly active substances that have unfortunately got into these foods. Exact analytical procedures have made it possible for residues, even at a concentration of 1 part per billion by weight, to be definitely ascertained. Generally, these are insecticides or pesticides, with chlorine-substituted hydrocarbons, such as lindane, aldrin, dieldrin, heptachlor, DDT or organic phosphorus compounds, etc. In no other class of substances are the advantages and disadvantages, the blessings and curses so intimately linked as in these chemical pesticides. For some people, they herald a better future as effective weapons in the campaign against hunger; for others, they are products of a diabolical spirit of invention pushed by greed for profit and power. One point: insecticides, no matter how poisonous they may be, are the products of chemical and biological research, and are per se neither good nor evil. The important thing is what man makes out of them; above all, how he applies them.

The problem confronting us here can best be illustrated by means of a concrete example, the much discussed DDT (= dichlor-diphenyl-trichlorethane)[14]. The fact that many insects are sensitive to the high toxicity of this contact poison was first observed in 1939 by the chemist Müller, in Basel. This substance, however, had been synthesized 67 years before in Strasbourg by the Ph.D. candidate O. Zeidler. One of the drawbacks of DDT is its lack of specificity; i.e. not only are pests affected, but it also possesses chemical stability, which means that it remains biologically active over practically unlimited periods. Repeated applications lead to accumulation, mainly due to its high fat solubility which greatly favors persistence in the fatty phase of vegetable and animal organisms. The world-wide use of this insecticide has had the result that also human beings, mostly via the natural feeding chains, have become increasingly contaminated with DDT. Owing to the marked fat solubility of this substance, every time it is transmitted to the next link in the nutrition chain, there is a further accumulation. This is the only explanation of the fact that in the adipose tissue of humans, amounts measuring from 2 to 10 mg/kg have been detected. However, the most heavily affected animals are the birds of prey, which stand at the end of the nutrition chain. In the organisms of these birds, most of the DDT accumulates to such a degree that disturbances in their calcium metabolism occur. They are no longer capable of producing normal eggs; either no eggs are produced, or only eggs with a very thin shell. The inevitable consequence is that the continuance of these species is in grave peril[15].

For the human race, a fateful conflict of interests now results. The highly industrialized nations are afraid of a further increase in contamination; they are taking steps (in Europe, Sweden was the first country to do so) to prohibit the use of DDT over wide regions or to ban it completely[16]. Uncertainty regarding the effect of these residues in the long run and genuine concern about the future of life in nature are probably the main motives behind these prohibitions. This development can be accepted by those in favor of a reasonable use of insecticides, because DDT began to be replaced several years ago by a new generation of more specific, easily broken down insecticides, and by novel biological methods of combating pests (e. g. the release of sterile males). The large populations of the developing countries have entirely different worries. They are indifferent to these slight residues in the adipose tissue (if such exist). What counts for them is the increase of food production. If they had to choose between insecticide-treated foodstuffs and no food at all — unfortunately a very realistic alternative — it would not be difficult for them to decide. Insecticides, especially DDT, have proved of inestimable worth in combating malaria and other diseases transmitted by insects, for example, spotted fever. The award of the Nobel Prize to Müller was partly based on the recognition that DDT has helped to eradicate malaria in a large number of countries [15, 17]. Mankind is confronted with the difficult task of arriving at a compromise which is acceptable. It will not be easy to achieve this goal. Most of the experts believe that it will not be possible to do without these highly effective insecticides, as long as man is determined to fight parasites and to achieve maximum crop yields. They vigorously demand that an end be put to indiscriminate spraying. Public authorities and manufacturers have to enforce that the application is highly specific and restricted to cases where it is indicated. A maximum of effect must be achieved with a minimum of pollution [15].

An Attempt to Assess the Situation

Which are the dangers threatening man and animal in the air we breathe, in the water we drink and in the food we eat? Before we take up the different possible kinds of damage to health, two preliminary observations must be made:

The ancient rule of Paracelsus "Sola dosis facit venenum" (= only the dose determines whether a substance is poisonous or not) is still valid; in the last analysis, it is not the dose that counts, but the total concentration accumulating in the organism or in the most sensitive organ or tissue. A critical state can be reached if over long periods minute amounts are absorbed and continuously stored. Plenty of evidence supports this rule: In certain prisoner-of-war camps, people have been tortured to death by having water poured into them. Long continued inhalation of pure oxygen can result in severe cell damage. Even common cooking salt, if given in excess, can endanger life, especially in the case of infants. Depending on the total dose, one and the same substance can be an element that is vital to life, a medicament or a highly toxic poison.

We have to free ourselves from the deep-seated prejudice that all substances which are naturally occuring are gifts of God, and to be regarded as benign; and that all substances furnished by chemistry are suspect. The most dangerous poisons which occasionally turn up in foodstuffs, for instance, botulinus toxin or aflatoxin, are formed by micro organisms. On the other hand, experience shows that synthetically produced medicaments and vitamins are by no means inferior in effect to substances laboriously isolated from natural products.

Three Dangers

The deleterious effect of all these substances can operate in different ways:

1. After a short or possibly longer period of time, a functional disturbance in an organ or tissue can develop. Such an acute or chronic toxic effect can in most cases be ascribed to more or less specific damages to enzymes, i. e. those molecules in the cells which catalyze and

control metabolic processes. The effect is especially disastrous when an enzyme of vital importance (e. g. the acetylcholinesterase necessary for nerve function) is blocked. This is the case with the organic phosphorus compounds (insecticides, nerve gas), even in minute concentrations. Depending on the dose and the degree of individual sensitivity, slight and vague complaints, or a manifest illness or sudden death may be the result of intoxication. The problem bound up with these effects stem from observations made on animals which can be applied only with certain reservations to human beings [18]. Even very slight alterations in the enzyme equipment or in the sensitivity of regulative processes may result in toxic effects, and the effectiveness of a medicament can vary very widely. In addition, it is a fact that people differ from one another not only in respect of appearance, constitution and blood group, but also in their enzyme pattern. This pronounced biochemical individuality is also the underlying reason why medical treatment can never be ready-made but always has to be adapted to every single individual [19].

2. Substances with a mutagenic effect are not primarily bad for the organism, but they cause in the germ cells a modification of the genetic information. In the germ cells, a change in structure of the DNA (desoxy-ribonucleic acid) in the cell nucleus occurs. Mutagenic substances appear in our natural environment in considerable number. Formerly, the chief concern of doctors was the ionizing radiation or radioactivity, with its damage to germ cells; now, the accent has shifted to the mutagenic substances [20, 21, 22]. Since mutations in man practically always result in malformations or disadvantageous metabolic defects, or are even incompatible with

life, such substances must, under all circumstances, be kept away from making contact with human beings.

It is extremely difficult to state with certainty whether a given substance is mutagenic or not for, here again, there are relatively large differences between species. Two such substances occurring in food are attracting a good deal of attention at the present time: caffeine, the stimulant so highly esteemed by man, as well as nitrite which is used as a 0.6 percent additive to pickling brine to treat meat products in order to retain an attractive red coloring. Both substances may produce chromosome breaks in cultures in vitro. This finding is not decisive for the alteration of the genetic material, but at any rate it raises suspicion. It has to be borne in mind that the effect in the organism is only slight under the usually prevailing conditions. It is probably much smaller than the spontaneous mutation rate. We can note with resignation: If man's food had been subject, from the outset, to the same strict regulations, as is now the case, the use of coffee would have had to be prohibited on account of the caffeine contained therein or the roasting of meat owing to the cancer-causing hydrocarbons produced in the process. Only further research on mutagenic agents, which is only beginning, will show whether there is any danger or not [20].

3. Man no doubt experiences his greatest dread when confronted with cancer which is the second frequent cause of death after cardiac and circulatory diseases. Research findings indicate that malignant growth is the result of a fundamental disturbance in cell harmony and is presumably to be ascribed to a failure of control over cell growth and cell mitosis. Many causes are capable of producing this disturbance,

mainly the cancerogenic agents. There are a large number of chemical substances with which cancer experimentally can be induced in animals. Some of them, for instance, 3.4-benz-pyrene, also occur in very slight concentrations in our natural environment, in particles of road dust, in smoke, including the smoke caused by many people[23]. The considerable increase in lung cancer, which is an established fact in all civilized countries, is no doubt bound up with this agent. Man is obviously prepared to expose himself to a calculated risk, in order to be "in". The legislators in nearly all countries demand that substances which have been ascertained to be cancerogenic must be kept away from contacting human beings. This is a requirement of the Delaney Amendmend, a measure put forward in the USA in 1958. Opinions differ about the interpretation, especially when it is a question of understanding what is meant by "under adequate experimental conditions". The rat and the mouse, which are used in large numbers for studies of cancerogenic action, display, like man, a certain tendency towards the formation of spontaneous tumors, i.e. they occur also in untreated control animals. If the cancerogenic effect of the substance to be investigated is slight, long-term and costly experiments are necessary and problems arise when it comes to their evaluation and interpretation. Thus the appearance of one single tumor in the control series and the diagnosis of the tumor as a benign formation or as cancer can cause the entire evidence to collapse like a house of cards. There are borderline cases where the investigator, if he really wants to be honest, must leave the question open as to the cancerogenic effect. I believe I am entitled to make the statement that, at the present time in the control of the action of new medicaments, new additives or other useful chemical substances, the greatest stress is laid on their eventual cancerogenic action.

Heredity and Environment as Decisive Factors

To what extent is there a risk that technology endangers the health of mankind? Many dangerous factors have been enumerated — the price we have to pay for all the comforts of our age — but it would be wrong to fall into pessimism. Positive aspects must be considered: Industrialization, which began a hundred years ago, has brought us additional sources of income and has increased prosperity. The average life expectancy has gone up by years; this is a proof that living conditions and medical attention have become much better. The young generation is also changed: they reach physical maturity sooner, grow taller and intellectually, they prove to be more critical; in the face of all life offers them, they are more experiment-minded and probably also more open-minded socially than previous generations. Man has at the present time means at his disposal to bring about far-reaching changes in his environment. It is also evident that man is not only determined by his heredity but, to a much greater extent than was formerly believed, by the environment. Environment acts on growth and appearance, but also in modifying metabolism. The protein-rich diet which we value so highly has a stimulating effect on hormone formation and enzyme synthesis[24]. The white man, under the impulse of external and internal stimuli, often boasts about his achievements and his enterprise. He does so with a contemptuous glance at all the throngs of people in countries where hunger prevails, who remain in passivity and are merely leading a dozing existence. This is unfair, because it is now a well-known fact that undernourishment during the first three years of life has irreversible effects on the development of the brain and blocks initiative[25].

Even the hereditary factors, the conservative components of man's constitution, can change. Biochemical research in paleogenetics has revealed that modifications in molecular structure and function have occurred in all phases of evolution. In the case of hemoglobin, for instance, 22 million years are required for such a structural modification (i. e. the successful retention of an amino acid substitution). Nature takes its time. Modern man, owing to hormone and metabolic control mechanisms, can adapt to the life in a technological age, but he does not have enough time to adapt genetically by mutation and selection. In this respect, man is at a great disadvantage compared to the rapidly multiplying bacteria and insects. The rapid succession of generations, combined with the appearance of resistent mutants, enables them to resist man's campaigns against them — no matter how efficient antibiotics and pesticides are.

Therapy:
New Legislation, More Research and Understanding, Additional Resources

What can be done? An improvement of the situation can be achieved by new legislation and regulations, by intensification of research, by technical measures as well as by suitable urban and regional planning. The potentialities and problems of legislative measures may be explained on the example of the problems of insecticide residues in food. In the "Directive on traffic in foods and articles of use" of the year 1936 there occurs the concise clause: "Foodstuffs must not contain deleterious substances"[26]. Since then, analytical methods have been developed which are a thousand times more sensitive than the old methods. The fiction of a "zero tolerance level" had to be abandoned[27]. This development was taken into account by a decision of the Swiss Federal Council of March 3, 1969 in the form of an amendment to Article 6 of the above Directive. This amendment contains a list of the legally admitted maximum concentrations. These limits are based on figures of WHO and FAO. For most highly effective insecticides, they lie in the range between 1 and 100 parts per billion[28].

The Cantonal Government of Berne, on March 11, 1969, issued a "Directive on the marketing and application of pesticides", especially of persistent chlorinated hydrocarbons. The use of such persistent chlorinated hydrocarbons is prohibited in all production plants, in premises and on materials which in any way are connected with the production, extraction, storage, processing, preparation, and sale of foodstuffs[29]. Previous experience has shown that, despite this clear prohibition enacted 1 1/2 years ago, it is very difficult not to overrun the tolerances. This puts the Cantonal authorities, who have to enforce the laws and regulations, in an embarrassing position. The Cantonal chemists are seeking to get as precise a picture as possible of the current contamination by continuously checking the products of numerous manufacturers and imported goods. Unfortunately, the overall picture that is emerging is not very good.

Before we resign, it must be borne in mind that probably several years are needed until the above-mentioned laws and directives begin to be effective. The tolerances have been fixed on the basis of experimental findings, with a safety factor of 1 : 100. This was done in the conviction that — when the health of mankind is at stake — only a high degree of safety may be regarded as "safe enough". The dilemma confronting us is as follows: Must we respect these severe norms ("prestige tolerances"), although it is very difficult to put them into practice, or may they be temporarily raised by, for example, lowering the safety factor from 100 to 30? These and other questions are now being presented to the scientist. They show that still greater effort to solve the residue problem will have to be made by all concerned.

Our defensive measures have to be based on legislation that really bears on all sources of immission, whether foodstuff residues or immissions in air and water are involved. It will not be possible to avoid severe punishment; after all, in the Middle Ages, well poisoning was one of the most reprehensible crimes. The cause is best served if public authorities and representatives of science and business work in close cooperation. The application of efficient measures will probably be welcomed by all sides[30]. The public authorities have the difficult task of reconciling two claims of fundamental importance: Man in the technological age has a right to pure air, clear water and a right to foods that are free of toxic substances. On the other hand, industry needs acceptable production conditions, for, after all, productive activity remains the basis of our prosperity.

The problems arising from the confrontation between technology and the natural environment call for intensified research, theoretical and practical, in all fields affected by this confrontation; they require cooperation by specialists in many fields. Research will produce the necessary knowledge only if experts in sciences, in agriculture and in the field of medicine work closely together. This combination of efforts has given rise to a new discipline, ecology, or environmental science[31, 32]. All disciplines which are likely to improve man's living conditions are to be encouraged, but top priority must be granted to ecology. New institutions are being created in the USA — often held up as a model in the field of science — and respectable sums are being made available for intensifying all efforts in the field of ecology and coordinating them[8, 33].

We cannot deny the fact that increased endeavors are indicated in our country. What is the use of scientific knowledge if it is not put into practice or if funds are lacking, or if there is no readiness to see the consequences? I wholeheartedly support a colleague of mine who recently declared that the first step in saving our environment is not increased research, but courage and the will to convert into action the available research findings. The realization of many projects demands insight, willingness to compromise and, above all, money. Public education and orientation must create in the body of the voters that degree of readiness to make sacrifices that is the prerequisite for raising the large funds necessary for ecological reform. It must be clear to everyone by now that it is not the growth of the gross national product that determines our standard of living, but the amount of financial resources which we are ready to spend for social purposes, for instance, for the protection of our natural environment. Man himself has to decide, in the last analysis, how much pure air, pure water, and freedom from noise and dirt are worth to him[8]. New guide-lines will have to be set up: Thus a central purification plant or a modern refuse disposal system are part of the image of a progressive community just as much as are modern schools or beautiful sports arenas. Man in the advanced industrial society must become accustomed to the fact that money must be spent not only for the acquisition of desirable goods but also on their disposal.

Man: Owner or Trustee?

The younger generation is taking a very critical view of us and our actions, and we will have to show them in a convincing way that the present responsible generation does not regard itself as the owner of the natural environment, but as a trustee who is seeking to hand it over to the coming generation in the same condition in which we found it[36]. Solemn warnings or romantic pessimism are not necessary to convince our fellow men of the urgency and significance of an up-to-date system of natural conservation. Facts and cautiously formulated forecasts as well as the now much-demanded direct-method orientation are perfectly sufficient. What do young people think of their parents and ancestors if whole classes of children or whole platoons of soldiers have to set about clearing lake shores, forests and watercourses of tins, bottles, motor-car parts, and many other forms of 20th century detritus? We often make it unnecessarily easy for young critics to put pressure on us if it is only a question of preserving still unspoiled recreation areas. It should no longer be necessary to awake the conscience for ecological problems by means of sensational news reports at regular intervals in the press and on the radio. In old times, man was on the defensive against nature and had to struggle to survive; modern man is now about to achieve complete mastery over the physical world. His role in the coming century, which will be still more dominated by technology, ought not to be that of a bold conqueror, but that of a wise, cautious regent.

The Role of the University

In the confrontation between technology and the natural environment, the university must not stand on the sidelines. As a consequence of its intellectual responsibilities, it has a crucial part to play. In teaching and research, full attention has to be devoted to the problem of the threatened environment. Thus the value of a research project does not depend alone on the quality of the investigators and on the scientific value of the project and its influence on other branches of sciences, but just as much on the social value of the study and its potential use for the economy.

At the university level, especially in the fields of medicine, the natural and the social sciences, there are many chances to train students to be not merely scholars, but also independent critical thinkers, people who not only excel in their specialized fields, but are also aware of their responsibilities towards the world around them. Biology teaching would convey half truths if it dealt only with molecular biology and ignored the background features of the struggle for existence of many species or even biotopes.

The highest aim of the physician is to maintain health. It should be the endeavor of all scientific men to maintain the natural world in a viable condition, to the extent at which this is still possible. Nature needs our protection: Exploitation through excessive use must be prevented. The balances disturbed by immission of all kinds have to be restored by corresponding corrective measures. We cannot turn back the wheel of history. Our aim must be: healthy human beings in a modified but not irreversibly imperilled natural environment. It is the highest duty of responsible people, not only to see to it, but if necessary to fight for it that all men — whether they live in highly industrialized countries or in developing regions — can participate in the technological progress of our age.

Economic prosperity, paired with social justice, are not the only achievement, but man must be able to live in an environment that makes life for everybody worth living.

References

1 a) Ehrlich, P.: Wir sind dabei, den Stern Erde zu ermorden. Ramparts Magazine 1969
b) Toward the Year 2000 in Daedalus, American Academy of Arts and Sciences 1967
c) Man and his future; a Ciba Foundation Volume; Churchill Ltd. London 1963
d) The next ninety years. Proceedings of a conference held at the California Institute of Technology, Pasadena 1967
e) Picht, G.: Mut zur Utopie; die großen Zukunftsaufgaben. Piper, Munich 1969
f) Graham, F.: Since silent spring. Houghton Mifflin, Boston 1970
2 Schaefer, K.: The environmental crisis; commentary on the 1970 European conservation year. Experientia 26: 672–676, 1970
3 Organisation mondiale de santé: Communiqué de presse du 21 mai 1970, concernant le système mondial de surveillance et de détection contre la pollution. Division d'information OMS, avenue Appia, 1211 Genève
4 Botschaft des Bundesrates an die Bundesversammlung über die Ergänzung der Bundesverfassung durch einen Artikel 24 septies betreffend den Schutz des Menschen und seiner natürlichen Umwelt gegen schädliche oder lästige Einwirkungen. 6. Mai 1970. EDMZ, Bern
5 The Biosphere. Scientific American, September 1970
6 Reflexions sur l'avenir du système de santé; contribution à l'élaboration d'une politique sanitaire. Rapport du groupe de travail sur la prospectivité de la santé. Editeur: Secrétariat général du gouvernement, direction de la documentation, 29–31, quai Voltaire, Paris 7e 1969
7 Gesundheitspolitik – heute. Symposium der Schweiz. Akademie der medizinischen Wissenschaften, Juni 1970. Benno Schwabe, Basel 1971
8 Environmental Quality. The first annual report of the Council on Environmental Quality; transmitted to the Congress, August 1970. Ed. by Superintendent of Documents, U.S. Govt. printing office; Washington D.C. 20402
9 Die Reinhaltung der Luft; Grundsätze zur Beurteilung lufthygienischer Probleme, herausgegeben von der Eidg. Kommission für Lufthygiene; Beilage zum Bulletin des Eidg. Gesundheitsamtes, 21 March 1970

10 Gilgen, A.: Luftverunreinigung und Gesundheit. 21 March, Zeitschrift für Praeventivmedizin 15: 217, 1970
11 Botschaft des Bundesrates an die Bundesversammlung zu einem neuen Gewässerschutzgesetz und Bericht zum Volksbegehren für den Gewässerschutz vom 26. August 1970. EDMZ, Bern
12 Ernährungsbericht 1969. Herausgegeben von der Deutschen Gesellschaft für Ernährung, Frankfurt a. M.
13 Sanders, H. J.: Food Additives. Chemical & Engineering News. Issues of 10 and 17 October 1966
14 Müller, P.: DDT; das Insektizid Dichlordiphenyl-trichloraethan und seine Bedeutung, Bde 1 und 2. Birkhäuser, Basel 1955 und 1959
15 Report, parts I and II, of secretary's commission on pesticides and their relationship to environmental health: Recommendations and summaries. Dept. of Health, Education, and Welfare, Washington D.C.; November 1969
16 Recent Reports on DDT:
DDT: Criticism, Curbs are on the upswing. Science 164: 936–937, 1969
Environment: Focus on DDT, the uninvited additive. Science 166: 975–977, 1969
Pesticide research: Industry, USDA pursure different paths. Science 166: 1383–1386, 1969
Insect control: Alternatives to the use of conventional pesticides. Science 168: 456–458, 1970
Agricultural pest control and the environment. Science 168: 1419–1424, 1970
Status of DDT-Restrictions in the U.S. Information Bulletin of the foreign agricultural service. Febr. 11th 1970
No total ban on DDT-use. Chem. & Eng. News 17. November 1969
A lesson of DDT, Editorial in Chem. & Eng. News; 24 November 1969
Pace quickens for DDT-restrictions. Chem. & Eng. News; 26 January 1970
17 Moore, N. W.: Implications of the pesticide age; complex ecosystems are being polluted in the name of progress. Ceres; FAO-Review 3: 26–33, 1970
18 Zbinden, G.: Drug safety: Experimental programs. Science 164: 643–647, 1970

[19] Williams, R. J.: Biochemical Individuality. J. Wiley, New York 1963

[20] a) Sanders, H. J.: Chemical Mutagens: the road to genetic disaster? Chemical and Engineering News. Issues of 19 May and 2 June 1969
b) Chemical Mutagenesis in Mammals and Man. Ed. by F. Vogel and G. Röhrborn. Springer, Berlin 1970

[21] Schubert, J.: Mutagenicity and Cytotoxicity of irradiated foods and food components. WHO-Technical Bulletin OMS, Genève 1970

[22] Huber, H., J. Halter: 13. Bericht der Eidg. Kommission zur Überwachung der Radioaktivität für das Jahr 1969 zuhanden des Bundesrates. Bulletin des Eidg. Gesundheitsamtes vom 5. September 1970

[23] Schaad, R., A. Gilgen: 3,4-Benzpyren im Staubsediment von Zürich. Zeitschrift für Praeventivmedizin 15: 87–96, 1970

[24] v. Muralt, A.: Protein-Calorie-Malnutrition; a Nestlé Foundation Symposium. Springer, Heidelberg (1969), see contribution Enzymes and Nutrition.

[25] Scrimshaw, N. S., J. E. Gordon: Malnutrition, Learning, and Behavior. M.I.T.-Press, Cambridge Mass 1967

[26] Verordnung über den Verkehr mit Lebensmitteln und Gebrauchsgegenständen (vom 26. Mai 1939), Stand 1. März 1972; Bundeskanzlei Bern

[27] Report on «no residue» and «zero tolerance»; Pesticide Residues committee; National Academy of Sciences; Washington D.C. 1965

[28] I Bundesratsbeschluß betreffend Änderung der Verordnung über den Verkehr mit Lebensmitteln und Gebrauchsgegenständen, vom 3. März 1969
II Verfügung des Eidg. Departement des Innern über Rückstände von Pflanzen und Vorratsschutzmitteln, vom 19. Mai 1969

[29] Verordnung des Regierungsrates des Kantons Bern über das Inverkehrbringen und die Anwendung von Schädlingsbekämpfungsmitteln, insbesondere von persistenten chlorierten Kohlenwasserstoffen. RRB vom 11. März 1969

[30] The chemical feast: J. Turner, R. Nader's study group report on the food and drug administration. Grossman, New York 1970

[31] Interdisciplinary research relevant to problems of our society; report submitted by the National Science Foundation, Washington D.C. 1970

[32] Odum, E. P.: Ecology. Holt, Rinehart and Winston, New York 1963

[33] Beispiel: Institut für oekologische Chemie (der Gesellschaft für Strahlenforschung, München) gegründet 1969

[34] Aebi, H.: Wissenschaftspolitische Probleme in Medizin und Biologie. Schweiz. med. Wochenschrift 100: 485–495, 1970

[35] Bloch, H.: Gedanken zu einer schweiz. Wissenschaftspolitik. Referat anläßlich des 50jährigen Bestehens des schweiz. Chemikerverbandes, publiziert im Chimia, 24: 321–327, 1970

[36] Präsidialansprache von E. Junod, Präsident der Schweiz. Gesellschaft für chemische Industrie, gehalten am 18. Juni 1970 in Interlaken (deutsche Übersetzung), bes. Seiten 4 und 12

Some of the Opportunities for Science in the Food Industry

W. B. Murphy

President, Campbell Soup Company, March 28, 1972

This talk begins with three observations by distinguished scientists.

The first is from a speech given on last March 6, which characterized the trace element research field as the most rapidly developing field of nutrition. It was said to have potential of major impact in nutrition research. The fact that it has a long way to go should stimulate the interest of younger scientists. The availability of new analytical methodologies which extend detection limits to the parts per billion range make the trace element nutrition field a science. Chromium is functional in the body at levels of 3—5 parts per billion. Cobalt, which in its active form, is present as a B$_{12}$ complex, is functional at less than 1 part per billion and certain other trace elements are expected to exert their essential effects at the part per trillion level.

The second observation is in a statement from another distinguished scientist written this past February 28, that "increasing concern with the hazards to chemicals in the environment coincides with progressive disillusionment over the benefits conferred by technology as a whole and chemical agents in particular. The drive toward avoidance of 'toxic' chemicals, the belief that pests can be completely controlled biologically, and the use of 'organic' food are symptomatic of a profound unease stemming from distrust of the safety evaluations of environmental toxicology." He also said, "understanding of orders of magnitude of environmental hazard is an essential step in evaluating the benefit vs. risk equation".

The third observation is from a talk given on last February 28. The scientist said, "The question is not 'Can the world feed itself?', but rather, 'Can the World afford to feed itself?' In other words, will the world environment withstand the pressure of the necessary modern technological changes in agriculture required to feed more and more people?" He also said, "we cannot sustain, and probably should not promote, on a world-wide basis, the U.S. approach to mechanized agriculture on cultivated lands... because... the average U.S. citizen today is using over 21,000 calories of outside energy per day for all purposes, including 10,000 calories of fossil fuel, and yet our food needs are only about 3,000 calories. Obviously, this drain on depletable resources — and the accompanying problems of pollution — cannot be continued." The observation was made that to grow a pound of wheat in the field requires about 1,500 pounds of water and it requires over a ton of water to produce a pound of bread. The scientist predicts serious water deficiencies by the year 2000. He also makes the statement that if the 3.7 billion people in the world had to depend on "organic" farming and chemical-free backyard food production, our land resource requirements would triple and our water requirements would be so great that many people would be without this essential element. He points out that "it is apparent from any ecological analysis that we are nearing the upper limits of world population growth and that a massive worldwide effort must be made to curtail the population explosion and bring people and resources into balance."

Now, after those thought-provoking statements, here is my message.

Over the past three or four decades, there have been many developments, much new knowledge and an almost geometric rise in activity in research and development having to do with food-related chemicals, food ingredients, food and food effects. Despite this large increment in research and in our knowledge, the further we go the greater the potential unfolds for future exploration. In so many instances, surface or partial knowledge indicates a clear answer, which becomes cloudy when one digs deeper.

It is not possible — certainly not possible for me — to list anywhere near the many opportunities for science in the food industry. What I shall try to do here is to outline three major areas for future scientific study that appear to hold promise of substantial benefits — a sort of overview of some of the possibilities as I see them. The three areas are, first, meeting the need for greater food resources, second, upgrading the quality of foods, and third, where we stand on food nutrition knowledge.

Meeting the Need for Greater Food Resources

It is obvious that after full exploitation of present food-producing lands, with continuing population growth, we are going to need additional growing areas. While there is quite a bit of unused agricultural land in the temperate zones, it will be imperative to bring large new growing areas into production and these must be in the tropics and in the oceans. This is a long-term matter. The world's arable lands are estimated at 6.6 billion acres. Something like 3 billion acres are in use in agriculture. Incidentally, in the United States we're losing about a million acres of good farming land each year, mostly to highways and urbanization. Maybe we need to have zoning to preserve our food-growing lands. The Rockefeller Foundation in their typical forward-looking manner has established research stations in two tropical areas for the purpose of developing knowledge to enable these areas to be utilized for food to a far greater extent.

It is strange that the parts of this planet which receive the greatest amount of the sun's energy have been far less useful to us for food production than the temperate zones which receive less such energy. The opening up of new tropical growth areas does not represent an impossible task, by any means, but it will involve a struggle because of special problems having to do with temperatures, lack of seasonality, predators, disease, and maintenance of soil physical conditions. Solving these problems will involve many disciplines including plant and animal genetics, soil chemistry, pesticide-fungicide-nematocide-virus disease research, plus water resource development, agricultural chemical application, and, in addition, the many activities that are needed to capitalize on new growing areas; namely,

finance, transportation, education, housing, and politics. Probably the best attack will be to utilize crops such as legumes, solanaceous species and others indigenous to the tropics.

We tackled an important crop in an area near the Equator. It took us about 7 years to lick the disease and predator problems via the genetic route and by use of widespread areas. Now we are stopped for political reasons. Politics may be the most difficult problem for food resource developments in the tropics.

It is also believed that there are many ways of taking advantage of the oceans for the production of high-quality protein foods. The potential for growing fish and shellfish is almost beyond our imagination. A visit to the mussel farms off the coast of Portugal and France show how the countless particles of shellfish spat can grow to full-size shellfish using the ocean's nutrients under controlled conditions.

Single cell protein development is in the distant future because of consumer disinterest in strange foods, but, I believe, will come as traditional food supplies become inadequate.

For the short term, the largest opportunity to increase our food resources is by higher productivity in existing growing areas. The opportunity is enormous and there are a number of ways to achieve this.

The average grower produces far less than the best grower. Here is an opportunity for the communication sciences, among other needs. For example: in our country and in others where we operate, the best vegetable grower attains yields per acre about double that of the average and about four times that of the lowest. According to data we have for several growing areas, the yields for corn and for soybeans are about 4 times greater for the best grower as compared to the poorest. We believe for many other crops the comparisons are similar.

The good growers maximize yields by optimum plant populations, fertilization, moisture supply, use of the best varieties and closely controlled application of predator and disease controls and other good cultural practices. But the best growers are not usually a majority.

The dissemination of the complex information needed for high production of crops, both plant and animal, involves many difficulties, probably of the same order of difficulty as that for the education of consumers on nutritional needs. This is one of the important reasons for high vs. low farm yields.

We need new breakthroughs in crop yields through new concepts in agricultural practices as, for example, new ways to supply nitrogen to legumes. We need to develop and exploit tissue cultures to point the way to better understanding of biological systems and to serve as a reservoir for potentially better production techniques.

The genetic development of higher-yielding varieties of many types of food ingredients is not a new subject but is still a largely untapped opportunity. The company I am with is working on several of the vegetables, the legumes, and some animals. We have achieved some excellent results and more will come. Several other companies are active here also.

We need to improve the efficiency of photosynthetic effects in plant growth by genetic means. By genetics also we need to reduce the weather effects and the effects of other unfavorable conditions. Because of the cost-reducing effects of mechanical harvesting, a higher degree of the uniform-maturing character of crops is needed for maximizing yields.

We should have predator controls by means that are supplemental to chemi-

cals. The development of disease-resistant and predator-resistant varieties and biological controls are promising possibilities. In the United States, crop losses due to insects, plant diseases, nematodes and weeds are equal to about $1/3$ of gross farm income. These losses are far greater in less developed countries.

We need to develop highly efficient and non-persistent chemical pesticides. Biological controls sound great but only a few are effective thus far. The use of attractants or hormones is probably for the distant future. Chemicals are now the most effective tools for many agricultural problems. Chemicals must be applied with intelligence and with restraint. I suppose it is a human tendency to consider that, if the prescribed dosage is good, an overdosage is even better. Also, there is sometimes an underestimation of the importance of the withdrawal period for control treatments. Here again we need skillful application of the communication sciences.

The whole area of microstructure, microaction, and microinterreaction is a fertile field for plant growing betterment. An understanding of genetic causes and effects and of the genes themselves is imperative since so much of our future progress depends on genetics.

The development of resistant varieties and of high-yielding varieties is a slow process but can be immense in impact. The opportunity for private, academic and government research in the matter of advances by the genetic route is so great that it should appeal to the bright young biological scientists. The conclusion here is that we need to get on with the job. There are too few good plant and animal breeders at work in the world today. If the youth of today, many of whom seem bored, aimless, and dispirited, could only perceive the excitement of the vast unexplored of the biological microworld —

the tremendous potential of the unknown truth and the undiscovered fact.

Those of us who live in localities where there is plenty of water are less prone to think of water limitations than those who live in the Southwest or in some parts of Central and South America or many parts of Asia and Africa, but the problem of water conservation, effluent disposal and water re-use deserve to be given high priority. The disciplines of limnology and hydrology are beginning to come into their own. Incidentally, for many food crops, it would be desirable to perform the first ingredient preparatory operations at the growing site so that much of the solid waste load does not accumulate at the processing plant, but is utilized as a sort of fertilizer for the farm. Some of this is now going on.

Last year, according to the U.S. Geological Survey, we used about 400 billion gallons of water per day. This is about $^{1}/_{3}$ of the fresh water available in terms of total streamflow. Streamflow is defined as the amount of fresh water available for our use and represents about 30 % of our rainfall and snowmelt evaporates and transpires. In the next 10 years, our fresh water usage is expected to double unless more rigorous conservation methods are employed.

Because water is cheap, tremendous amounts are used through absence of the economic need to conserve it. For example, irrigation ditches being unlined and open to the air take more water than needed because of seepage and evaporation. As demand increases, the cost of conserving water will be accepted as necessary.

The drip-technique of water supply in agriculture holds much promise — here is a water application method for plants that can increase water utilization in large measure by reducing run-off and evaporation.

Last year irrigation for agricultural purposes in the United States took about 35% of the total water withdrawn form the available supply; industry took about 58 %, and household use was about 7 %.

Our water problems are really economic problems rather than those of absolute supply, since the re-use possibilities are there but at a cost.

The opening of new lands for food growing via irrigation as exemplified by the Snowy Mountain project in Southeast Australia is a model for study and replication.

In summary, with respect to food resources, it appears that the rapid spread of advanced food-growing technology and its allied activities in existing food-producing areas holds the greatest promise for immediate food resource gains and the potential is enormous.

Upgrading the Quality of Foods

The achievement of better-quality foods is a second major opportunity for food sciences and technology. I mean here more nutritious, more attractive food ingredients and foods with greater utilization properties generally achieved via the genetic route. I also mean a better balanced essential amino acid content and better availability among some of the most important protein foods, especially the vegetable proteins, and the development of higher protein levels. While we are talking about better-quality foods, we must include the matter of better preservation methods, which include better containers.

Vegetable proteins have been long used even though they are not yet as good quality as they need to be. It is believed that through genetic development, vegetable proteins can be made to have balanced essential amino acid content with excellent availability. The legumes as a source of protein are important, but as consumed today, are short on several essential amino acids, especially methionine, and they need to be improved as to protein availability. But there are some varieties of legumes that are quite high in methionine content and better than average in availability. The door is open to producing higher-quality vegetable proteins through breeding.

The treatment of food from the raw state to that which is ready for the consumer has many areas for improvement. One of the most important is the protection of the nutrients in the raw ingredient, both by conservation and by stabilization. This now is by means that hardly could be called sophisticated, but much can be achieved through the development of more closely controlled processing and the elimination of delays. Involved here will be greater precision of machinery and of process action and a higher level of instrumentation. Better storage, faster delivery and harvesting at nutrient peaks all have an important bearing on food quality. The control of contaminating organisms is a highly important field requiring first, improved analytical methodology and the development of means to inactivate the organisms. This can have a major effect on food safety. Contaminating organisms in food ingredients can be inactivated by high frequency electronics.

It does not take a mental giant to conclude that food acceptance is an essential food requirement. I think all mothers of young children will agree heartily. There is too little we know about the whole subject of food acceptance in the areas of temperature, appearance, texture, and flavor chemicals.

Where We Stand on Food and Nutrition Knowledge

On the matter of food and nutrition knowledge, I am going to talk in general about how much we know and, then, the problem of getting adequate knowledge to consumers.

Every year additional essential nutrients are discovered. Furthermore, we have an enormous amount to learn about the signifiance of nutrients in relation to the body systems. We know that an essential nutrient can also be a toxic substance and we also know that arbitrary standards of acceptance are just that — arbitrary. To say that 0.50 parts per million is all right and 0.51 is not all right is a lazy way of setting a standard. Cumulative effects and variable reactions as well as combination effects compound the difficulties and challenge the investigator.

The task of the toxicologist is complicated by the fact that what is usually required is an analysis, not of the acute effects of large doses, but of the effects of very small doses accumulated gradually, of variabiltiy of response within a large population and of the effects of other environmental variables and of disease. Toxicologists, aware of their own limitations and with a responsibility for helping to protect the public health, have to lean on rather crude and cumbersome procedures to avoid any uncertainty about the safety of products they stamp with approval. But even excessive caution cannot guarantee safety if the substance of fundamental biological knowledge is inadequate. Results of extensive testing on animals may not be applicable to man. Extrapolation of animal data to man must be done with the utmost care and caution. The conversion of the large Pine Bluff, Arkansas chemical warfare facility to a testing station for long-term, low-level toxicological feeding studies can be a major step forward for assistance in measuring toxic effects.

I would like to pass on a warning concerning very low levels of substances as they are measured in the laboratory. It has been our experience, when exquisite testing is involved as in the case of parts per million or per billion, the testing must be done under sterile conditions since contamination can take effect in the act of making the test.

It is hoped that there is dawning a new era of scientific appraisal in the food field. If it could just be generally recognized that either too much or too little of any nutrient can be hazardous and that there are complexities in the human system that make simplistic conclusions dangerous — we will have gone a long way to introduce a better balance into our deliberations.

As to food knowledge, I think we must find some way to educate human beings on how to eat well. A sequel to that is how to get them to do so. Here we have a need for the communications science and the disciplines of psychology, philosophy, and psychiatry. Somehow or other, we must get common sense into our attitudes. We must find some way to stop the penchant of some scientists to rush into print for the sake of rushing into print and to insulate the fad from adoption until it has been proved. This country has had a long history of food fads. They seem to come and go endlessly. Generally, these are products of ignorance, superficial knowledge, or just plain publicity-seeking — usually hooked up one way or another to money-making. Also, some way should be found to stop the superficial setting of standards that are in the political area rather than being sound nutritionally. I do not think we need a lot of laws for these things — we just need a high degree of objectivity

on the part of consumers, government officials, educators, and industry.

Finding a way to tell consumers broadly how to eat well is about as frustrating a task as I can imagine. There are more than 50 and probably more like 75 essential nutrients — maybe even 100 before the identification task is completed. The minimum of 50 includes the eight essential amino acids plus arginine and histidine that are essential for infants; the 15 essential vitamins plus others not yet discovered; the 19 essential minerals including copper, selenium, nickel, tin, vanadium, zinc, etc. plus others not yet known as to vital importance — I hear that lead will be on the list shortly; that makes 44 plus; add linoleic and arachidonic fatty acids — that's 46; add water, 47; then add carbohydrates for calories, bulk, and metabolism — that's 48; then other fat for additional energy and metabolism purposes, and other protein for additional nitrogen — that reaches 50. But since we're intelligent creatures, we'd better add psychological acceptance and physiological acceptance as essential to nutrition. That totals 52 essential nutrients and nutrient factors.

A word of caution about protein. There is a tendency today to use the term protein without qualification. We all know that there are good, fair, and poor proteins. To label something "protein" is almost akin to using the word "posie" in the botanical field. Here is an unhappy predicament that ought to be faced up to by those who are responsible for nutritive labeling. Because it complicates the nutrition education problem does not excuse its being swept under the rug. The feeding of greater quanties of inferior quality protein in an attempt to remedy the lack of quality is not a good answer, especially for young children and certain vulnerable groups.

To eat well would be to know what has been consumed and what the body needs in all of the essential nutrients, and those needs vary with age, sex, race, allergies, eating habits, mental attitude, food last eaten, drinks last drunk, etc., etc. The average human being doesn't have the capability of keeping track of its nutrient intake in terms of individual essential items, so some other practical system must be developed. The systems of the essential four food groups or essential seven food groups are very crude and in my opinion not very satisfactory. I have no good answers here. My own eating habits consist of sticking with a wide variety of properly selected and well-prepared traditional foods and making sure that the protein is high-quality protein, and not eating too much of any major nutrient. I stick with the traditional foods because that's what my ancestors consumed and because the substitutes are too chemically simple to gain my confidence. Any traditional meat, vegetable, or grain food is a highly complex combination of a great many chemical constituents — sometimes constantly changing. Synthetic or man-made products intended to take the place of the complex meats, vegetables, and grains have a long way to go to be comparable.

Incidentally, the key to proper nutrition could hinge on nutrients that scientists have not succeeded in isolating as of this date. The only aspect of this in which one can be confident, rests in the belief that decades of experience indicate that they are present in traditional food staples.

I've mentioned some of the opportunities for science in the food industry including ingredients, agriculture, water, foods, education, and information. This is an exciting part of our economy and our life — and along with medicine, possibly the most exciting.

And now — despite all of the possibilities that have been mentioned for food supply, for food quality, and for food knowledge, I would like to return to a point made earlier. There is a limit to this earth's capacity in terms of population and food supply. As intelligent human beings, we must take this inexorable fact into account if our descendants are to have a good life.

Some FAO Activities and Attitudes Concerning Pesticides

E. E. Turtle

Plant Production and Protection Division, FAO, Rome

Foreword. The following is an outline of a talk given by Dr. Turtle at a meeting of experts organized by the Soil Resources Development and Conservation Service of FAO from 25 to 28 January 1972 in Rome and supported by the Swedish International Development Authority (SIDA). The report on the meeting entitled "Effects of Intensive Fertilizer Use on the Human Environment" has been published as FAO Soils Bulletin No. 16.

Introduction. Some explanation is needed for the inclusion of a talk on this subject in a Symposium primarily concerned with fertilizers. I understand that it was felt that future activities and studies relating to fertilizers and the environment might benefit from a knowledge of experience in the pesticides field.

At the outset we can accept that there are a number of similarities. Some of these can be listed as follows:

Both fertilizers and pesticides are *agrochemicals* being used in increasing amounts and which are essential to increasing agricultural productivity throughout the world.

Both are responsible for environmental changes of some kind.

Both are subject to public discussion and criticism on the latter grounds.

In both cases much of the criticism is not based on objective study of the available facts.

However, there are important differences. The most striking is that pesticides by their very nature are toxic whereas fertilizers generally are not. Some pesticides are toxic to only a very narrow range of species; but most of them are more or less toxic to a wide range, which often includes man and other non-pest species. As a result, over the years, there have been numerous poisonings and fatalities from pesticides. The fact that these incidents have often followed failures to take simple precautions — amongst packers, distributors and users, has led to stricter supervision and control by authorities in many countries. Similarly, the occurrence of residues in foods has raised doubts, stimulated much research and led to the introduction of tolerance levels and other restrictions. In short, it has long been accepted that new pesticides should only be introduced after critical assessment of the possible risks and that subsequently

they are only used under adequate technical supervision.

This leads to a description of the FAO Pesticides Program since it is designed essentially to ensure safety in the introduction and use of pesticides.

FAO Pesticides Program

This does not refer to a program designed to increase the use of pesticides: there is no such program! FAO's activities in the plant protection field only lead to any increases in use of pesticides insofar as studies and advisory activities of the respective biologists, at headquarters or in the field, lead to recommendations that pesticides rather than other control methods are appropriate for the particular conditions. The FAO Pesticides Program commenced in 1959 and followed a period during which a number of fatalities had occured among field applications. The accidents were largely due to *unsatisfactory practices* in the use of some recently introduced pesticides of relatively high acute toxicity, and it was felt that governments needed guidance. In 1962 an intergovernmental conference on the *use of pesticides in agriculture* was held in Rome and FAO subsequently established a program aimed to raise standards in supervision and use of pesticides. Groups of experts were formed to advise on specific aspects of the subject.

One group of experts dealing with the official *control of pesticides* has drawn up advisory papers and prepared a list of standard specifications to help governments in the inauguration and administration of schemes for supervising the introduction and use of pesticides. Such schemes are normally based on the principle that pesticides should not be used unless they are first registered with the authorities. Advice has been provided from Headquarters and consultants have made visits to various countries requiring assistance on this subject. Furthermore, the advice has always been that Departments of Health and of Agriculture should both be involved.

Another group of experts which meets jointly with a WHO group of experts on an interdisciplinary basis, has been concerned with *residues of pesticides in foods.* Each year it considers the available data on the occurrence of residues of particular pesticides in foodstuffs. Its task is to assess the data critically and, when the data are adequate, to make recommendations for tolerances, or acceptable residue limits which are considered suitable for general adoption for commodities moving from one country to another. When the group finds that the data are inadequate, attention is drawn to the deficiencies and to the kind of research that needs to be done in order to remedy them. Some recommendations are made on a temporary or conditional basis pending the conduct of further research to clear up doubts about specific points which are listed. The reports of this group summarize the data considered. Subsequently, the recommendations for maximum acceptable levels or tolerances form the basis of inter-governmental discussions which are organized under the joint FAO/ WHO Codex Alimentarius Commission with a view to building up a series of internationally agreed tolerance levels. These particular activities are concerned with residues in food. Nevertheless, in 1969 the group expressed some uneasiness regarding the distribution of DDT in the environment with consequential occurences in food; and it recommended that "its use should be restricted to circumstances in which there are no satisfactory alternatives".

The third group of experts set up for the 1962 Conference has been concerned with the increasing occurrence of strains of pests that are *resistant* to the pesticides formerly used. This latter problem needs to be tackled not only to ensure a continuous basis for controlling pest species but also to overcome the tendency of farmers to apply more pesticides to overcome resistance when it occurs.

Another expert group has been concerned with promoting research programs for the control of major pests by the integration of all *methods of pest control,* including biological and chemical methods, based on ecological considerations. Particular attention has been paid to the development of integrated procedures for the control of pests of cotton because this is a crop on which excessive amounts of pesticides have often been applied, sometimes only to provide assurance against possible attack by pests. The general conception of this part of the program, therefore, is to make sure that pesticides are only used where they are really needed.

Aid to Developing Countries

FAO is responsible for projects designed to improve competence in the control of pests in many countries. These projects normally are broad in conception, containing major inputs in entomology, plant pathology and similar biological subjects. Some include assistance in setting up procedures for regulating the introduction of new products. Others are substantially concerned with the ability to measure pesticide residues. These are mainly concerned with residues in foods but the facilities developed are being employed for making measurements in various other environmental materials such as soil and water. But resources for this work are limited at present and it is hoped to increase this activity, since data on the actual occurrence of residues are essential to any rational consideration of their importance in any particular situation. The need for measurements is by no means confined to foods and materials of terrestrial origins. The Fisheries Department of FAO is very interested in measuring the occurrence of residues in fish and water and the FAO Technical Conference on Marine Pollution here in Rome (December 1970) paid much attention to monitoring of the occurrence of residues in the marine environment.

Attitudes on Some Important Points

We recognize that pesticides have sometimes been applied without sufficient study of needs or of alternatives; also that some organo-chlorine pesticides are more persistent than is needed to control some of the pests against which they are used and that their residues have become distributed widely in the environment. Nevertheless, many widely reported statements on this subject cannot be sustantiated when examined objectively and in the light of the facts. For this reason, it is essential to assess such information objectively and critically on an inter-disciplinary basis before deciding on any necessary action.

In deciding on action such as the withdrawal of a cheap and widely used product furthermore, it is necessary to take into account the various likely consequences, such as any new hazards or any economic or — on the DDT question — any social effects which might follow. FAO is following such a policy, while encouraging research and other activities leading to the adoption of satisfactory alternatives. With its multiple responsibility for the promotion of productivity in agriculture, for technical assistance in developing countries and for the maintenance of standards of quality in food and with its background of experience in assembling experts on a multi-disciplinary basis, in my view FAO is well suited to provide objective information and advice of this kind.

In summary, the activities of FAO in the pesticide field have been largely directed to the encouragement of higher standards of scientific and technical supervision of introduction and use. Alternative methods of pest control are encouraged and any excessive uses discouraged wherever possible. By increasing the technical and scientific input, we are confident that it will continue to be possible to continue to make full use of pesticides in agricultural production throughout the developing world.

DDT-Chlorophenothene:
The Situation in the Federal Republic of Germany

H. P. Tombergs

Bonn-Bad Godesberg, Postfach 490

A law has been enacted regarding the use of DDT (BGB I) in the Federal Republic of Germany.

This law decrees a comprehensive ban on the handling of DDT and DDT-formulations as well as the banning of transacting food of animal origin (such as meat, fat, milk, and milk products, etc.), and certain consumption goods (cosmetics), in case the tolerances, for the determination of which the law contains an enabling authority, are exceeded.

For food of plant origin, tolerances are already fixed for DDT-residues, namely in the "Tolerance Regulation for Plant Protection" of November 1966. These tolerances are going to be reduced to lower values.

Furthermore, DDT-tolerances for animal feed and drugs, e. g. in bases of salves, such as wool fat, are given in the feed law and the drug law, respectively.

This is a comprehensive regulation, especially as the world "transaction" (Verkehr) includes: production, import, export, selling, buying, and use. Three exceptions are provided from this ban: "to combat lice, bedbugs, and pharaoants"; the exceptions are not valid for the use in man or vertebrates, and in animal housing, in food production.

The fact that DDT has reached an extraordinary dispersion due to its relatively high resistance and as a consequence of exaggerated use and misuse in organisms and environment, and the way that causes an increase in higher organisms, are too well-known to be mentioned here in more detail.

Much emphasis ought to be put on the contents of DDT in body fat and lipoid-containing organs of man and in mothers' milk.

Although the amounts of DDT in commerce in the Federal Republic of Germany are decreasing year by year (1970: 150 tons) — there has been no DDT-production in the FRG for several years — human fatty tissues contain 3.3 ppm (0.7—0.9), and mothers' milk 3.8 ppm DDT (1.5—8.3), on fat basis.

Even if one could not determine unequivocally whether these quantities were hazardous to human health, hazards as regards exposure of organs by effecting the metabolism (hormones, liver-enzymes), especially in cases of disease, malnutrition, and famine, should not be excluded. Upon evaluation of the need of DDT in pest control, the Federal Government concluded that such a potential hazard cannot be accepted. Therefore, the banning law provides for protection of man and his environment. Furthermore, the Federal Government has authorized the competent ministry of health to observe the possible exceptional use, and to investigate possible hazardous consequences of DDT-substitutes.

The Federal Government believes it has taken an important step in the sector of environmental and health protection by enacting this law and the regulations to follow it.

Welcoming Speech

U. Lohmar

Member of the German Federal Parliament, at the Second International Symposium "Chemical and Toxicological Aspects of Environmental Quality"
Munich, 27/28 May 1971

The discussions and talks which you will have at this international symposium during the next two days are bound to attract considerable attention. This is

guaranteed first of all by the scientific
stature of the experts attending here from
international and German research institutions, and also by the subject matter
itself. I am not a natural scientist. I started out as a lawyer and sociologist and
then went into politics. As a member of
Parliament, who is particularly responsible for economic policy, I hope to receive from your sessions valuable information about the practical steps politicians can or should take to aid science
in maintaining or restoring the environment so as to serve man best. Surely,
scientific progress does not also mean intoxication of our *natural environment.*
Therefore every activity in science and
economic production must raise the question of whether or not man's natural environment in the industrial society is pre-
served and safeguarded. If this is not
the case, it must be the duty of Government and Science to take all necessary
steps against it. And this is our problem
in politics: Since we are not experts ourselves, we can only learn from you, the
natural scientists, how to cope with such
dangers. The example of *environmental
protection,* which is of real concern to all
of us, demonstrates that politics and
science can no longer be treated separately but that each must rely on the other
to a considerable degree. This is why I
gladly accepted the sponsorship of this
Congress. Politics sometimes has other
rules than in science. Public funds never
meet the sum of all requests put before
the Government. This involves the problem of determining priorities for the
political tasks and expenditures. Since
the German and international public has
become more and more interested in, and
even alarmed by the subject of environment, there have been hints enough in

our country that by spending more money
on environmental protection we will have
to spend less money on education or on
science in general. Basically, I think this
is an error, because we are dealing here
with two aspects of one and the same
problem. Who says that there are no
other fields in which Government spending could be cut in order to make funds
available for the protection of the environment? But this is — and I am now
dealing with one of the characteristic features of politics — not only a question of
objective priorities, but also of the political and at the same time scientific
awareness of the policy makers, namely
for which purpose which funds can be (so
to speak) released. Normally it takes quite
a while until politics have come to grips
with the entire scope of a new problem.
As far as the question of education is
concerned, it took as much as a whole
decade here in Germany. When it comes
to the protection of the environment, we
can afford this time lag even less. For this
reason I am once again repeating my
request: please consider it your task to
indicate clearly and precisely what will
have to be done for which reasons and
with which means, in order to cope with
the problems of environmental quality.
Only then will politicians be able to
switch tracks, and perhaps economics will
then cooperate more than is sometimes
the case. As far as I am concerned, I will
gladly promise to do what I can in order
to draw the attention of the Federal Government and my colleagues in the *Bun-
destag* to such substantiated proposals or
requests as will be made by your Congress, and to urge them to take action.
With these thoughts in mind I now wish
you — or better I wish us — a fruitful
and rewarding symposium.

New Technologies for the Alleviation of Environmental Pollution

H. Gershinowitz

The Rockefeller University, New York, N.Y. 10021, USA

Summary. A review of existing and probable technologies indicates that many environmental problems can be solved or alleviated by wise use of technology. The prospects are more favorable for air pollution and solid waste disposal than for water pollution. It is pointed out that technology is not a panacea. It must be used with discretion and with wisdom. The need for remedial technology should be based not only on local or short term considerations but also on broad ecological principles. The misuse of technology should be prevented by punitive and restrictive legislation, by fiscal policy and by economic analysis which takes into account all costs to society of damage to the environment. In spite of the dangers inherent in the ever increasing use of technology, only through wise use of more and better technology can mankind survive and improve his condition.

Zusammenfassung. Eine Übersicht über bestehende und mögliche Technologien zeigt, daß viele Umweltprobleme durch den geschickten Einsatz der Technologie gelöst oder gemindert werden können. Gegen Luftverschmutzung und Müllbeseitigung bestehen bessere Möglichkeiten als gegen Wasserverschmutzung. Technologie ist kein Allheilmittel, sie muß mit Vorsicht und Überlegung angewandt werden. Die Notwendigkeit, die Umweltqualität durch technologische Heilmittel zu verbessern, darf nicht nur auf ortsgebundenen oder kurzfristigen Beweggründen, sondern auch auf umfassenden ökologischen Grundsätzen basieren. Mißbrauch von Technologie sollte durch Bestrafung und einschränkende Gesetzgebung, durch Steuerpolitik und wirtschaftliche Analysen verhindert werden. Dies würde alle Unkosten, die die Gesellschaft durch Umweltschäden tragen muß, umfassen. Trotz der Gefahren, die der wachsende Gebrauch von Technologie mit sich bringt, kann der Mensch nur durch überlegte Anwendung fortschrittlicher Technologie überleben und seine Lebensbedingungen verbessern.

This paper is a general survey, with some specific examples, of the rapidly growing field of the application of *technology* to the alleviation of environmental pollution. The amount of work under way is enormous and impossible to deal with in a short paper. For that reason, most of the references cited are themselves reviews of progress in limited areas, so that the reader may explore in more detail the developments mentioned.

To speak of *alleviating pollution* by technology is anathema to many people today. To them, technology is the origin, not only of environmental pollution, but of most of today's social problems. Without technology cities would presumably not exist. Industry would not be fouling the atmosphere and the water with its waste products. Applications of technology would not be reducing man to the role of attendant on machines, making his life full of meaningless repetitive motions. The products of industry would not degrade man's tastes and his use of leisure time. The other extreme is to consider technology as the source of all good, as the panacea for all of our ills.

Both of these extremes are dangerous. The first is romantic nonsense, ending with everyone a noble savage, racked by disease and at the mercy of the elements. The second could lead to Orwell's *1984* or to Huxley's *Brave New World*, with crime made impossible by constant electronic supervision of every act of man and unhappiness eliminated by chemotherapy.

Anyone who believes that technology is the sole *cause of environmental deterioration* should travel through rural India or rural Mexico. I have approached villages in India in the early evening and have seen a pall of smoke and dust hanging over them, indistinguishable at a distance from the pall which hangs over our Western cities. When one gets closer, one realizes that the sources are different. The odors make one aware that the haze comes from the smoke of burning cow dung used for fuel, and from the dust raised by people, bullock carts and camels, moving along the unpaved streets of the village.

A devastated landscape is not always the consequence of the misuse of technology. The barren countryside of Yucatan, with cactus as its only commercial crop, is eloquent testimony to the results of burning off the forests and planting crops until the soil is exhausted.

Even nature itself is not a respecter of purity of the environment. A single *volcanic eruption* can put into the atmosphere a cloud of particles lasting for years and exceeding, at least up to now, all of man's debris. On a global basis, nature without technology is still the *major source of contaminants of the atmosphere*. Natural processes are the sources of two-thirds of the *sulfur compounds*, ninety-nine percent of the *nitrogen compounds*, and ninety percent of the *hydrocarbons* which are found in the atmosphere[1]. I specified on a global basis, because our *pollution problems* of today are with excessive or undesired local concentrations, largely arising from man's activities. But even the specification of what is too much is often subjective. The haze over the Great Smoky Mountains of Tennessee, which is the consequence of reactions in the atmoshere of *terpene hydrocarbons* which evaporate from pine trees is considered beautiful and romantic, while the haze over our cities caused by hydrocarbons from *automobile exhausts* is almost universally considered undesirable.

As usual, the truth lies somewhere between the two extremes. Technology is neither the source of all evils nor the panacea for all ills. What is of primary importance is the interaction of people with technology and the interaction of technology with natural processes.

Even if the growth of population should eventually be halted, even if the reasons or motives which lead people to assemble in larger and larger urban agglomerations could be diminished, it would take many years before the way of life as we know it now could be changed so as to reduce or eliminate the undesirable effects of civilization. We must use all the resources at our disposal to reduce man's impact on the environment. Technology is not the least of these resources.

Up to now the *benefits of technology,* which are an important part of our Western civilization, have far outweighed the bad effects. Many of the latter have been due to unwise or improper use of technology. Our problem now is that we have reached a critical stage in our relationship with nature. In many local situations the scale of man's interference with natural processes has become of the same magnitude as the processes themselves. The self-purifying mechanisms of nature can no longer cope with these interferences. Many of the things we are doing to nature are undesirable; many may even be unnecessary; most can be eliminated by technology used wisely.

We must be very careful, however, to make sure that new technologies, developed to alleviate the disturbances to the environment, do not themselves introduce new problems. This means that we must carry on a continuing process of *technology assessment,* of anticipating, estimating, and evaluating, the effects of new technologies on the environment and on human society itself. After adoption, new technologies must be monitored, not only to make sure that anticipated problems are under control, but that unanticipated problems do not arise. This is a matter of much concern in the United States[2,3]. Formal procedures for carrying out technology assessment have already been proposed in legislation now under consideration by the U.S. Congress. There is much debate about the means which should be used. The need is apparent.

The danger is that too much control and too early analysis will stifle initiative and innovation and make it impossible to do the good that can result from well planned and properly used technology.

Not only must we examine the possible consequences of new technology, we must re-examine the *economic factors* which determine our choice of technologies. This does not imply that we need a new economics. Economists assure us that the means are available for including both direct costs and externalities[4]. What is needed is a way in which to evaluate the externalities, a way in which to distribute the costs of protecting the environment. *Taxation* is, of course, one of the most obvious ways in which to do this. It is not at all certain that it is the best way. In the United States, for example, taxes have been proposed by the federal government which would be based on the sulfur content of fuels used by power plants and on the lead content of motor gasolines. Such taxes are opposed in principle by some who maintain that these taxes are merely a *licence to pollute,* and that governments must establish *standards for effluents* and must impose severe penalties on those who violate the regulations.

It may very well be that 'new' technologies for protection of the environment may indeed be old technologies, which have not been used because they were economically not competitive with other technologies when considered on a basis which did not include total *costs to society,* but which may become attractive when the costs of making sure that the processes and the products are not harmful to man and his environment are included.

After this rather lengthy introduction, let us now look at some of the major sources of environmental pollution and examine the extent to which we might expect technology to eliminate or reduce that pollution.

Although the *pollution problems* of the air, the *water*, and the *land* are often inextricably linked, so that alleviation of one contributes to the aggravation of another, it remains both convenient and desirable to examine them separately. In the first place, the individual's perception of the atmosphere is more direct and more universal than his perception of water and land. Air is all around him wherever he is. Air is more homogeneous than water or land. What affects one man affects his neighbors as well. It is probably for these reasons that much of the furor about the condition of the environment centers around the air of our cities.

A second reason for considering air, water, and land separately is that the technologies needed to deal with them are often quite distinct. I shall, therefore, take them up in that order, but always remembering, and reminding you occasionally, that one does not solve the real problems of environmental pollution by putting wastes where they cannot be seen or into other media.

Air Pollution

It is my opinion that the problems of *air pollution* are less serious and more susceptible to alleviation than those of the water or the land. If we exclude the few effects of global occurence but as yet of uncertain importance, such as the *increase of carbon dioxide* and of particulate matter, the problems of air pollution are problems of cities or of limited areas with high concentrations of industries, such as the Ruhr in Germany, the Ohio River Valley in the United States, or the Midlands in England. Except for relatively short periods in a few locations, the air is in constant motion, both horizontally and vertically. There are many mechanisms, both physical and chemical, through which the atmosphere cleanses

itself. Sunlight provides unlimited amounts of energy, and the abundant supply of oxygen insures that eventually those substances which remain in the atmosphere will be oxidized to their most stable thermodynamic state. We must remember, however, that *rain* is one of the most important mechanisms, and that rain brings back to the land and the water such harmful substances as sulfur dioxide and minute particles of lead and pesticides. The goal, therefore, is the removal or elimination at the source of undesirable substances, whether they remain in the atmosphere for long or not.

The principle sources of urban air pollution are: 1) *automobiles,* which produce *hydrocarbons, nitrogen oxides, carbon monoxide*, and *particulates;* the last of these include *lead, rubber,* and *asbestos;* 2) *electric power* generating plants, which produce *sulfur oxides, nitrogen oxides,* and *particulates;* 3) *domestic heating,* which produces *sulfur oxides, nitrogen oxides, and particulates;* and 4) *incineration of garbage,* which produces *particulates* and increasing amounts of *hydrogen chloride* and other *corrosive substances.* Control of pollution from domestic heating is primarily a matter of regulation rather than technology. Cities which have imposed restrictions on the allowable sulfur and ash content of fuels for domestic heating have achieved substantial reductions in air pollution.

Incineration is a more complex problem and I shall deal with it when I discuss solid wastes.

Large *central power plants* present a major problem. *Soft coal* is the principle fuel for these plants. Most soft coal is high in sulfur content and also contains large amounts of non-combustible inorganic materials, part of which is carried off with the combustion products as very small particles called *fly-ash*. Coal is not just carbon which contains some impurities.

It is a mixture of highly condensed aromatic hydrocarbons, many of which have sulfur, oxygen, and nitrogen as part of the ring structures. The sulfur cannot be removed by physical processes but only by chemical reactions which break apart the molecules. The two most studied processes for *desulfurization of coal* are high pressure hydrogenation and partial combustion. In both of these, technological advances have been substantial but large scale application has been inhibited by the high cost of the available processes. It is now certain that, despite the increasing use of nuclear energy for the generation of electric power, fossil fuels will have to be the major source of energy at least to the end of this century. The world reserves of petroleum are too small, and their value for other uses to great, to allow their use for central power plants except under exceptional circumstances. Consequently, renewed attention is being given to the *desulfurization of coal*[5]. Since that is such a difficult and costly process, the removal of sulfur oxides from the combustion gases is under consideration as an alternative. A large number of processes have been proposed, and many carried through the pilot plant stage, but none has as yet achieved wide acceptance. The problems are not so much technical feasibility but of economic profitability when judged by the usual criteria. However, there is a growing realization that *cost-benefit analyses* which take into account all the externalities in addition to the plant costs, as well as the need to reduce the *emission of sulfur oxides* in many places regardless of costs, will make it necessary and even economically attractive to use the available and rapidly improving technologies.

There is a paradox involved in the desulfurization of coal which may make the removal of sulfur oxides from stack gases the more attractive process. Large central power plants produce a major part of the particulate matter in urban atmospheres. One of the most effective ways of *removing particulates* is by electrostatic precipitation. It has been found that this process is much less efficient when low sulfur coals are used and that it is very difficult to reduce the particulate content of the combustion products to an acceptable level.

The complexity of these problems of sulfur and particulate removal has caused an increasing amount of interest in the long-distance *transmission of electricity*. Power plants are located in cities because it has been cheaper to transport the fuel to the cities than to transport the electricity from the mine to the cities. While slow and steady progress is being made by increasing the voltages of overhead transmission lines, the most exciting developments are in underground power transmission[6]. In most cities, electric power is already transmitted underground, and the decreasing public tolerance for unsightly overhead transmission lines through scenic areas is making it very probable that more and more underground transmission lines will be required. At present, underground transmission is ten to twenty times more expensive than overhead lines. The stimulus for research and development is the knowledge that in theory the efficiency of a cable can be much greater than that of bare wires. Cables insulated with compressed gases are already in use and are being improved, but the great hope for the future lies in *cryogenic cables*. At the temperature of liquid hydrogen, aluminium has a resistance only one-thousandth of its magnitude at room temperature.

In the long run, the development of much more efficient methods of transmission of electrical energy over long distances will be an essential prerequisite to solving the problem of site selection for power plants, not only for fossil fuel plants but for nuclear plants as well. I have mentioned the more prominent current problems, but many more are arising as the amount of generating capacity increases. *Thermal pollution,* which I shall discuss in more detail later, is already a danger and will become more important as the number of nuclear plants increases, because up to now these plants operate at lower temperatures and are less efficient thermodynamically than most modern high temperature fossil-fueled plants. Here too we can anticipate help from technology. At least two methods are being developed for conversion of chemical or thermal energy to electrical energy without going through the Carnot cycle, the *fuel cell* and *magnetohydrodynamics.* For both of them, application is a matter of competitive cost rather than basic principles of technology.

I am quite confident that as far as generation of electrical energy is concerned, technology can make it possible for us to have both electric power and a satisfactory environment. I feel, however, that I must emphasize the fact that technology alone will not be enough. The need for novel technology must be proven, and the pressures to apply the technology must be applied by the people, by their elected representatives, and by the courts of law. I wish that I could have as much confidence in human nature as I have in the capabilities of technology.

The automobile as a source of urban air pollution presents a more difficult problem. This is not because new technologies may not be able to alleviate the severity of *air pollution* caused *by automobiles,* but because the basic problem is not air pollution, as serious as that is, but the over-all problem of *intra-urban transportation.* In many large cities, you can walk a mile faster than you can cover the same distance in a motorized vehicle. The emission of hydrocarbons and carbon monoxide by automobiles has already been greatly reduced and will be reduced further in a few years. There are no technological barriers to producing *lead-free gasolines* or to making engines which will run well on such gasolines. But often, when I sit in a taxi, blocked by traffic, in New York or in Paris, I wonder about the possibility of a complete blockage of motion in the city, caused by the interlocking of automobiles at a few key intersections, resolvable only by backing cars away, one by one, from the periphery. The problem for technology here is not so much the pollution free motor vehicle, but a better system of *intra-urban transportation.*

A great deal of attention is being given to the improvement of conventional means of transportation. *Battery driven cars* will probably soon be in use in some cities, in spite of their many disadvantages. Attempts are being made to woo the commuter from his automobile by providing more efficient and more rapid train service from suburbs to the inner city. But, in my opinion, the really exciting developments are the more unconventional proposals. Much thought is being given to moving sidewalks for the centers of cities. An American scientist has proposed a combination of a moving sidewalk together with a network of small, overhead, cable-pulled cars, moving about twenty miles per hour[7]. The inventor estimates that the system could operate profitably at a fare of twenty-five cents. In New York City, buses and subways now cost thirty cents.

What is now needed to solve *urban transportation* problems is daring, vision, and initiative on the part of municipal authorities. When these qualities have been made manifest, technology will rush in with possible solutions.

There are other problems of air pollution from aircraft, pollution from pesticides which vaporize or are spread as aerosols, pollution from industrial processes. For each of these there appear to be technological solutions, but a prerequisite to their development and application is an assignment of priorities, a recognition and acceptance of the need for increased costs and regulations setting criteria and standards.

Water Pollution

I wish that I could be as optimistic about *water pollution* as I have been about the air. The problems of water pollution are more critical than those of the atmosphere for several reasons. In the first place, the processes of *regeneration* and *self-purification* are much slower. In the second place, the deleterious, or lethal concentrations of pollutants are often smaller than those in the air. Thirdly, the sources of contaminants are more diverse and more complex than those which provide the major part of air pollution. Finally, the natural processes which are affected by contamination of the water are more sensitive in themselves and more essential to the propagation of life in general than are those which take place in the air. The marshes, wetlands, and estuaries are breeding places for many varieties of *marine life* and *water fowl*. It is these areas which are most threatened by the activities of man.

It is for all these reasons that the role of technology in alleviating environmental problems is not as clear for water as it is for air. This is not universally true, however. Take the case of thermal pollution from *electric power generating plants*. Increasing the temperature of water is harmful to living organisms mainly for two reasons. It reduces the amount of oxygen which can be carried in solution, and it changes the rate of metabolic processes, making it impossible for many organisms to survive or to reproduce. Two solutions for thermal pollution have been proposed, one technological, the other biological. In the first, cooling towers or other heat exchange systems can be used to reduce the temperature of the water which is returned to the environment. This may be feasible in some geographic locations, but in others it would only contribute to air pollution, causing fogs. To me, the most promising solution to thermal pollution would be a combination of technology and biology, the use of the available heat in a beneficial way rather than to reduce it to a harmless level. An enormous amount of work is now being done on the *biological effects of temperature*. I quote a recent article in this field: "More papers on this subject were produced in a single year in the sixties than in all the years from 1900 to 1920[8]."

Man himself could benefit directly from *constructive use of waste heat*. In many parts of the more industrialized parts of the world the waters used for recreation are too cold for swimming. Judicious choice of location for discharge of hot water which would not interfere with other biota would greatly increase the opportunities for recreation along the coasts of California, of the United Kingdom, and of Scandinavia, for example. As far as other living organisms are concerned, schemes have been proposed for changing the local fauna and flora to those more suited to tropical or subtropical conditions. Warm water used to irrigate greenhouses could make possible economically attractive year round cultivation of sensitive crops in northern Europe. Part of the work needed to achieve this would be technological and part biological, so here too technology will be important.

The *chemical contaminants of water* are more difficult to deal with. They originate primarily from three sources, municipal wastes, industrial wastes and by-products, and agricultural wastes and run-offs. For *municipal wastes,* the technology is almost all available. The problems are mostly economic and political. What goes into the air affects the people who put it into the air. What is put into the water affects mainly those downstream. It is difficult, therefore, to convince municipalities, and the taxpayers who support the activities of municipalities, that they should expend large amounts of money to improve the condition of their neighbors. What is required is legislation by larger governmental units to compel towns and cities to purify their wastes. This is already happening in many countries. In addition, regulations are being made to prevent the introduction into sewage of substances which are hard to remove by conventional treating methods, *phosphate-containing detergents,* for example.

There is an additional problem, at least in the United States. The teaching and practice of sanitary *engineering* has tended to be a bit old-fashioned. The emphasis is still on the removal of human excrement and other organic matter and of the micro-organisms which thrive on these and which are harmful to human beings. Little attention has been paid to synthetic organic chemicals, to laundry products, or to the wastes from small industries which enter the municipal sewage systems. Nor has enough attention been given to modern chemical engineering, which can improve the efficiency and decrease the cost of water purification. These weaknesses are now generally recognized and great progress has been made in the last ten years[9].

Industrial wastes and by-products have always been recognized as hazards or, at least, as undesirable by the industries which produce them. For a long time it has been realized that the costs of eliminating health hazards and reducing environmental pollution must be taken into account in the economics of the production and the manufacture of materials and chemicals. What has changed is the magnitude of the potential costs. It has been customary to consider such costs as marginal. In some cases already, and in many cases in the future, the costs of maintaining the quality of the environment may make it impossible to operate in certain geographic areas and, eventually, make some substances more expensive than competitive, less polluting materials. There are special papers devoted to the economics of environmental pollution in this volume, so I do not wish to dwell too long on this issue. I do want to make one additional point, however. One should be cautious about identifying conservation of resources with preservation of the quality of the environment. Total *re-cycling* is often recommended as the ultimate solution to pollution. It is indeed, but it may not be the most feasible, either technologically or economically, and it may not be even the best method for some time to come. After all, it is the most abundant elements which contribute the most to our wastes. Recovery of waste products is already causing a glut of some substances. Calcium chloride is one example and sulfur may soon be another.

All of these factors pose a great challenge to technology. Even where economic incentive is lacking, restrictive legislation will make it necessary to reduce effluents from industrial plants. In principle there do not seem to be any technological barriers. To reduce theory to practice is still an enormous task.

In order to provide food for the undernourished peoples of the world and for the rapidly increasing population, more and more *fertilizers* and *pesticides* will be needed. More and more cattle and other animals will be fed in central locations. The problems already encountered which result from these practices will be magnified and become critical unless improved methods are developed for controlling *wastes* and *run-offs* from *agriculture* and *animal husbandry*. The pesticide problem is recognized and on the way to control, but the problems of *excess nitrogen* and *phosphors* leading to *eutrophication* of waterways, of *oxygen depletion* and contamination of rivers and lakes caused by the excrement from poultry farms and cattle feeding lots have received much less attention.

The technology of *sewage treatment* is just as applicable to chickens as it is to humans. Once the problem is acknowledged, sanitary regulations will be imposed and treatment of the effluents will be made compulsory.

The problem of *run-off of fertilizers* is a more difficult one and technology is less applicable. The increases of yield which result from fertilization and the inefficiency of uptake of fertilizers by plants, together with the vagaries of the weather, often lead the farmer to over-fertilize. More conservative practices and careful monitoring of the waterways in agricultural drainage basins are the most probable ways in which to reduce the magnitude of this problem.

There is one source of environmental pollution which in my opinion is not receiving enough attention; that is, the loss of substances during *transport*. In recent years, many of the most serious incidents of water pollution have been caused by leaks from railroad cars, tank trucks, pipelines, and ships. Many of these incidents have been *accidents*. Some of these accidents have been due to faulty design, but many have been due to carelessness, negligence, and even willful disregard of orders and regulations. There is too much tendency to say 'accidents will happen' and to concentrate on ways of cleaning up after accidents. I remember a slogan which was posted in our laboratories, 'Safety Is No Accident'. Technology must be developed to make transport more accident proof, but, in addition, by education and by penalties, those in charge of transport must be made personally responsible for avoidable accidents. The nuclear powers take great precautions to prevent accidental discharge of atomic weapons. It is not unreasonable to take precautions against the accidental release of materials which can poison large numbers of plants and animals or despoil the oceans and shorelines.

Solid Wastes

Solid wastes provide the clearest examples of the inter-relationship of environmental problems. If one incinerates them, there is a risk of contaminating the air; if one buries them or dumps them at sea, the water may be polluted; if one leaves them on the surface, one accumulates masses of unsightly, and often unhealthful, rubbish. There are many problems here for technology, ranging from reducing the amounts that are generated, improving the methods of collection, and, ultimately, disposing of them.

The amount of solid waste per capita may not be the most accurate index of affluence, but it certainly is the most visible. Paper and other cellulose products do not pose much of a problem since they are easily combustible, biologically degradable, and — where suitable collection methods are available — economically worth recycling. They should be separated from other wastes in order to make effective use of these procedures.

This is not so much a matter of technology as of consumer education and the provision of suitable methods of collection.

The disposal of plastics and of non-returnable beverage containers, both plastic and metal, presents much more serious problems. A substantial amount of work is being done on the development of *degradable plastics* and on safe methods for the *incineration* of chlorine-containing plastics. One of the major problems arises from the prevalence of small scale *incinerators* in individual buildings. These were originally encouraged by municipalities, particularly in the United States, in order to reduce the amount of garbage which must be collected by trucks moving from house to house. While it is technologically feasible to design and build efficient incinerators, it is very difficult to insure that they will be operated properly. They are used for only a few hours a day and the individuals who are responsible for them rarely have the training or technical competence which is required.

Small incinerators are now being discouraged or forbidden and methods of collection of wastes are being improved. Equipment for the *compaction of solid wastes* has been developed which greatly reduces bulk and makes it more susceptible to mechanical handling.

If one were building a new city, one would probably provide for *grinding or maceration* of all solid wastes at the source and their transport as slurries through a sewage system, separate from sanitary and storm sewers. The slurries could then be treated effectively and efficiently at large central installations and the water could be re-cycled into the collection system.

An alternative to incineration is *controlled pyrolysis*. This process also is most useful when a prior segregation of materials has been made so that specific products of value can be made under specified conditions of pyrolysis.

It does seem that many of the most serious current problems of pollution due to solid wastes can at least be alleviated by technology. The development of this technology is going slowly, however, and the alternative in the minds of many concerned persons is restrictive regulation. As of March, 1971, fifteen of the states of the United States had under consideration legislation which would forbid the use of non-returnable beverage containers. In some of these states, the legislation would also forbid the use of packaging materials which were not biologically degradable. Here we have a real challenge to, and incentives for, chemical technology, biological technology, and transport technology.

Conclusions

The possibilities for alleviating environmental pollution by means of new technologies are very favorable. I have given only hints of the accomplishments and prospects in this short paper. In many cases it is not so much a matter of new technologies but the abandonment of old technologies. In others, it is a matter of looking again at technologies available but not used, in the light of new economic factors or restrictive and punative legislation. The one point which I hope that I have made clear is that despite the evils which technology has brought with it, only through better and wisely applied technology can mankind survive and improve its condition.

References

[1] The Standford Research Institute: Sources, Abundance and Fate of Gaseous Atmospheric Pollutants. A report prepared for the American Petroleum Institute. 1966

[2] The National Academy of Sciences: Processes of Assessment and Choice. A report prepared for the Committee on Science and Astronautics, U.S. House of Representatives. Supt. of Documents, U.S. Government Printing Office. Washington, D.C. 20402, 1969

[3] The National Academy of Engineering Technology: Processes of Assessment and Choice. A report prepared for the Committee on Science and Astronautics, the U.S. House of Representatives. Supt. of Documents, U.S. Government Printing Office, Washington, D.C. 20402, 1969

[4] Coffman, R. B.: Economics, ecology and political decisions. Bull. Atomic Scientists 28: 2–4, 1971

[5] Squires, A. M.: Clean power from coal. Science 1969: 821–8, 1970

[6] Rose, P. H.: Underground power transmission. Science 170: 267–73, 1970

[7] Avery, W. H.: A report presented at the meeting of the Am. Assoc. Adv. Science. Dallas, Texas 1968

[8] Cairns, J. Jr.: Thermal pollution: A cause for concern. J. Water Pollution Control Federation 43: 55–66, 1971

[9] Holcomb, R. W.: Waste water treatment: The tide is turning. Science 169: 457–59, 1970

received: May 1971

Recent Air Pollution Problems

P. C. Blokker

Stichting Concawe, The Hague 2012, Netherlands

Summary. A review is given of recent air pollution problems, mostly new aspects of pollutants known for many centuries. First an evaluation of the problems in general is given. It is emphasized that the core of community air pollution problems has been and will be the products generated in energy production. Because of the growth of population and increasing standard of living in underdeveloped countries, air pollution could become a very great problem if no rigorous actions are taken.

Further, the new aspects of specific pollutants are discussed. For sulphur dioxide the new aspect is that the drift of this pollutant over great distances may have serious consequences. For nitrogen oxides the main concern is their capacity to react photochemically with unsaturated hydrocarbons from auto exhaust, which gives rise to the formation of different kinds of short-lived free radicals, semistable compounds, and other secondary pollutants such as ozone, organic hydroperoxides, and peroxyacyl nitrates. Several investigators consider ozone and photochemical oxidants as the potentially most hazardous of all pollutants. For lead, the main fear is that at increased levels some interference with certain enzyme systems might occur. For carbon monoxide, concern is mainly for the response of people suffering from diseases associated with impaired oxygen saturation of the blood and of people with certain types of anemia. For carbon dioxide, concern centers on the possible effects on the temperature of the earth's surface. Finally, the great uncertainty with respect to possible effects of artificial asthmatic and carcinogenic agents is discussed.

Zusammenfassung. Der Beitrag gibt eine Übersicht aktueller Probleme der Luftverschmutzung, besonders neuer Gesichtspunkte über Verunreiniger, die größtenteils schon seit Jahrhunderten bekannt sind. Zunächst wird eine allgemeine Bewertung der Probleme gegeben. Es wird betont, daß Produkte aus der Energieerzeugung schon immer Kern der Probleme der Luftverschmutzung waren und daß sie es auch weiterhin sein werden. Sollten keine harten Maßnahmen ergriffen werden, so würde die Luftverschmutzung durch Bevölkerungszunahme und Steigen des Lebensstandards in den Entwicklungsländern zu einem großen Problem werden.

Weiterhin werden neue Aspekte spezifischer Verunreiniger behandelt. Bei Schwefeldioxid ist dieser neue Aspekt die Befürchtung schwerer Folgen durch seine Verteilung über große Entfernungen. Bei Stickstoffoxiden ist das

Hauptproblem die photochemische Reaktionsfähigkeit mit ungesättigten Kohlenwasserstoffen aus Autoabgasen. Dadurch entstehen verschiedene kurzlebige, freie Radikale, metastabile Substanzen und andere Sekundärverunreinigungen wie Ozon, organische Hydroperoxide und Peroxiacylnitrate. Verschiedene Wissenschaftler betrachten Ozon und photochemische Oxidantien als die potentiell weitaus gefährlichsten Verunreiniger. Bei Blei besteht die Gefahr, daß bei hohen Anteilen eine Wechselwirkung mit bestimmten Enzymsystemen entstehen könnte. Kohlenmonoxid schadet hauptsächlich Personen, die unter einer Krankheit leiden, die die Sauerstoffsättigung des Blutes beeinträchtigt, oder solchen, die an Anämie leiden. Kohlendioxid verursacht Besorgnis durch die Beeinflussung der Temperatur der Erdoberfläche.

Abschließend wird die große Ungewißheit hinsichtlich der Auswirkungen künstlicher asthmafördernder und karzinogener Substanzen behandelt.

Introduction

From the title one might think that recent problems have something to do with the appearance of pollutants not known or seldom occuring in the past. With the possible exception of some pesticides this is only true for a small number of pollutants, which are not very important for the environment or at most are of significance in a few locations, such as vapors of new organic materials. Real, important new problems, however, are caused by pollutants which we have known, or at least were present in considerable quantities, over many centuries, such as *smoke, dusts, carbon monoxide, carbon dioxide, sulphur oxides, nitrogen oxides, hydrocarbons,* and *evil-smelling gases.* But in the past these pollutants were only of *local importance,* such as the coal smoke in the mid 13th century in Nottingham and some other towns. In the present century a few serious incidents have occurred, which could conceivably be considered as being only of local importance such as Donora in 1948, Poza Rica in 1950 and the Meuse Valley in 1930, and on a larger scale the disaster in London in 1952. After 1940 with the large increase in population and crowding, many more cities began to show evidence of *air pollution,* fortunately less serious than during the London disaster. This kind of pollution was caused by gases or substances which had been present for a long period in the atmosphere and the only reason for their present importance is the higher concentrations in which they are now occuring.

The causes of this, and this is also true for pollution of water and soil, are the increase in population density, the still greater increase in energy consumption per capita and the increase in pollution-causing products resulting from the economic growth in certain areas of the world. This has caused such an enormous increase in the *gaseous, liquid,* and *solid wastes* (the latter two often being converted into gaseous waste) that quite a number of people, including many scientists, consider it as a very serious future threat to our environment and even to human life. Others are more optimistic and trust that advanced technical knowledge will provide the tools to find solutions for the problems. Nevertheless, these problems are of great concern to nearly all scientists, who are directly or indirectly involved with the biosphere.

Biersteker[1] has stressed that human beings are hard at work improving their *micro-environment* but that the *macro-environment* has deteriorated. He and many others state that the ecological systems which determine the environment are often vulnerable and pollutants may cause loss of the ecosystem's stability by diminishing the number of species. The problem with air pollution, in common with water pollution, is that effects may become widespread. The capacity of air and oceans to receive pollutants without adverse consequences is very large but not inexhaustible.

One difficulty with air pollution is that breathing of air cannot be avoided. In one respect, the situation with air pollution is more favorable than with persistent pollution of water (e. g. by heavy metals, chlorinated hydrocarbons) viz. that the lifetime of most air pollutants is relatively short (estimations are 4 days for sulphur dioxide, 2 years for carbon monoxide).

Even if the level of air pollution were considered to be tolerable at the present time, there is no doubt that it would become a very great problem if no rigorous action were taken. The growth of population and increasing standard of living in underdeveloped countries could cause pollution to increase rapidly, mainly because of increase in energy production, industries, motor cars etc.

Evaluation of the Problems in General

In order to face these problems and to be able to solve them it is necessary that we have quantitative knowledge of the hazards.

While reasonable knowledge exists on acute effects of many single pollutants

our knowledge of the joint effects of several pollutants together, of reactions between pollutants and of long-term effects on ecology and human health, is very limited. This lack of information understandably has led to the questionable solution that concentrations of pollutants should not rise much above the natural background. But is the natural background really a good yardstick and moreover, which *background level* is really natural? The difficulty with the concept of natural background is shown in the case of lead. Patterson[2] estimates the natural level in air as being $0.0005 \ \mu g/m^3$, whereas in remote mountain areas the present level is 0.12[3]; by means of speculative deductions Patterson estimates the natural lead level in blood to be 0.25 $\mu g/100$ ml, whereas the level in New Guinea aborigines and Mexican Indians is about 20 $\mu g/100$ ml[4,5]. Recently Stokinger[6] has given excellent comprehensive toxicologic evaluations of potential human health hazards from man-made and natural pollutants for air, water, and food. For *air pollution* similar reviews in 1968 have been given by Stokinger, Coffin[7] and Goldsmith[8]. The recent evaluation by Stokinger is restricted to *long-term health effects* of concentrations near to present or envisaged pollutant levels. He considers respiratory *irritants (ozone and oxidant)* and *carcinogens* (partly from air) as the most important air pollutants with a general geographic distribution and *asthmatic agents* and *asbestos* with a more localized distribution. Goldsmith[9] stresses the importance of non-disease effects, such as some respiratory reactions, sensory irritation, reactions to odors, central nervous system reactions, altered responses to high demand, and biochemical changes.

In the following survey problems of air pollutants that are only of local importance (e. g. fluoride emission) or that are of decreasing importance are not discussed. The latter refers for example to smoke and gases from the burning of coal. *Coal smoke,* the chief atmospheric pollutant for centuries, still is a major contributor to air pollution in many urbanized areas, but the great improvements obtained in British and American cities have demonstrated that severe smoke pollution is preventable, although the cost may be large.

The core of community *air pollution* problems has been and will be the products generated in energy production. The nature of these products are closely related to the kind of fuel and the way in which it is used. Therefore, if in the future fossil fuels were largely replaced by nuclear energy, totally different kinds of air and *water pollution* problems would have to be faced.

Recent air pollution problems, often old problems with new aspects, are summarized below.

Problems with Specific Pollutants

Sulphur Dioxide

Pollution by sulphur dioxide emitted by domestic and industrial *burning of fuel* and by smelting industries has been a problem for many decades. It has been resolved in a number of locations but is still very important. The new aspect is that some scientists believe that the drift of sulphur dioxide from one country to another may have serious consequences. This especially refers to the drift of sulphur dioxide from Central and Western Europe to northern countries, where it is feared that the decrease of pH in rainwater by dissolved sulphur dioxide ("acid rain") may have adverse effects because of the low buffer capacity of some rivers, lakes, and soil. Other investigators believe that this is exaggerated. Elaborate studies are to be carried out in different

countries under auspices of the OECD (Organization of Economic Cooperation and Development).

In 1965 the estimated *amount of the SO2* emitted by pollutant sources in the northern hemisphere was 136 million tons, equivalent to 68 million tons of sulphur, whereas the natural sulphur emission (from H_2S) is estimated at 62 million tons[10].

It is estimated that in Europe about 400,000 hectares (1,000,000 acres) of forest is affected by air pollution, mainly by SO2. The survival of roughly 1/3 of this forest area is threatened[11].

In several areas in the world the very sensitive *epiphytic flora* are declining, mainly due to the presence of SO2. Around various larger towns in Europe and America many epiphytic lichens and mosses are extinct (so-called epiphytic deserts)[12]. To our present knowledge this is not a serious phenomenon because they have as far as we know no effect on the food chain, but it is a sign that the *ecology* of vast areas can be changed by very low concentrations.

According to Stokinger, no health effects may be anticipated from SO2 per se at any conceivable community level, but through interactions with other air pollutants, e.g. adsorption to fine particles, catalytic oxidation to SO3 and formation to sulphates, acute effects may be produced and even long-term effects cannot be excluded at the present state of knowledge.

Technically it is possible to remove most of the sulphur from liquid fuels or to remove SO2 from flue gases, but the costs for removing sulphur from residual fuel and flue gases are high. This has up to now, hampered practical application and measures have been confined mostly to the use of fuels of naturally low sulphur content and by use of high chimneys. As more fossil energy is used, the pressure to remove sulphur and (or) sulphur dioxide will steadily increase.

Nitrogen Oxides, Ozone, Oxidants

Nitrogen oxides, the total *emission* of which is estimated by Robinson and Robbins[10] as roughly 36 % of that of sulphur dioxide, would appear to be exonerated as community air pollutants of health concern at present levels, but there is concern with regard to their capacity to react photochemically with unsaturated hydrocarbons from auto exhaust which gives rise to the formation of different kinds of short-lived free radicals, semi-stable compounds and other *secondary pollutants* such as *ozone, formaldehyde, organic hydroperoxides,* PAN *(peroxyacetyl nitrate)* and other peroxyacyl nitrates. Stokinger[6] and others consider ozone and photochemical oxidants as the potentially most hazardous of all pollutants, mainly because in animal studies long term effects of exposure to concentrations not far above those sometimes found in photochemical smog areas have been observed, such as emphysema, increase in *pulmonary cancer* incidence and an acceleration of the aging process. Stokinger[7] gives a number of arguments for the similarity of action of ozone and the much less potent nitrogen dioxide to ionizing radiation, e. g. their capability of initiating free radical chain reactions, of producing *chromosomal aberrations,* and of causing *premature ageing* of animals repeatedly exposed.

Notwithstanding the large amount of work that has been devoted to problems connected with formation and mode of action of *photochemical smog,* much more research is needed. For instance the development of a large excess of ozone during severe smog periods in Los Angeles usually signals an ensuing rapid decline in eye irritation. Is that because the olefins are nearly exhausted and the ozone is oxidizing the irritating intermediate compounds as has been suggested[13]?

The formation of oxidants has regularly been found in the Netherlands and in West Germany, although originally it was thought that sun irradiation would not be strong enough in these areas. Further, how can the so-called *"Open Air Factor" (OAF)* found by Druett and May[14] be explained? They found that a mysterious component of open air, present even in the dark, has germ-killing ability (E. coli were sometimes killed within half an hour). OAF disappeared very rapidly in any form of enclosure or contact with walls; it could not be SO_2, NO_2, formaldehyde or ozone because these revealed no germicidal action at realistic levels. The authors suggest that it might be a reactive product formed in the dark from ozone and car exhaust.

To keep the *"oxidant" levels* in meteorologically unfavorable areas below maximum levels (values between 100—300 µg/m^3 for 1 hour calculated as ozone have been proposed), limitation of the concentration in auto exhaust of either unsaturated hydrocarbons or nitrogen oxides or both is indicated. Up until now all measures have been towards the first, by improving combustion either by better combustion in the cylinder itself or by afterburning. This system, already compulsory in California, has the tendency to increase the concentration of nitrogen oxides.

Also decreasing the concentration of nitrogen oxides in *auto exhaust* is being considered. However, laboratory investigations indicate that with effective hydrocarbon control, moderate reduction of oxides of nitrogen may not further reduce smog effects. In fact some laboratory data indicate that moderate reduction of oxides of nitrogen emissions may negate some of the benefits gained from hydrocarbon control[15, 16]. In order to achieve clear-cut benefits beyond those from hydrocarbon control, the concentration of oxides of nitrogen should be drastically reduced. It seems doubtful whether, with the present uncertainty of supporting data, this is warranted. This would probably necessitate catalytic devices for decomposing nitrogen oxides and omission of lead from the fuel to avoid poisoning of the catalyst.

Lead

The information published to date does not permit any definite conclusions concerning an identifiable hazard to human health at the prevailing levels of *lead pollution* in the atmosphere; the general opinion among toxicological authorities is, that there is no valid evidence of an adverse effect of present air levels.

However, since there remains some doubt whether at increased levels some interference with certain enzyme systems, e. g. in porphyrine synthesis might occur, an increase in the lead levels in the atmosphere is considered undesirable.

A great many research projects into the lead problem are underway. The objectives that seem to be of most importance[17] are systematic investigations of the lead content of skeletons from different historical periods, the study of effects of lead on specific enzymes (such as delta-amino levulinic acid dehydrogenase), the study of the limits of the capacity of the animal body to excrete lead absorbed by various routes and a continous study of the health of people suffering from a defect in their excretory capacity (such as certain renal diseases).

A gradual diminution of lead in gasoline leading to *unleaded gasoline* is technically feasible but if equal fuel performance were required, this would lead to very high investment costs and an increase in price of gasoline with a new set of combustion problems to deal with.

Carbon Monoxide

Stokinger[7] draws attention to the fact that the ratio of hazardous (toxic) levels to current urban air levels is smaller for CO than for other pollutants in urban atmospheres, particularly when comparison is made with SO_2. Levels of 50 ppm of CO in air and 4—5% of carboxy hemoglobin in blood have been recorded. Concern for these levels, found in stagnant traffic, involves not so much the direct effect to healthy individuals, but mainly the response of people suffering from diseases associated with an impaired oxygen saturation of the blood and of people with certain types of *anemia* which may be associated with an excessive metabolic production of carbon monoxide, such as in certain hemolytic anemias. Another seemingly remote possibility might be that CO-induced *hypoxemia* of long duration could accelerate the development of degenerative vascular diseases *(aterosclerosis)*. Further, a decreased reaction to sensory stimuli has been observed in individuals exposed to high CO concentrations. For all these reasons the American MIC value of as little as 20 ppm CO for 8 hours and 70 ppm for 1 hour has been set.

By far the main source of CO is motor exhaust. To reduce the present concentrations better combustion or afterburning can solve the problem, as with the unsaturated hydrocarbons in the exhaust.

Carbon Dioxide

Carbon dioxide normally is not considered a pollutant. This probably is because CO_2 is a natural constituent of the air at a fairly high concentration (300 ppm) and is an integral factor in the life cycle of the earth. Nevertheless changes in *atmospheric CO_2* could be very important for man's environment. As long ago as the middle of the 19th century Tyndall suggested that the glacial periods may have been due to a change in the atmospheric CO_2 content. In 1896 this hypothesis was effectively advocated by Arrhenius[18]. Based on measurements by Langley, Arrhenius showed that an increase in the content of CO_2 and (or) water vapor would increase the temperature of the earth's surface because it would increase the absorption of the long-wave back radiation of the earth. The geologist Chamberlin[19] does not reject this hypothesis but shows that insufficient information on actual changes in the depletion and enrichment pattern of CO_2 during and between the glacial periods exists to prove the hypothesis. At present this problem is still unsolved.

Since the end of the 19th century an artificial source of CO_2, fuel usage by man has become important[10]. Arrhenius has mentioned this already. Investigations at certain locations in the northern hemisphere have shown that the CO_2 content of the atmosphere has increased from the 19th century base of 290 ppm to 320 ppm viz. with 10%. This shows that the balance between the release of CO_2 and the consumption by photosynthesis in plants and by the very important subsequent formation of calcium carbonate in oceans, has been disturbed and some investigators suggest that this could cause a gradual melting of the polar ice caps.. Revelle[20], for example, states that a 25% increase of CO_2 would increase the absorbed and back radiated energy by 2% and that, if half of this energy were available (greenhouse effect) to melt the polar ice caps, the ice would disappear in about 400 years, resulting in inundation of coastal areas (melting of the Antarctic icecap would result in a 20 m rise in sea level) and in a change in the *world climate*.

According to the latest analysis by Manabe and Wetherald[21] a change in the CO_2 content from 300 to 600 ppm would increase the temperature of the earth's surface by 2.4° C.

McCormick and Ludwig[22] have indicated that a possible increase in fine *particulate material*, some of which is caused by man-made pollution, may have the effect of increasing the reflectivity of the earth's atmosphere and hence to reduce the amount of radiation received from the sun and to have a cooling effect. This effect could reverse any warming trend due to CO_2. Also an increase in cloud average could have a cooling effect because of interception of the incoming radiation. Whether the CO_2 warming theory or the submicron particles cooling theory are right, the possibility that the abundant pollutants (CO_2 and submicron particles) will have a much greater and worldwide effect than the pollutants we are normally dealing with should not be overlooked.

Asthmatic Agents

After 1950 it was discovered that besides the natural *asthma-promoting agents*, such as many types of pollen and microorganisms, artificial agents are sometimes emitted causing outbreaks of allergenic character or bronchial irritation, e. g. "Yokohama asthma", later characterized as bronchitis in Japan, asthma-like phenomena in New Orleans and Minnesota in the USA [6, 8]. The precise nature of these allergens is still unknown; they are associated with particulate fractions probably originating from processes like grain milling, burning or roasting of organic material.

Carcinogenic Agents

These substances, part of which can reach us via polluted air, are giving much concern, mainly because there is a higher incidence of lung cancer in urban areas than in rural ones without knowing what the urban "factor" really is and to what extent it is due to air pollution.

The greater concentration of known *carcinogens in urban* air compared to that in non-urban air is not considered sufficient to account for the urban factor as such[8]. Goldsmith explains that the available evidence has failed to support the hypothesis that exposure to community air pollution is a causal factor affecting lung cancer rates, but Stokinger[7] states that there is increasing evidence favoring the implication of air pollution as an important factor in lung cancer. When surveying the literature with the often conflicting facts and statements, it is quite evident that much more research has to be carried out before any real conclusions can be drawn. The reason for the great difficulties in identifying the cause of the urban factor is many sided. Partly it is inherent to the gaps in knowledge of cancer itself, partly it is due to insufficient knowledge on the type and amount of carcinogens, cocarcinogens, cancer promoters (which even can be found among seemingly inert particles), and cancer inhibitors; insufficient knowledge on all kinds of interactions; and the fact that carcinogens are ubiquitous in nature and are synthesized there (e. g. for plant growth regulation), and can reach us in different ways, viz. via water, air, and food. All this is complicated by a lack of knowledge regarding the effects of the increased circulation of viruses in congested areas, the effects of urban crowding, as such and the great variability in susceptibility of individuals, etc.

The number and types of carcinogens and cancer promoters found experimentally are increasing annually. To quote an example, *asbestos*, causing a special type of *lung cancer* (mesothelioma), has recently drawn much attention. Some investigators think it likely that *polycyclic aromatics* and certain trace metals are essential to *mesothelioma*. The sources from which asbestos can reach the human body are not exactly known, nor is the degree of exposure necessary to induce the tumors, but in view of the known brief occupational exposures of man in many of the reported cases, it has been suspected that the reasons of these cancers might be from non-industrial environments such as air pollution[7]. Therefore, further study is needed, also in finding improved methods in identification of asbestos fibers within tissues, ascertaining which varieties of asbestos are harmful, and in collecting airborne particulates.

References

[1] Biersteker, K.: De ruimtelijke ordening in algemeen medisch perspectief. (Town and country planning in general medical perspective.) Tijdschr. Soc. Geneeskunde 48: 470–479, 1970

[2] Patterson, C. C.: Contaminated and natural lead environments of man. Arch. Environm. Health 11: 344–360, 1965

[3] Dept. of Publ. Health, State of California: Lead in the Environment and its Effect on Humans. March 1967

[4] Goldwater, L. J., A. W. Hoover: International study of "normal" levels of lead in blood and urine. Arch. Environm. Health 15: 60–63, 1967

[5] Kehoe, R. A.: Contaminated and natural lead environments of man. Arch. Environm. Health 11: 736–739, 1965

[6] Stokinger, H. E.: The spectre of today's environmental pollution – U.S.A. Brand: New perspectives from an old scout. Am. Ind. Hyg. Ass. J. 30: No. 3 195–217, 1969

[7] Stokinger, H. E., D. L. Coffin: Biological effects of air pollutants. In: Air Pollution I, ed. by A. Stern. Academic Press pp. 446–546, 1968

[8] Goldsmith, J. R.: Effects of air pollution on human health. In: Air Pollution I, ed. by A. Stern. Academic Press 547–616, 1968

[9] Goldsmith, J. R.: Nondisease effects of air pollution. Environm. Res. 2: 93–101, 1969

[10] Robinson, E., R. C. Robbins: Sources, abundance and fate of gaseous atmospheric pollutants. Stanford Research Institute. Prepared for API, February 1968 and Supplement Report June 1969.

[11] Bossavy, J.: Information sur les dommages causés par la Pollution de l'air aux Plantes et aux animaux dans les pays européens. In: Air Pollution: Proc. of the first European Congress on the Influence of Air Pollution on Plants and Animals, April 1968. Centre for Agricultural Pub-

lishing and Documentation, Wageninden pp. 15–26. 1969

[12] Barkman, J. J.: The influence of air pollution on bryophytes and lichens. In: Air Pollution: Proc. of the first European Congress on the Influence of Air Pollution on Plants and Animals. April 1968. Centre for Agricultural Publishing and Documentation, Wageningen pp. 197–209, 1969

[13] Chambers, L. A.: Classification and extent of air pollution problems. In: Air Pollution I, ed. by A. Stern. Academic Press 1968, p. 16

[14] Druett, H., K. May: The open air factor. New Scientist 579–581, 13 March 1969

[15] Environmental Protection Agency (Air Pollution Control Office, Washington D.C.): Air Quality Criteria for nitrogenoxides. Jan. 1971

[16] Romanovsky, J. C., R. M. Ingels, J. R. Gordon: Estimation of smog effects in the hydrocarbon-nitric oxide system. J.A.P.C.A. 17: 454–459, 1967

[17] Kehoe, R. A.: Toxicological appraisal of lead in relation to the tolerable concentration in the ambient air. J.A.P.C.A. 19: 690–700, 1969

[18] Arrhenius, S. A.: The influence of the carbonic acid in the air upon the temperature of the ground. Phil. Mag. 41: (V), 237–276, 1896

[19] Chamberlin, T. C.: An attempt to frame a working hypothesis of the cause of glacial periods on an atmospheric basis. J. of Geology 7: 545/584, 667/685, 751/787, 1899

[20] Revelle, R.: Atmospheric carbon dioxide. In: Restoring the Quality of our Environment, Report of the Environmental Pollution Panels. President's Science Advisory Committee. The White House, Nov. 1965

[21] Manabe, S., R. T. Wetherald: J. Atmos. Sci. 24: 241, 1967

[22] McCormick, R. A., J. A. Ludwig: Science 156: 1358, 1967

Drinking Water and Waste Water Problems

L. Mendia

Department of Sanitary Engineering, University of Naples, Italy

Summary. Drinking water and waste water problems are discussed regarding the hazard to human health by biodegradable and non-biodegradable wastes, as well as the ecological interferences. Parameters especially affecting water pollution, are social-economic characteristics of the communities, their hygienic conditions, waste water treatment, and physical, chemical, and biological properties of the various waste waters. As far as the consequences of pollution are concerned, the protection of human health is discussed together with technological and economic possibilities.

Zusammenfassung. Trink- und Abwasserprobleme werden bezüglich der Gefährdung der Gesundheit der Bevölkerung durch biologisch abbaubare und nicht abbaubare Abfälle diskutiert, aber auch die ökologischen Wechselwirkungen behandelt. Parameter, welche die Wasserverunreinigung besonders beeinflussen, sind sozial-ökonomische Eigenschaften der Gemeinden, ihre hygienischen Bedingungen, die Abwasserbehandlung, physikalische, chemische und biologische Eigenschaften der verschiedenen Abwässer. Was die Konsequenzen der Verunreinigungen angeht, wird besonders auf die Sicherung der Gesundheit und technologischen sowie wirtschaftlichen Möglichkeiten hingewiesen.

The existing correlations between waste disposal systems, drinking water and human health have been the prevailing subjects of *traditional hygiene* for almost one hundred years. Today, more than ever, they constitute one of the main problems of the environmental crisis, which, unfortunately, is of universal character.

While geographical conditions add morphological differences to the problem they do not change the substance of it.

The *drinking water crisis* involves, in the same dramatic way, the developing countries as well as the industrialized ones.

In the last few decades, due to the demographic explosion, industrial development, the continous and indiscriminate introduction of new products in the domestic and agricultural economy, and the transformations introduced in the living standard of all types of human communities, the problem of *drinking water pollution* has reached new and larger dimensions.

As a matter of fact, pollution by *sewage* has gradually increased as a consequence of growing amounts of waste from industrial and agricultural activities.

Today, it is necessary to develop a defensive action not only against organic (biodegradables) and biological pollutants which still remain the main source of danger to *public health*, but also against the non-biodegradable wastes and inorganic compounds, which can exert a direct toxic effect and even a very alarming, accumulative action. The most dramatic aspect for the future of water resources is constituted by the progressive increase in this kind of pollutants as connected exclusively with the industrial development.

For many years this argument has been taken into consideration by international organizations, scientific associations and scientists and, therefore, it is doubtful whether we shall be able to make a contribution to its study with this brief report. However, following the experience and knowledge acquired, we shall try to outline some of the arguments to be considered for the conservation of the environment, and, in particular, water deficiency: the ecological challenge which seems to characterize our epoch.

The fundamental terms to be considered are, on the one hand, the water supply and, on the other, waste disposal. The problem has to be solved, on the whole, as a unitary system, in view of the existing connections between the two single sub-systems: *water supply* and *sewage disposal*.

Evidently each of the problems (sub-systems) has to be solved first in the "optimal" way, according to the specific parameters which characterize it. Subsequently, each of the solutions reached should be verified, considering the influences on the quality of the water assigned to drinking purposes.

This assertion, which may seem obvious, means that when environmental health situations have to be faced, it is necessary to consider, as a complete whole, all the components of that specific problem (technical, geographical, economical, social, legal, administrative, etc.), not only in the light of the present, but particularly with regard to future developments. The water supply concerns the management and administration of water resources. These, in turn, condition the development of urbanization, agriculture and industries. Wastes, which can seriously degrade the quality of water supply, result from various activities. Consequently, it is easy to assume that the conservation of drinking water has priority over the numerous elements to be considered for regional development. Therefore, it is indispensable that other activities be planned and developed on the basis of this philosophy. The sources, vectors and, therefore, the consequences of *water pollution* depend on certain groups of *factors*.

In the first group, we should include those inherent to the sources of the pollution. For example: Ia) socio-economic characteristics of the communities (creating the wastes): urban, agricultural, artisan, industrial, mixed; Ib) hygienic conditions and health situation of the communities (endemic disease zones; epidemic disease episodes); Ic) waste water disposal systems (static, dynamic, techniques for the treatment of domestic and urban sewage, techniques for the treatment and pretreatment of industrial waste waters; Id) physical, chemical, and biological characteristics of the different types of waste waters originating from the area (for the industrial area it would also be worthwhile to have knowledge of the raw materials and technological processes adopted, at least for the major activities); Ie) hydrological and hydraulic characteristics of diluting bodies, dilution and self-purification capacities.

Among the factors which function as the *vehicle of pollutants* from the sources, we should consider: IIa) the soil, i.e. the geological and hydrogeological characteristics of the region involved (in cases where the waste disposal is effected by dispersion or spraying); IIb) hydraulic sources used for drinking water supply: underground (conditions of protection of the ground water), surface waters (rivers, lakes, artificial reservoirs).

As far as the consequences of pollution are concerned, in this specific case we should consider those factors which characterize the quality of drinking water sources (water quality criteria).

Such factors could be summarized as follows: IIIa) *public health safety;* IIIb) *palatability;* IIIc) *technological* and *economical convenience.*

The first aspect, which obviously is of predominant importance, has to be considered from the biological point of view (pathogenic agents, parasites, vectors) and from the toxicological point of view. This latter action has to be examined by evaluating the direct dangers (poisoning) as well as the accumulative capacities (particularly underhand) of some pollutants (heavy metals, biocides, hydrocarbons, etc.).

The subject, due to the numerous implications of chemical, physiological, biological, ecological, technological, economical, and analytical nature, is under the control of the health authorities all over the world and codified by international standards through the action of the World Health Organization.

As far as palatability is concerned, the main characteristics to be considered are: temperature, turbidity, color, odor, taste; the optimum values can, due to the local situation, sometimes surpass the suggested limits, in view of the factor of flow availability.

It should be pointed out – even though obvious – that the problem of the *community water supply* is urgently to be solved in terms of quantity as well as of quality.

In this light, the argument of the "technological and economical convenience" finds its appropriate allocation. By such a definition we mean the evaluation of the technical possibilities and economic validity of submitting the raw water to possible treatment processes necessary to correct the original characteristics of them, so that they reach the quality standards needed for drinking water (not only in terms of safety but also considering the domestic economy, the safeguard of the supply and distribution system).

Among these well-known difficulties, we shall mention: the incrustations, corrosions, sediments, biological growths; the excessive consumption of soap; the difficulties of cooking food, the adulteration of flavors and tastes of beverages and foods; the inconveniences of washing clothes, etc. The treatment processes for correcting the characteristics which can provoke such inconveniences (hardness, instability of bicarbonic equilibrium, suspended solids, degree of mineralization, iron, manganese, plankton, etc.), influence, of course, the cost of water and therefore the *economy* of the whole *water supply* system. It must be noted that the above-mentioned inconveniences should be correlated mainly to the natural quality of some waters, more than to the pollution, unless to particular local situations connected with industrial waste disposal. But, in view of the principle that the problem of water resources has to be analyzed and solved as a whole, we assume it worthwile to also consider the above-mentioned aspects (IIIc).

At this point, we should like to give an example of a general character. In the case of a *community water supply* program, the priority choice is toward the use of *underground water*. Nevertheless, if the supply of such waters should impose considerable expenses for treatment (as well as for their adduction) the result is a heavy economic burden and, therefore, should be rejected as an initial step. Consequently, as a second step, we should be induced to consider the possibility of using *surface waters* available in the region. Evaluating this opportunity, it arises that the use of surface waters imposes not only some types of treatment (i.e. clarification, disinfection) but also a number of limits and criteria on the management of the river basin, in view of pollution control (exclusion of certain activities, screening of the type and size of further human and industrial installations, high water quality criteria, etc.). Therefore, before defining the water source to be developed, it is necessary to establish an *economical balance* which considers not only the effective costs of the necessary treatment to be adopted for the different situations (here the costs for the adduction of the water itself are included) but also the "sacrifices" that each solution imposes on the community in order to protect the quality of the waters against the urban and/or industrial development consequences. Obviously, the weight of the "sacrifice" will be varied following the predominant economy of the region. A typical example is represented by the difficulties in the management of artificial reservoirs. In this case, due to the very slow hydraulic regimen and the relatively small reservoir capacities, strict criteria of protection are necessary not only for the reservoir but also for the river supplying it, for a long distance upstream.

From this short outline, we may see how numerous the variables characterizing the problem are (those we pointed out represent only a few examples) and how, therefore, the optimal solution of it should be reached in different ways, according to the particular situations. Among the main conditioning elements we can mention:

1) the geographical and economical situations (developing countries, industrialized regions, arid zones, tropical regions, climates, etc.),

2) the water supply sources (underground, surface, etc.),

3) the main sources of pollution (biological, chemical),

4) the nature of the various water supplies used in the region, the amount of flow requested for each case (drinking, agriculture, industrial services).

On the basis of the points outlined it becomes obvious how the approach to the problem can vary, passing from developing countries, of tropical or sub-tropical regions, to an industrialized continental region.

From the numerous investigations and studies carried out during the last few years, it emerges how diseases connected with water for direct (ingestion) or indirect (contact, vectors, shortage, etc.) reasons still represent one of the main *sources of mortality* and morbidity in many countries. However, it is encouraging to consider that these dangers and damages should be reduced to a minimum and, in some cases, should disappear completely. We have, in this sense, the positive experience of the sanitation action developed in the last century all over the world. This action has mainly faced such problems as: the improvement of the environmental situation (land sanitation, malaria and vector control), the diffusion of the use of hygienic facilities at home, the construction of sewage systems, the final wastes (liquid and solid) disposal, the health education of populations, and the construction of aqueducts and water treatment plants.

If we consider the *history of environmental hygiene* during the last few decades, we can observe that many diseases, once endemic, have disappeared from large regions and countries *(malaria, ty-phoid fever)* thanks to the development of *sanitary engineering.*

On the contrary, in some regions of Asia, for example, due to the primordial systems of sewage disposal and water supply, cholera is still endemic, representing a potential source of epidemic.

This vast exemplification means that damages to hygienic nature connected with the water supply can be eliminated by developing opportune programs of *environmental sanitation.* The difficulties are not technical ones, because of the present status of the art of sanitary engineering, but mainly of economical nature. The basic difficulties are concerned with the peculiar political and economical situation of the countries where these programs of sanitation are more urgently requested. We want to mention the developing countries, many of them characterized by a recent political evolution and/or an emerging economy. As an example, we should like to emphasize the technical, economical and cultural assistance that international agencies give for *projects of environmental sanitation* to countries in need of this kind of help. Particular mention must be given to the programs of the World Health Organization in the specific field of community water supply. This optimistic view of the traditional horizon of water supply has been jeopardized during the last two decades. As a matter of fact, new dangers to *public health* have arisen in connection with the *water quality.* These came from the pollution induced by the use of new products introduced in the agricultural and domestic economy (biocides, detergents) as well as by the development of industrial and artisan activities.

The industrial development is characterized on the one hand by an increase in the requested flows and on the other by an increase in *waste waters.* In the case

of a water body where there are two concurrent events, the quality level of water supplied by it could be seriously compromised. In fact, the increase in the flows of water withdrawal reduces the diluting and self-purification capacities of water body, at the same time when the waste water flows have been increased.

In addition to this aspect of quantitative origin, we should point out the importance of the problems connected with the nature of the chemical pollutants. As mentioned before, these pollutants, no less than the biological ones, present a serious problem for human health.

The main worries come from the trace level, at which they can still exert their damaging action (accumulative capacities of certain human organs), the technological difficulties of their removal, and the economic implications of the politics of control of the sources of these pollutants *(preventive action).*

The present alarm for the ecological crisis is mainly related to these types of new pollutants. The management of water resources, as well as the water classification and the water quality criteria, have to be re-examined or planned, taking into consideration the particular aspects of the pollution, i. e., toxicity (direct or induced), interference in biological processes of purification, and eutrophication phenomena.

Consequently, it is necessary, on the one hand, to find new (non-conventional) hydric resources in order to satisfy those requests which do not need water of a particular quality, saving high-quality sources for drinking purposes, and, on the other hand, to develop a stronger action for the *pollution control* of *natural water.*

The first program would find its solution through the use of brackish and marine waters, the re-use of sewage effluents for municipal industrial and agricultural purposes, the recycling of the industrial effluents.

As far as the planning of pollution control programs is concerned, care must be taken that the errors committed in many industrialized countries are not repeated in industrializing and developing countries. Industrial development does not necessarily mean an environmental degradation. On the contrary! the present ecological philosophy has the opposite effect.

Disposal and Utilization of Solid Waste

T. Teeuwen

The Foundation for solid Waste disposal, Utrechtsweg 223, Amersfoort, Netherlands

Summary. After a classification of the different categories of solid waste, a prognosis is given of the amounts of waste to be disposed of in the Netherlands during the coming thirty years.

The principal methods for treatment and transport of municipal refuse and normal industrial refuse are mentioned and, in particular, the environmental, planological, and economic aspects of the different methods are discussed.

A comparison of costs between alternative methods makes conclusions possible for future planning of treatment methods.

An outline is given of existing systems and systems to be developed in the future on the principle of re-use and recycling.

Zusammenfassung. Nach einer Klassifikation verschiedener Kategorien von festen Abfallstoffen wird eine Prognose aufgetragen von den zu beseitigenden Abfallmengen in den Niederlanden während der nächsten dreißig Jahre. Die wichtigsten Methoden für die Verarbeitung und den Transport des städtischen und industriellen Mülls werden erwähnt, und besonders die milieuhygienischen, planologischen und ökonomischen Aspekte der verschiedenen Methoden werden vorgetragen.

Aus einem Kostenvergleich zwischen den alternativen Methoden lassen sich Schlußfolgerungen ziehen für die zukünftige Planung von Beseitigungsmethoden.

Eine Übersicht wird gegeben sowohl von existierenden als auch von zukünftig zu entwickelnden Methoden auf dem Prinzip von Wiedergebrauch und Rezirkulation.

Introduction

The problems of *solid waste* elimination do not (yet) manifest themselves everywhere in the same way. They are mostly felt in those areas where technical development has progressed furthest and where, as a consequence, the consumptive spending is highest.

The data mentioned in this outline, refer to circumstances as they are in the Netherlands.

As already shown in many figures and graphs, the technical development and its use of primary energy and raw materials has a literally explosive character. It is a fact that each raw material and each finished product reaches the waste stage after a certain period of use. Characteristic for these is, furthermore, that the lifespan of goods and products is getting shorter, which amplifies the cumulative character of the waste-problems.

Besides a growth in quantity of wastes there is a change in quality. The removal of wastes has become so great a problem, because elimination of waste matter is not possible. Even with the most stringent elimination treatment, one can only speak of conversion of one matter into another, and it often happens that certain unfavorable or dangerous characteristics then remain in the residual products.

Categories and Quantities per Category

Experience teaches us that correct and efficient treatment of waste matters requires separating them into different categories.

The largest category — from the viewpoint of volume — is that of *municipal refuse*.

This includes domestic refuse from homes, shops, offices, public utility services and institutions, and smaller factories of local importance; also market and street refuse, gardenwaste etc. All these types of waste are normally collected by the public cleansing services.

In the Netherlands, the quantity of municipal refuse in 1970 amounted to approx. 270 kg per inhabitant per year.

Besides the municipal refuse there is an important category consisting of *industrial wastes* that can be treated in the normal way.

This category can mainly be compared with municipal refuse, it consists principally of lightweight products: packing materials, paper, plastics etc. In 1970 this quantity equaled approx. 25% by weight of the municipal refuse.

Special types of industrial wastes include *wastes from agrarian industries* and from agricultural and horticultural industries, markets, etc. In the Netherlands half of these are horticultural wastes and half market refuse.

Besides these, there are the *wastes from the bio-industry, stockbreeding and slaughterhouses* (e. g. excessive dung). Because of the fact that these wastes generally contain a very high degree of water it is difficult to find a market for it in the agricultural area.

A similar group is formed by *sludge from water-treatment plants* for *domestic waste water*, and corresponding *industrial waste water*.

In Holland approx. 35% of the communities have one or more treatment plants for domestic waste water at their disposal. In a number of cases it is possible for industries to be connected with the existing sewage system of the town, in order to have the industrial waste water purified together with the domestic waste water.

One category, which deserves full attention from the viewpoint of environmental protection is *waste matters* that are dificult to eliminate form the chemical and petro-chemical industries, laboratories, research institutes, as well as from plants using chemicals, *oil products, lacquers, acid-baths*, etc. in their processing or manufacture.

The waste-matters in this category are in many cases liquid or semi-liquid. This group includes among others, the volatile, explosive, poisonous waste matters, which are generally aggressive and very dangerous. In 1970, the quantity of the combustible part was stated as being approx. 50,000 tons per year. The quantity of these wastes in relation to the other wastes, is very small (about 1%).

The waste matters mentioned up to now all have in common the fact that they are mainly organic in nature and in view of the environment more or less polluting. Besides these, there are non-polluting wastes.

To this group belong the heavy wastes, such as building-and demolition waste, rubble and other wastes that are easy to dump, and mainly inorganic in nature, originating from urban and industrial areas. To these can be added the wastes from the mining industry and certain wastes from the steel industry, such as foundry-sand.

Quite a different kind of category is created by *discarded cars* and other large metal objects.

Finally, there are still some special categories, such as: destruction material, dead animals, pathogenic wastes, nuclear waste, for which — because of their dangerous condition — quite different disposal methods are necessary. In most countries elimination of these wastes is legally controlled.

Some information will now be given about the principal treatment methods of municipal refuse and industrial wastes that can be compared with municipal refuse, thereby giving some attention to environmental quality and planological and economic aspects.

The most popular way of treatment is still dumping. In fact, by dumping of wastes one means the final storage of waste-matter in the countryside. It is clear that, if the wastes are indiscriminately taken to open plots, this would be the simplest and cheapest way of waste elimination, but from a viewpoint of environment, certainly the most objectionable.

Another method of dumping is the *controlled tipping method* which can be understood as the method, by which environmental as well as aesthetic and social objections to dumping are conclusively fought and prevented. One condition for applying this method is that there must be a suitable site, which can be considered a valuable rural area in the future from the urban planning standpoint.

In order to prevent pollution of surface and groundwater it may be necessary in certain cases to isolate the dumping site entirely from these waters by installing a drainage system, or a ditch. It might be necessary to install a water treatment plant for treatment of the water before draining it off into surface water.

Before starting a dumping project it will be necessary to know exactly — through studies and research on the spot (geological and geo-hydrological research) — which of the measures mentioned are applicable.

In order to obtain acceptable working conditions on the site, raw refuse should be compressed by a bulldozer and covered daily with a layer of covering material. A good working method requires the waste matter supplied in small quantities at a time, in such a way that the bulldozer can handle and compact it immediately after unloading on the spot.

The site where the fresh wastes are dumped within the tipping area should be kept as small as possible. To give the site a high degree of regularity (which will be of great importance for the future use of it), the compressible and the non-compressible wastes should be carted and dumped separately. Also de-watered sludges can be treated under certain cirumstances. The working-method is greatly dependent on the groundwater level, the kind of soil and the climate.

Composting of Domestic Waste and Incineration or Dumping of the Non-Compostable Wastes

The conversion of municipal refuse into compost has been practiced on a large scale in the Netherlands for some time.

It is a treatment in which compostable organic matter is fermented in a microbiological way.

There are well-known systems which pulverize the refuse mechanically before fermentation and then it is stripped of non-compostable substances such as metals, glass, hard and soft plastics, pieces of wood, etc., by magnets, breakers, ballistic separators, and sieves.

After the preliminary treatment the remaining waste matter is fermented in heaps for 4 to 5 months with addition of water; aeration aids the fermentation process.

Another method, developed in Holland and still used on a large scale, is the "van Maanen" system in which waste matter is subjected to a fermentation process in its original size. Here too, the conditions are made as favorable as possible (aerating and watering) in order to obtain good fermentation. After about half a year the fermented waste (compost) is separated by sieving from the remaining non-compostable waste matter. The "van Maanen" system can only be applied in more or less isolated areas — the fresh refuse attracts vermin and birds, because of the permanent presence of great quantities of fresh waste matter.

This is contrasted with the composting plants where the waste matter is first pulverized: after pulverizing, it does not attract any flies and rats and also reduces the degree of disfigurement in the country-side.

Through changes in the structure of domestic refuse, composting has become more difficult in the last 15 years. On the one hand more and more attention had to be paid to the finishing treatment of the compost in order to obtain a good product.

On the other hand, more and more demands were made on the quality as in Holland the market for compost underwent change: from the field of agriculture, via the field of horticulture into the direction of recreation and gardens (public parks).

Besides the problem of good quality, the non-compostable part of domestic waste continually increased, namely from approx. 10 to 15% in the early fifties, to approx. 55% in 1970.

In addition to the domestic waste, it also became necessary to remove a growing amount of bulky waste and industrial waste.

Given all these factors, composting methods only offer a solution for diminishing a part of waste matter.

As a consequence, the number of municipal composting-plants in Holland have been greatly reduced in the last few years.

Of the original 16 plants, only 6 are left now. The difficulties experienced by these municipal plants exist in a lesser degree with the plants operating on the "van Maanen"-composting system, because here the waste matter is not subjected to a preliminary pulverizing process, which makes better quality control possible.

In the existing van Maanen plants, there are larger dumping areas, so that the composting method here can be combined with dumping of the residues and the non-compostable waste matter.

The Most Radical Treatment: Incineration

This method offers the possibility of effectively eliminating a great part of the waste matter in a quick and hygienic way.

The combustible part of the waste is practically entirely converted to carbon dioxide and water. In most furnaces, domestic as well as bulky refuse can be incinerated, even though the bulky pieces must be reduced in size in order to be admitted to the furnace via the normal fill shaft (hopper) for domestic waste.

The same applies to combustible industrial waste. Combined incineration of municipal and industrial wastes offers advantages from a technical viewpoint, namely:

— a greater quantity of waste matter enables changeover to bigger incinerating-units, hence lower specific costs;

— domestic waste has a homogenizing effect on the other waste matters, which yields favorable result with filling, predrying and incineration; the lower caloric value of domestic waste (1500 to 2000 kcal/kg) also acts favorably on incineration, in a sense that the melting point of the slag, is reached less quickly than in exclusive burning of industrial waste, which, as a rule, has much higher caloric value.

Besides *flue gases* arising in the process of incineration, there are also residues, such as slags and *fly ash,* which need to be eliminated into water, or into soil.

The quantity of residue in the present structure of waste matter in Holland is approx. 8 % of the original volume.

From a certain treating capacity it can be advantageous to convert the slag into a product that can be used as filling material in road construction, or in public parks. In this conversion, iron is removed from the slag; the iron can be sold. In this way the quantities of incineration residue can be reduced to only a fraction of the original weight. Incineration does not offer a universal solution. For the waste matter mentioned above — such as building and demolition waste, rubble, non-combustible industrial waste, etc. — this treatment is not possible. These waste matters have to be treated in a different way.

Dependending on local conditions, requirements must be established for flue gases.

A high degree of gas purification in waste incineration can only be achieved with an electrostatic dust collector or by means of a wet gas precipitator.

Limiting the air pollution is, to a large extent, a question of good furnace design and allowing enough time for adequate final combustion of the gases.

In order to purify flue gases from an incinerating plant, it is necessary to cool them down from approx. 900° C to at least 350° C. Consideration should be given to whether there are advantages in attempting heat recovery. It may be possible that the heat can be transfered via a steam or hot-water boiler to an installation for direct heating, or that the heat can be used for industrial purposes.

Power generation is a possibility for utilizing heat, which deserves special consideration in larger plants. Examples in Holland are the incineration plants in Rotterdam, the Hague, and Amsterdam.

In contrast to some years ago — in cases where no power is generated — it is no longer necessary to make use of a steam boiler; a much simpler and cheaper form of flue-gas cooling, namely spraying water into the flue gases, will suffice.

For this, the water mist is sprayed into the flue gases by means of a number of atomizers.

The gases are cooled by evaporation of the water; furthermore it appears that the *electrostatic dust collector* is more effective due to the presence of a great deal of vapor in the flue gases.

A special method of *flue gas cooling*, in which the heat of the flue gas is used, is use of the heat for drying sludge from water purification plants.

This combination of waste treatment is very interesting from the viewpoint of heat economy. It is, however, still a problem whether this method of direct contact of flue gas and sludge can be realized without the flue gas developing unpleasant odors.

The installation of a heat exchanger between gases and sludge would mean an extra investment with the chance that the expected advantages will not materialize.

Certain burnable waste matters cannot be burned together with the municipal refuse, because they are not fit to be burned in the normal incinerator furnace. This applies to most combustible chemical wastes. *Incineration* of these materials must be done in installations especially designed, observing the necessary precautions and under special conditions. The *liquid waste* matters are sprayed into furnaces, designed for the purpose, with the aid of pressurized atomizers.

The solid chemical wastes and the liquid and semi-liquid or low melting wastes cannot be incinerated in a furnace with moving grates. For these wastes, rotating drum type furnaces are used.

Extra measures for cleaning the flue gases are necessary when certain poisonous compounds appear in the flue gases in too high concentrations. This generally requires very expensive preparations.

It is a well-known fact that certain materials, appearing in municipal refuse, such as some plastics, also reproduce harmful gases when incinerated (hydrochloric acid, hydrofluoric acid). The concentration in flue gases from municipal refuse is, however, so slight, that only a high increase in concentration would necessitate extra measures for the purification of these gases; first by installing a higher chimney and second by implementing a wet gas purification system.

It is probable that such steps may be necessary in the future.

Other Methods of Waste Elimination

In combination with the treatment methods mentioned above, there are a number of additional techniques for the purpose of transforming waste matter into a condition in which it can be treated in an easier, or a more acceptable way.

Examples are: compressing, pulverizing, or breaking of wastes, prior to controlled tipping, composting, or incineration. These methods will not be further discussed here.

Planological Aspects in Waste Elimination

The *planological aspects* are an important factor in the problem of *dumping*, because with this form of waste elimination a demand is made on the availability of land area in the country side.

The use of space in the various treatment methods is shown schematically in Fig. 1. In this figure, six columns are shown of which the highest — in the middle – shows the demand on space with the controlled tipping method.

The letters in this column represent the following categories already mentioned above:

A) Municipal refuse, density after compressing by bulldozer: 600 kg/m³.
B) Normal industrial waste, mainly corresponding to municipal refuse.
C) Agrarian waste, converted into 50% dry solids.

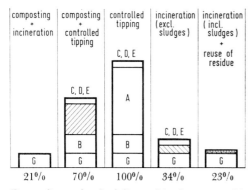

Fig. 1. Spaces for final disposal in the countryside with treatment methods

D) Waste from the bio-industry, converted into 50% dry solids.
E) Sludge from water treatment plants, converted into 50% dry solids.
F) Building and demolition wastes, etc.

(Categories G and H, mentioned above, are not given). The use of space in controlled tipping is taken as 100%.

With the composting method, the demand on space for the non-compostable waste and the residues of composting amounts to 67% of the original.

If the composting method is combined with the incinerating-method, the demand on space is reduced to 15%.

The demand on space for incineration, without further treatment of the residue and without combined incineration of sludge amounts to 29%.

When sludge is also incinerated and the residue used again, the demand on space is approx. 17%.

It must be obvious that in the application of each method a certain demand is made on rural space. One great difference is whether the wastes are more or less contaminating from the viewpoint of environmental quality.

Generally, it will be easier to find space for inorganic wastes — demolition waste, rubble — than for organic wastes. The former can possibly be dumped into water; in this case measures will have to be taken against water and soil pollution.

In densley populated urban areas in Holland there are no more possibilities for dumping.

The incineration method offers a solution for combustible wastes.

For non-combustible wastes it is already necessary to find a dumping site at a relatively great distance from its origin.

Economic Aspects

If a choice has to be made among alternative methods of waste elimination,

which is especially the case with municipal refuse and combustible industrial wastes, cost aspects can settle the matter, after planological and environmental aspects have been taken into consideration.

As a rule, controlled tipping, even if this takes place under stringent measures for prevention of environmental contamination, is considerably cheaper not in the neighborhood, but at a great distance from the place where it is produced; extra transportation is necessary to utilize cheaper tipping methods.

Especially in areas with a greater quantity of waste, the cost of local incineration will not differ much from the costs of a satisfactory tipping method at a great distance.

Costs of Treatment

In Fig. 2 the costs per ton of waste matter are shown for a number of alternative treatment methods for municipal refuse and combustible industrial wastes, in relation to the quantity annually treated.

The following methods are distinguished:

A) controlled tipping,
B) controlled tipping at a large plant in an isolated area equipped for supply by rail, eventually combined with composting by the "van Maanen" system,

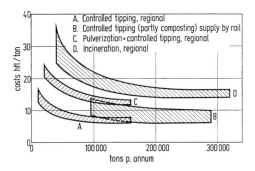

Fig. 2. Treatment costs for town refuse and combustible industrial wastes

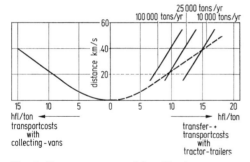

Fig. 3. Transport costs with collecting vans and tractor trailers

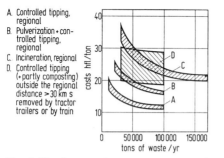

A. Controlled tipping, regional
B. Pulverization + controlled tipping, regional
C. Incineration, regional
D. Controlled tipping (+ partly composting) outside the regional distance > 30 km s removed by tractor trailers or by train

Fig. 4. Total costs for transport, transfer and treatment

C) mechanical pulverization of wastes, prior to controlled tipping, and
D) incineration.

Costs of Transport

In Fig. 3, the *costs* per ton *of transported waste* as a function of the distance are shown for two different types of transport. Vertically the distance is shown in kilometers.

The left side of the graph shows the costs of transport in collecting vans; the right side shows the costs for tractor trailer combinations, including costs for transfering wastes from the collecting van to the trailer.

The latter costs depend on the quantity of waste to be transported annually. If the quantity is small, the costs of transfer are high.

The choice between collecting van transport and tractor trailer transport depends on the distance, as well as on the annual quantity to be transported (transferred). For example, a comparison shows that with an annual capacity of 25,000 tons, transported by tractor-trailer is to be considered only for distances of more than 20 km.

Total Costs for Treatment and Transport

In Fig. 4, the total costs per ton for treatment and extra transport (from collecting area to treatment site, excluding collecting costs) are shown as a function of annual capacity for a number of alternative elimination methods. Here too, the costs refer to the elimination of municipal waste and combustible industrial waste.

The following methods are given:

A) controlled tipping within the region; transport in collecting vans,
B) pulverization and controlled tipping within the region; transport in collecting vans,
C) incineration within the region; transport in collecting vans, and
D) removing the wastes to a large controlled tipping plant, outside the region (possibly with partial composting);
transport by tractor-trailer or rail.

From this comparison of costs the following conclusions can be drawn:

1) The utilization of treatment methods that are justified from the viewpoint of environmental health, require — for reasons of costs — regional co-operation.

2) For the time that within the region — or at a short distance from it — controlled tipping can be done in an appropriate manner, this method yields the lowest costs.

3) The pulverizing method combined with controlled tipping is always more expensive than controlled tipping of non-pulverized material and has, from the viewpoint of environmental health, rather small advantages over the latter.
4) Removal of waste by means of long distance transport to plants where controlled tipping is possible (possibly combined with composting) is attractive for economic reasons in smaller regions, which have no dumping space of their own. For larger regions the incinerating method should be considered.

The limit, depending on the transport distance and the manner of transport, is 50,000 to 80,000 tons annually which corresponds to 125,000 to 200,000 inhabitants.

Use of Wastes

Re-Use via the Waste Stage

Incinerating, composting and controlled tipping are methods, in which some types of reuse are already put into practice. The waste matters or residues are generally used in a different way or for a different purpose than their original use.

With incineration the following examples can be mentioned:
1) heat recovery for industrial purposes, regional heating, generating electricity;
2) use of ashes and slags after iron removal, breaking and sieving as, for instance, hardening and filling material in the construction and upkeep of roads and paths.

With *composting*, part of the residue is utilized as compost in the fields of agriculture, horticulture, recreation, and public parks.

Also, with the controlled tipping method, the residue can partly be used again (in a cultural and technical sense) for landscaping and land reconstruction. A special type of reuse takes place through natural recirculation, in which the entire biosphere is involved: the *carbon cycle*.

A great part of the wastes come from natural products originated from this cycle (products from nature: wood, wool, natural rubber, leather).

With the incineration and composting of wastes, the organic matters involved go through part of this cycle: with incineration through very quick direct conversion of organic matter into carbonic acid and water, with composting via slower biological destruction processes in the soil, in which case carbonic acid and gaseous hydrocarbons are also created. Whether there is a question of recirculation in a practical sense, depends on the natural (period) time of formation of the original matter.

The cycle for living organic matter — wood, animal products, wool, natural rubber etc. — can, from the human point of view, be considered as closed.

This case is contrasted with fossilized organic mattes, e. g. oil, coal, gas, and the synthetic materials made from it. In this case, the cycle has a very long span (some millions of years) and the original materials must be considered as not being self-renewing. Many inorganic substances are products of even slower single growing-processes. Here it is especially necessary that an effort be made to find a sensible means of reuse. With processing treatments, such as incineration, composting and pulverizing (the latter being preliminary treatment for, among other things, controlled tipping) magnetic metals can be easily recovered.

Especially with *incineration*, relatively clean scrap iron can be obtained. The recovery of non-magnetic metals — copper, tin, zinc, aluminium — can for the present composition of municipal refuse, only economically be done by hand sorting.

Although *hand sorting* is frequently done in composting plants, it is generally considered a labor unworthy of man, which in future methods, certainly must be avoided.

Experience with composting plants clearly shows that mechanical removal of certain components from municipal refuse, with some exceptions, cannot be realized economically.

In other words: for efficient reuse of certain wastes it is desirable (if not necessary) to keep the wastes separate at the source in order to make separate collecting and further treatment possible.

Examples are:
— steel manufacture from used cans and other ferrous metals,
— manufacture of cardboard from old paper and cardboard, and
— manufacture of new glass from old glass.

These examples refer to waste reuse in essentially the same direction as the original. Here a regeneration process is always necessary to prepare the waste for reuse.

Furthermore, there are possibilities where wastes are reused, after treatment, in a different capacity from the original.

Examples are:
1) The use of *grinded glass in bitumen* for road construction to improve the friction of the surface.
2) The manufacture of alcohol from cellulose-containing materials.
3) Fermenting of completely compostable waste (e. g. agrarian waste).
4) The manufacture of gaseous and liquid hydrocarbons from *automobile tires*, via a dry distillation process.

Besides certain components from municipal waste, as well as with several industrial wastes, reuse is possible in the same or in a different direction of use, for instance with wood, metal, fiber waste, etc. Especially with chemical wastes, there are many possibilities for reuse for the same purpose as the original one or for a different one, for example:
1) Recovery of chlorine from hydrochloric acid formed during chlorination processes in the plastic industry (PVC).
2) Recovery of acetone and phenol from tar formed during the diphenylalpropane formation process (used as a basic material for resin fabrication).
3) Recovery of scarce metals such as copper, nickel, silver, cobalt, chromium, used in the electroplanting industry.
4) Reuse of acids and bases used for washing processes in the same or other processes.

With all these types of waste, all kinds of technical, physical, chemical, or biochemical processes are necessary to make the original waste fit the new purpose.

In these cases it is particularly necessary to keep the waste matters for reuse separate at their source.

Re-Use of Material Without Reaching the Waste Stage

Finally, there are a great many cases where reuse of goods and products takes place more than once — before the definite waste stage is reached.

Examples are, all kinds of *packing material*, for which a deposit must be paid: glass packing, wooden boxes, tanks, and barrels. These goods are continuously recycled without being treated or regenerated. At the most cleaning must be performed between uses.

Many possibilities for reuse arise in which goods or products find a second or third use via the second-hand trade. In most cases, each successive use means a degradation vis à vis the previous one, until the product is degraded to waste.

Reasons for Re-Use of Wastes

There are two main reasons for reusing wastes and for stimulating natural recirculation processes in waste treatment, namely:

1) Re-Use for Environmental Control

A growing number of materials, e.g. many inorganic materials and organic/inorganic compounds, are entering the environment which, above certain concentrations, are undesirable because they are injurious to human, animals, or plant health.

Examples are all sorts of metals such as chromium, mercury, zinc, lead, tin, cadmium, and others; furthermore, all kinds of organic matter and synthetic products such as chlorinated hydrocarbons, elements from paints, and so on.

The way in which these substances enter the environment can be varied. Sometimes they enter the environment in concentrated form due to accidents (collisions) spilling, etc., and often in low concentration as components in waste matters. For instance, dangerous or injurious substances appear more often in small quantities in municipal refuse, as part of paints and dyes, coatings, medicines, galvanized products, empty packing materials, electrical and electronic apparatuses, etc.

2) Re-Use for Conservation of Raw Materials

Fossil organic compounds and a great number of inorganic substances (among others metals) are formed in the earth's crust through a growing process that can be considered unique. It goes without saying, that these materials must be carefully controlled. An attempt must be made to limit these materials to a minimum in wastes.

Reuse of many products must be encouraged to the maximum, both for environmental protection and economy of raw materials.

In most cases regeneration is only possible when these substances are kept separate from the other waste materials.

In most cases, regeneration after mixing with other wastes is impossible or excessively expensive (requiring too much energy) and therefore meaningless. For liquid wastes it is of great importance that they be kept as concentrated as possible.

Initiating and establishing these regeneration processes will require a great deal of work in terms of organization, standardization, legislation, and regulation — and, of course, in the education of human beings.

For most processes the techniques are well known. It is mostly a question of giving priority to investment in these techniques.

In *economic calculations* in this field both the social costs and the savings due to the fact that less raw materials are used must be considered.

Organic Chemicals (Except Agricultural, Animal Health Chemicals and Food Additives)

N. A. Iliff

Shell Centre, London S. E. 1, England

Introduction

One of the striking features of the chemical industry is the tremendous growth, over the past twenty years, in the *production of organic chemicals.* In 1950 world production (outside the Communist countries) totalled only 7 million tons. By 1970 it had grown to 63 million tons and it is estimated that in 1985 it will total 250 million tons. Even if thereafter the growth rate declines, it is clear that by the year 2000 production will be very large indeed.

Some figures for individual countries are as follows:

	Millions of tons of organic chemicals	
	1970	1985 (estimated)
U.S.A.	27	90
Japan	9	45
West Germany	6.5	28
U.K.	3.5	12

Along with the growth in volume there has been a considerable extension in the range of organic chemicals. Many are recent discoveries; some that are now commonplace commercial products were, no long ago, little more than scientific curiosities.

These developments arose largely from changes during the last few decades in the basic raw materials of the industry, and in the processes used. Formerly, in the aliphatic field, the base materials were natural products such as carbohydrates and fatty oils, and coal was the main basis of aromatic compounds. Fermentation processes were commonly a key step towards aliphatic compounds, while in the aromatics field, coal tar was the base material most readily available.

Increasingly, petroleum has displaced other *basic raw materials;* as regards volume of production, synthetic methods have overtaken fermentation. Petroleum is an abundant and readily available raw material, and from it can be obtained economically (and the more so the larger the scale) the building blocks for chemical synthesis, e. g. olefins, aromatic hydrocarbons. Thus there has been a considerable research incentive to devise synthetic methods for producing existing chemicals, and to produce new compounds and materials derived from them.

As a result, the organic chemicals industry is highly versatile and has a degree of flexibility that was previously unknown, and synthetic processes are possible in plants with very high capacities. In 1930 almost no petroleum was used as a chemical feedstock, but by 1950 it was the basis of 40% (by weight) of all organic chemical production. By 1970 this had risen to 88%, and it is estimated that by 1985 as much as 98% of all organic chemical production will be based on petroleum.

Outlets for Products
of the Chemical Industry

The increasing *volume of organic chemicals* being produced carries, in itself, important implications for environmental conservation. However, even more important is the increasing *range of compounds available*. More and more of these are substances which do not occur in nature and their fate in the environment as well as their possible effects upon it require particular attention.

The majority of organic chemicals (perhaps 75%) are further processed, either in their place of manufacture or elsewhere. Over two thirds of the latter are used to synthesize *end-products,* such as plastics and resins, synthetic fibers, synthetic rubbers and surface coatings, which are no longer "chemicals" in the accepted sense. They are relatively inert, but because their final disposal afterwards may present environmental problems, organic chemical producers are increasingly concerned with the *ultimate fate* of their *products*. They feel a sense of responsibility, but nevertheless can only give advice, being unable themselves to exert direct influence or control.

Of the remaining 25%, consisting of chemicals used as such, the greater part is further processed within industry itself to produce, either alone or in admixture with other materials, such products as *solvents* (aliphatic, aromatic, chlorinated, e. g. aerosol propellants), glycols (e. g. anti-freeze, brake fluids, cosmetics), and detergents.

Arising from these types of end-products, which contain "chemicals" as such, the "best guess" seems to be that at the present time a total amount of *up to* 20 million tons of manufactured organic chemicals *may* be entering the environment annually, out of a world total of about 60 million tons of organic chemicals pro-duced. Of this "up to" 20 million tons per annum, probably about half is accounted for by solvents, such as methanol, ethanol, higher alcohols and their derivatives, glycol ethers, chlorinated solvents and chlorofluorocarbons, and aromatic solvents. *Detergents* (active material) amount to about 1.5 million tons, and *organic pesticides* about 1 million tons. Of the *gaseous base chemicals* (e. g. lower olefins, butadiene, acetylene, and synthesis gas) used as manufacturing "building blocks", it is estimated that up to 1 million tons per annum enters the atmosphere (2—3% of the total used by industry). The remainder (roughly 7 million tons per annum) is made up of miscellaneous organic chemicals of many kinds, including many compounds, the end use of which has not been properly quantified — a task that would involve an immense amount of detailed analysis.

Man-Made Chemicals in the Environment

Chemicals can affect the environment in various ways depending on their nature, the way they are used, and the way in which they are disposed of after use.

At manufacturing plants the danger of pollution is always present, simply because the chemicals are there in quantity. Leaks of gases and volatile chemicals may give rise to atmospheric pollution which can create a nuisance because of odors, irritant effects etc. In some cases *damage to vegetation* could be caused, for instance, by ethylene. Problems of liquid effluent disposal from chemical works exist, and uncontrolled discharges into streams or rivers can give rise to severe pollution. However, at its most material, there is a financial incentive to the firms themselves to avoid losses, so good housekeeping can in itself be good business. In general the manufacturing record of the chemical and oil industries is good and improving, with increased effort devoted to minimizing and controlling pollution, in research, design, technology, and operation. Broadly, therefore, the *problems of pollution* by plants *manufacturing organic chemicals* are localized, and in general the same situation applies at locations in which chemicals are further processed to "non-chemical" end-products.

In processes in which chemicals are used *as such* (i. e. are not further converted), escape to the environment is, in general, likely to be more widespread. This applies especially to processes in which the chemical concerned cannot be economically recovered for *re-use:* thus while, for example, in solvent extraction every effort is made to recover the solvent (if for none other than economic reasons), in a process such as wool scouring the detergents used will appear in the liquid effluent, which may be difficult to deal with.

In the same way in factories in which lacquered articles or materials are made, e. g. motor car bodies or lacquer coated wires, the solvents used in the lacquers may ultimately find their way into the atmosphere, and the same certainly applies to paints used for instance in the household. Within the factory it may be more a question of the *workers' safety* and protection rather than one of environmental conservation: it can be thought of as becoming the latter when it passes outside the "factory fence".

In addition to the above, there are of course chemicals whose use depends wholly on their being directly injected into the environment, e. g. pesticides.

Losses of chemicals in transit, whether as intermediates or final products, are another possibility, though a great deal has been done by the chemical industry (internationally and on a cooperative basis) to minimize *spillages* and their effects by specifying containers and providing

simple printed instructions on the action to be taken in emergency.

The tonnages so far given do not include *lubricants* or industrial oils. While not specific chemicals, they are usually highly refined hydrocarbon products and contain special additives; they are therefore worthy of mention.

Current annual consumption of lubricating and industrial oils worldwide is on the order of 20 million tons per year. Investigation in one particular industrialized country indicates that at least 10% of lubricating oils and industrial oils are unaccounted for, after use by the public or by industry. It seems likely that they have gone into the environment. The average figure in some countries may be considerably higher than 10%.

It therefore seems reasonable to assume that, worldwide, over 2 million tons per year (and possibly up to 5 million tons per year) lubricating and industrial oils appear in the environment either in water courses or the soil. Some comments on the biodegradability of hydrocarbons will be made later.

Naturally Occurring Organic Chemicals in the Environment

Although the annual amount of man-made organic chemicals (excluding lubricants) which enter the environment may perhaps be as much as 20 million tons a year, this total is very small in comparison with the enormous tonnages of organic compounds naturally produced.

Over the ages, degradation and emanation cycles have become established, through which an equilibrium seems to be maintained. Although there is a great deal of knowledge about the detailed mechanisms involved, the way in which many of these cycles operate is obscure. What is clear is that some of them are on a massive scale; for instance it has been estimated that swamps and other natural sources emit as much as 1600 million tons of *methane* into the atmosphere each year. Even cattle, which emit methane equivalent to 7% of their energy intake, must contribute a world total of several million tons! It is estimated that the world's atmosphere contains a total of 4800 million tons of methane and it is evident that a balance situation exists.

Another high emission is that of *terpene-type hydrocarbons,* emitted from forests and other vegetation, amounting to an estimated 170 million tons per year. It is believed that these polymerize in the air, and presumably they are eventually eliminated from the air by rainfall, or deposition in aerosols.

Degradation of Chemicals in the Environment

It is probably reasonable to believe that the same kind of processes that apply in these cases apply also to the *degradation* of most of the *solvents* produced by the chemical industry, and that the action of ultraviolet light, oxygen and water vapor in the air are of prime importance.

However, for many chemicals *biodegradability* is probably the most important process contributing to their elimination from the environment. The extent to which pollution by such chemicals presents a danger to the environment depends on the rapidity, and extent, to which they are biodegraded by micro-organisms.

Micro-organisms that are able to degrade organic chemicals occur in soils, sewage systems, and other aqueous habitats, and with a few exceptions they are aerobic. Heavy *pollution of streams and rivers* with materials which eliminate the dissolved oxygen from the water prevents the biodegradation of chemicals which would otherwise be readily degradable,

and would disappear under aerobic conditions. The bacteria in sewage systems are able to degrade a very wide variety of chemicals, particularly if given time to adapt and acclimatize themselves to a new type of chemical. In the sea also, degradation processes continue; it could indeed well be that the discharge of raw sewage into the sea may be beneficial in promoting the growth of micro-organisms capable of degrading chemicals that enter the *marine environment.* This is of course provided that the sewage is kept free from non-degradable and potentially toxic substances, particularly those which are concentrated by fish such as certain heavy metals and organochlorine compounds.

Micro-organisms in general, or particular species in a microbial community, have an amazing ability to *adapt* themselves to new environmental conditions and nutrients. They are readily susceptible to genetic modifications and enzymic reorientation as a result of changes in environment or availability of nutrients. It seems probable that, either by mutation and natural selection or by non-genetic adaptation, one or other micro-organism can, under appropriate conditions, destroy any *biologically* synthesized organic molecule. In some cases an individual organism can decompose compounds within a large group of structurally unrelated molecules.

In general, therefore, those products of the organic chemicals industry which are also found in nature do not give the same cause for concern, from the environmental aspect, as those which are truly "synthetic" i. e. no not occur naturally. Thus chemicals that were formerly produced by fermentation processes would be expected to cause little concern, but many of the compounds and materials now being produced by the chemical and allied industries contain structures that are not known in nature. Whether or not they are biodegradable depends on whether micro-organisms can adapt themselves to these chemicals as nutrients. In very many cases they do so, but certain chemicals show extreme recalcitrance. Such resistance is not surprising, since even in nature there are certain materials that are highly resistant, e. g. soil humus.

Chemical Structure and Biodegradability-General

It is worthwhile to review the evidence on biodegradability and to attempt to relate it to molecular structure.

In the atmosphere it is hardly possible to relate molecular structure with the degradability of manufactured chemicals. There is the massive background of the *methane cycle,* involving unknown reactions in the atmosphere, to be taken into account; of the total *hydrocarbons in the air* only about 15% are from man-made or specific sources. In addition to the naturally occuring methane and terpene hydrocarbons, ethane, ethylene, butane, pentane, acetaldehyde, acetone, methanol, ethanol, and benzene, together with a surprisingly high concentration of *n. butanol,* have been reported. The extent to which these are products of the biosphere, breakdown products or industrial pollutants can hardly be assessed. *Chlorofluorocarbons* are present in the atmosphere, but these are almost certainly man-made and very stable.

In soils and water, degradation is mainly microbiological.

The biodegradation of organic chemicals proceeds mainly, though not always, through oxidation by *aerobic bacteria* in such habitats as sewage systems, soil and aquatic environments. In attempting to relate chemical structures with degradability, it is important to recognize that, while generalizations may be made, there

may be exceptions in relation to particular microbial ecosystems. The next section gives a review, in general terms, of what is known (on a fairly empirical basis) of the relationship between biodegradability and chemical structure.

Specific Examples of Relationship between Biodegradability and the Chemical Structure

The *degradability of the alkanes* depends on the extent of chain branching and on the chain length. Degradability decreases with increased branching; an alkane containing a quaternary carbon atom is particularly difficult to degrade. With increasing chain length, degradability of alkanes increases up to about fourteen carbon atoms, after which there is a progressive decrease. Broadly, these rules apply also to other compounds that may be considered as derivatives of the alkanes, including the side chains of alkylbenzenes.

Alkenes seem to be at least as readily degraded as the corresponding alkanes, though the position of the double bond may have an influence. The fact that octene-1 is comparable with n. alkanes as a growth substrate for bacteria, whereas 1,7 octadiene does not sustain bacterial growth, suggests that the terminal double bond is refractory towards bacterial oxidation; it seems therefore that the attack on octene-1 proceeds from the saturated end of the chain.

The resistance of *cyclohexane* (as compared with n.alkanes) to degradation is noteworthy. However, in an "acclimatized" sludge and in the presence of a readily degradable hydrocarbon, it presents no problems. (Possibly bacterial symbiosis or co-metabolism is involved, and such possibilities should not be overlooked in other cases).

Information on chloro substitution of alkanes is rather scanty, but 1-chloro mono substitution seems to render degradation easier, at least in some cases, and there are also indications that increasing chain length has a similar influence.

The introduction of oxygen into alkanes as a hydroxyl group, or as aldehyde or carboxyl, increases degradability. Thus alcohols, aldehydes, and fatty acids (also their glycerides and methyl esters) are readily degradable, and again the rule on chain branching applies. In the lower ketones there seems to be a tendency for the keto group to retard oxidation, but again increased chain length counteracts this.

Substitution of the OH in the carboxyl group of *fatty acids* by NH$_2$, to give an amide, reduces degradability, and nitriles seem to be even more resistant.

In general, the *insertion of an ether* linkage in an alkane causes resistance to degradation. Thus diethyl ether and dioxan are comparable in degradability with, say, tert.-butylbenzene. Diethyl ether is much more resistant than the isomeric n-, sec-, and iso-butanols.

Benzene in most cases is not readily attacked but this may be, in laboratory practice, because it is reasonably water soluble and fairly toxic. Under conditions of appreciable dilution, benzene was found to sustain bacterial growth at a high rate, comparable to the rate observed with normal paraffins.

A substituent in the ring commonly, but not always, increases degradability. Mono alkyl substitution does this to an extent depending on chain branching and chain length, and the fused ring in naphthalene renders the latter readily oxidizable. However, while toluene is more readily degradable than benzene, further methyl substitution causes a decrease in degradability, and as a group the polymethylbenzenes are quite resistant, the extent of this depending on the number and positions of the substituent methyl groups in the ring.

Phenol is readily degraded by bacteria once the latter are well "acclimatized" and it seems, therefore, that hydroxyl substitution of the aromatic ring promotes degradability. Similarly the aldehyde and carboxyl groups increase degradability, and benzaldehyde and benzoic acid readily degraded. Benzyl alcohol is rapidly oxidized, which shows the effect of -CH$_2$OH as a substituent (bearing in mind the effect of a methyl group substituted in benzene and the effect of OH on alkanes, this might be expected).

Nitrobenzene seems to be biologically inert, the NO$_2$ group apparently adding resistance. One NH$_2$ substituent in benzene increases the susceptibility to oxidation, but a further NH$_2$ or a methyl group confers resistance to degradation. Thus the phenylene diamines and toluidines are more resistant than aniline.

Chlorine substitution *in* the *benzene* ring appears to impart some resistance to structures that would otherwise be more readily degradable. In the case of phenol this applies particularly if a chlorine (or other halogen) atom is meta to the hydroxyl. Some similar but more complex effects result from chlorine substitution in benzoic acid and phenoxy alkanoic acids. Resistance due to *chlorine polysubstitution* has been observed with both straight chain and ring structures — e. g. *organochlorine insecticides, chlorinated biphenyls.*

The effect of structure on *synthetic detergents* is worthy of special consideration since problems of degradability which have occured in the past in the "domestic" field have been for all practical purposes solved; in the industrial field success seems to be on the way.

An earlier and widely used base material for domestic detergents was propylene tetramer benzene sulphonate. The alkyl chain was highly branched and therefore this material was very resistant to degradation and a marked improvement resulted from a change to linear alkylbenzene sulphonates. The linear alkane sulphonates are slightly more readily degradable than the linear alkylbenzene sulphonates. The sulphates and ethoxysulphates of linear primary alcohols are even more biodegradable than the linear alkane sulphonates but extensive and complex branching in the parent alcohol reduces this.

In ethoxylates, contrary to the case of alkyl benzene sulphonates, the replacement of branched by linear alkyl phenol only slightly increases degradability. However, primary alcohol ethoxylates are readily degradable to a high degree, although again this is reduced by complex branching in the parent alcohol, and also decreases with increasing numbers of ethylene oxide units per molecule.

Although not detergents, it is appropriate to mention the polyethylene glycols at this point. These degrade only slowly and replacement of ethylene oxide units by propylene oxide confers added resistance. In concluding this section one must make the general, but absolutely fundamental, point that degradability of a chemical does not necessarily imply its complete transformation into carbon dioxide and water. Non-degradable residues may be formed from the parent compound. This proviso lends weight to the urgent need for more research to establish such facts, the need for which is emphasized under "Conclusions" below.

Some Practical Examples in the Environment

It may perhaps be of interest to consider the possible *fate in the environment* of three of the major man-made organic chemicals that enter it — *ethylene, acetone* and *chlorofluorocarbons* being taken as examples. It must be strongly emphasized that any quantitative data is highly speculative (a fact which might be described as unfortunate, and which should be rectified).

It seems likely that, world-wide, up to 500,000 tons a year of ethylene are lost from chemical plants, i. e. 2—3 % of the total processed. Ethylene is a natural product emanated from vegetable sources and it might therefore be excepted that natural cycles would provide for its elimination. Fractions of a part per billion* have been found as normal constituent in "clean" air, e. g. at Port Barrow, Alaska. (Though this is not absolute proof of natural origin).

Research into the chemistry of Los Angeles smog has provided considerable insight into the way in which, inter alia, *olefins* are involved, and it would seem reasonable to surmise that similar reactions might apply generally, and to more widely dispersed olefins, including ethylene, the final elimination step being through rainfall or surface deposition of aerosols containing the breakdown products. It seems probable that the lifetime of olefins in the atmosphere is short.

Some 700,000 tons per annum *acetone* is used as a solvent (again world-wide), and it is possible that the majority of this enters the environment. Probably most of it enters the atmosphere through evaporative losses, but part may originate in aqueous solution, say, in trade effluents. Acetone is, in fact, found in air as a normal constituent, nearly 1 part per billion having been found at Port Barrow, Alaska. Since it is a compound that can be produced naturally by fermentation processes, it seems certain that degradation

---

* billon = thousand millon (10^9)

processes exist, which keep the level at an equilibrium. In the air acetone may undergo photolysis which could lead to its elimination, but it seems reasonable to believe that most of it is returned to earth by rainfall; oxidative degradation by micro-organisms will then result in its elimination. The same mechanisms can also apply to synthetic ketones such as methyl ethyl ketone and methyl isobutyl ketone.

A type of compound which does not occur in nature are the chlorofluorocarbons widely used in *aerosol propellants*. Of the 700,000 tons of carbon tetrachloride produced annually worldwide, about 95% is used for the manufacture of chlorofluorocarbons. These compounds are now found in the atmosphere as normal constituents. They are very stable and their presence, and any possible built up, needs watching; there seems to be no evidence that such highly halogonated structures can be attacked by micro-organisms. (Nevertheless fluorinated organic compounds can, in fact, be synthesized in nature, e.g. fluoracetamide has been found in certain plants, so that degradative processes involving fluorinated compounds cannot be discounted entirely; in fact microbiological growth has been observed at appreciable rates with 1-fluoro hexane).

This section has discussed three large-scale products about which a fair amount of statistical and technical data exists — regarding production, end use, degradation, etc. It could, however, well be that smaller tonnage products — many of which have complicated structures not found in nature — represent a more serious gap in our knowledge of the fate of organic chemicals in the environment

Conclusions

Increasing attention will have to be paid to *degradation-resistant synthetic chemicals*. Some of them, per se or as partial degradation products (which may be more resistant than the original), may enter the environment even in small quantities and arrive (ultimately) in the sea. In the course of this there is always the possibility that they may enter into the food chain and produce unforeseen results. There is a vast field for objective scientific research into the nature of these substances, their amounts and distribution, their toxicology, and possible adverse effects.

Such research has a place for specialist expertise of many kinds, and for scientists of various disciplines, chemists, microbiologists, ecologists, and physicists, to name but a few. Chemicals in the environment has become an extremely emotive subject, and in this field the need for scientific objectivity among those carrying out the research is imperative if real progress is to be made. Too readily has publicity been given to ill-founded but well intentioned "scares", resulting in the diversion of research effort to these from areas that are basically far more rewarding. There is also need for far more information on the "logistical" side — the quantities of man-made organic chemicals entering the environment, their relationship between those of natural origin, and their fate. This, together with the research mentioned above, demands monitoring systems of far greater sophistication and international operation than we have at the moment. Man will only win the battle against pollution by substituting objectivity for emotion, and cooperation for partisan attitudes.

Inorganic Chemicals in the Environment — with Special Reference to the Pollution Problems in Japan

M. Goto

Department of Chemistry, Gakushuin University, Toshima-ku, Tokyo, Japan

Summary. Data are indicated concerning environmental pollution by various pollutants. Special emphasis is put on mercury, catalysts, plastics' stabilizers, and pigments. The data are mainly available for Japan. Furthermore, diseases caused by environmental chemicals are discussed, such as the Minamata and the Itai-Itai disease as well as an intoxication caused by arsenic in milk products.

Zusammenfassung. Es werden Daten angegeben über die Umweltbelastung mit verschiedenen Verunreinigungsstoffen. Besonders eingegangen wird auf Quecksilber-Katalysatoren, Stabilisatoren von Kunststoffen, Pigmente. Die Daten gelten besonders für Japan. Darüber hinaus werden durch Umweltchemikalien verursachte Erkrankungen wie die Minamata- und die Itai-Itai-Krankheit und eine Vergiftung mit arsenhaltigen Milchprodukten diskutiert.

Environmental study and protection are matters of urgent necessity in every industrialized country. Table 1 shows the amount of pollutants discharged in Japan in 1967 and the amount estimated for 1975. The amount will increase approximately twice in 8 years.

Table 2 shows the degree of water pollution of the Tamagawa river (Tokyo) in 1951 and 1965. The water is still used as drinking water now.

The analytical data of suspended particles in the air in the different cities and those of sediments in Tokyo Bay are given in Tables 3 and 4.

In this article, some problems of environmental pollution due to inorganic chemicals included in chemical processes are

Table 1 Amounts of wastes discharged in Japan

Pollutants	1967	1975	ratio
SO$_2$	≡1,658,000 tons	≡ 3,671,000 tons	2.2
CO	≡7,584,000 tons	≡18,287,000 tons	2.4
Hydrocarbons	690 tons/day[1]	≡930 tons/day	1.4
Industrial Waste Water (BOD load)	5,792 tons/day[2]	12,823 tons/day	2.2
Municipal Waste Water (BOD load)	3,075 tons/day[2]	3,610 tons/day	1.2
Municipal and Industrial Wastes	1,163,700 tons/day	2,907,900 tons/day	2.5
Plastics	1,166,000 tons/day	5,100,000 tons/day	4.4

[1] Value for 1969; [2] Values for 1965

discussed, and some actual examples are referred to.

Mercury: World production of mercury is currently approx. 9,000 tons per year. The consumption of mercury in Japan in 1968 was 1,266 tons. This amount was used for manufacturing caustic soda (714 tons), inorganic chemicals (336 tons), catalysts (103 tons), pesticides (24 tons), and others (89 tons). Caustic soda industries are the important consumers of mercury (56%). In Japan in 1970 95% of soda production is based on the mercury process which uses mercury as electrodes. Hence, the products (caustic soda, chlorine, and hydrogen), salt water

Table 2   Water pollution of the Tamagawa River in Tokyo

Pollutants	1951	1965	ratio
Cl^-	7.64 ppm	29.61 ppm	3.87
SO_4^{--}	10.31 ppm	37.99 ppm	3.68
$\equiv NO_3^-$ (as N)	1.75 ppm	2.062 ppm	1.17
$\equiv NO_2^-$ (as N)	0.0149 ppm	0.162 ppm	10.87
Ammoniacal $\equiv N$	0.0249 ppm	4.863 ppm	195.30
$KMnO_4$ consumption	3.69 ppm	20.37 ppm	5.52
No. of bacteria (per cc)	700	265,286	378.98
Population	673,000	1,167,000	1.73

Table 3   Analytical Data (average) from Some National Air Monitoring Stations in Japan (1968)

	SO_2 ppm	total	Be	V	Cr	suspended particles $\mu/m^{3(1)}$ Mn	Fe	Co	Ni	Cu	Zn	Pb	NO_2 ppm	NO ppm
a	0.022	220	<0.0005	0.017	<0.040	0.15	4.5	<0.008	0.05	0.068	<1.1	<0.30	0.037	0.051
b	0.054	175	<0.0005	0.030	<0.040	0.21	3.5	<0.008	0.06	0.123	1.2	0.43	0.058	0.037
c	0.084	420	0.0005	0.119	0.048	0.72	14.7	0.011	0.17	0.395	2.4	0.89	0.048	0.052
d	0.029	297	<0.0005	0.075	<0.040	0.40	5.3	<0.008	0.09	0.088	1.8	0.74	0.021	0.059
e	0.083	231	<0.0005	0.069	<0.040	0.72	7.5	<0.008	0.13	0.149	2.0	0.70	0.020	0.015
f	0.027	236	<0.0005	0.022	<0.040	0.22	4.2	<0.008	0.07	0.058	<1.1	<0.30	0.023	0.017
g	0.046	202	<0.0005	0.028	<0.040	0.31	6.6	<0.008	0.06	0.078	<1.1	<0.30	0.022	0.034

Average Pb concentration for 1969: a, Sapporo; b, Tokyo; c, Kawasaki; d, Osaka; e, Amagasaki; f, Use; g, Kita-Kyushu.

Table 4   Analysis of the bottom sediments of tthe Tokyo Bay near Kawasaki, Industrial District (August, 1970)

spot No.*	COD ppm	total Hg ppm	Me- Hg ppm	Et- Hg ppm	Cd ppm	Pb ppm	Zn ppm	As ppm	CN ppm	Cr ppm	org. P ppm
1	14,100	13.2	0.024	ND	5.8	86	14.5	20.2	ND	12.0	ND
2	4,100	13.5	0.009	ND	3.1	91	5.8	16.3	ND	5.9	ND
3	6,400	8.1	0.014	ND	4.0	107	100.6	12.7	ND	10.5	ND
4	5,400	10.9	0.018	ND	40.4	365	3.3	42.5	5.9	7.5	ND
5	6,200	17.8	ND	ND	9.5	249	0.7	29.4	ND	10.1	ND
6	10,700	4.5	0.010	ND	5.3	233	4.9	11.7	ND	8.0	ND
7	1,000	4.8	0.002	ND	4.6	149	30.3	30.3	1.0	21.6	ND
8	8,100	6.4	ND	ND	4.6	229	24.8	33.0	ND	20.0	ND
9	9,900	4.3	0.006	ND	3.5	141	18.0	17.4	ND	11.9	ND

* Numbering from the mouth of the river Tamagawa to the river Tsurumigawa. Kanagawaken Rep.

Table 5 Production of catalysts in Japan

Usage	1969	1970
Petroleum refining	3,998 tons	4,860 tons
Petrochemical manufacturing	3,208	3,398
Polymer synthesis	1,769	2,836
Inorganic chemical production	1,788	2,159
Gas manufacturing	3,940	3,620
Oil and fat processing	1,125	1,269
Medicine and food manufacturing	444	625
Miscellaneous	485	58
Total	16,756	18,826

soda. High mercury concentration was found in the salt water mud, which was discarded in the open sea until quite recently. The caustic soda production in Japan by the mercury method in 1970 was 3,171,000 tons, and approx. 650 tons of mercury were lost in the manufacturing process in Japan in 1970.

Catalysts More than 70% of modern chemical products are produced using catalysts. About 90% of the products from petrochemical and macromolecular industries involve catalysts (metallic oxides, chlorides, alloys, etc.). The production of catalysts in Japan in 1970 was 18,826 tons, and appreciable amount of them was lost in products, washing water and high temperature distillates. Cheap catalysts are discarded as liquid or solid wastes.

mud and washing water are usually contaminated with mercury. It was estimated that 150—260 gm of mercury are usually lost for production of each ton of caustic

Table 6 Production of the stabilizers for vinyl chloride resin in Japan

Compounds	1965	1970
$3PbO \cdot PbSO_4 \cdot H_2O$	6,801 tons	1,2234 tons
$2PbO \cdot (C_{17}H_{35}COO)_2Pb$	1,456	2,933
$(C_{17}H_{35}COO)_2Pb$	5,090	9,251
$2PbO \cdot PbHPO_3 \cdot 1/2H_2O$ $PbO \cdot 2PbCO_3 \cdot H_2O$ $PbSiO_3 \cdot mSiO_2 \cdot nH_2O$	1,552	3,560
Total	14,899	28,348
$(C_{17}H_{35}COO)_2Cd$	1,072	1,199
$(C_{11}H_{23}COO)_2Cd$ $[CH_3(CH_2)_3CH(C_2H_5)COO]_2Cd$	1,614	5,335
Total	2,686	6,534
$(C_{17}H_{35}COO)_2Ba$	1,706	3,885
$(C_{11}H_{23}COO)_2Ba$, etc.	328	740
Total	2,034	4,624
$(C_{17}H_{35}COO)_2Ca$	1,075	3,134
$(C_{17}H_{35}COO)_2Zn$	438	1,142
$(C_{11}H_{23}COO)_2Sn(C_4H_9)_2$ $(C_{11}H_{23}COO)_2Sn(C_8H_{17})_2$ $(C_{17}H_{35}COO)_2Sn$, etc.	3,131	5,843
Total	24,263	49,255
Vinyl Chloride Resin	482,973	1,161,053

Only expensive catalysts are recovered with a loss of 5—6%. The production of catalysts in Japan in 1970 is given in Table 5.

Stabilizers: The production of plastics in Japan in 1970 was 5,130,000 tons (vinyl chloride resins: 1,161,000 tons, polyethylene: 1,305,000 tons, polystyrene: 668,000 tons, polypropylene: 581,000 tons). Vinylchloride products are processed with heat stabilizers whose chemical formuls and productions in Japan are listed in Table 6. Lead compounds are cheap and effective stabilizers (57% of stabilizers are lead compounds); they are used for manufacturing pipes. For the production of transparent films, cadmium, barium, and organic tin compounds are also used as stabilizers. Those stabilizers are ap-

parently the source of inorganic pollutants.

Pigments: Pigments are used as paints and also for coloring plastics, paper, and china. Chemical formulae and production of important pigments are listed in Table 7. Metallic pollution due to pigments often occurs in factories and in everyday life. Furthermore, almost all plastics are colored with metallic pigments, and the pollution from them must also be carefully examined.

Gilding: Gilding process can be accelerated by increasing the concentration of metallic compounds. Important metallic compounds for gilding are as follows: $CuSO_4 \cdot 5H_2O$, $Cu(BF_4)_2$, $Cu_2(CN)_2$, Cu pyrophosphate, Cu sulfamate; $NiSO_4 \cdot 6H_2O$, $NiCl_2 \cdot 6H_2O$, $Ni(BF_4)_2$, $NiSO_4 (NH_4)_2SO_4 \cdot 6H_2O$; $CoSO_4 \cdot 7H_2O$, $CoCl_2 \cdot$

Table 7 Production of metal pigment in Japan

Compounds	1965	1970
$2(Na_2O \cdot Al_2O_3 \cdot 2SiO_2) \cdot Na_2S_2$	795 tons	1,406 tons
TiO_2	98,124	163,218
$TiO_2 + Sb, Ni$	–	–
$Cr_2O_3, Cr_2O_3 \cdot 2H_2O$	–	–
$K_2O \cdot 4CrO_3 \cdot 4ZnO \cdot 3H_2O$ / $ZnCrO_4 \cdot 4Zn(OH)_2$	949	–
Fe_2O_3	19,885	47,012
Fe_3O_4	73	–
$FeK_3[Fe(CN)_6]_3 \cdot nH_2O$	2,599	3,186
$CoO \cdot nAl_2O_3$	–	–
$Co(AsO_4)_2, Co_3(PO_4)_2$	–	–
ZnO	35,683	61,589
$ZnS \cdot BaSO_4$	2,984	2,218
$CdS \cdot nCdSe$	–	207
CdS	–	407
$CdS \cdot HgS$	–	–
HgS	–	24,186
PbO	–	38,025
Pb_3O_4	8,352	13,755
$PbCrO_4$	8,555	16,651
$PbCrO_4 \cdot nPbMoO_4 \cdot mPbSO_4 \cdot xAl(OH)_3$	1,062	2,719
$2PbCO_3 \cdot Pb(OH)_2$	1,161	1,468

Table 8   Numbers of patients and deaths caused by pollutants (August, 1970)

Disease	No. of Patients	No. of Deaths
Bronchitis	1,566	
Yokkaichi	533	
Osaka	822	unknown
Kawasaki	211	
Minamata Disease	116	52
Minamata	75	46
Niigata	41	6
Itai-Itai Disease	96	100

- Cd 0,4~1 ppm in rice
- Cd >1 ppm in rice
::::: water pollution

Fig. 1   Environmental pollution in Japan

$6H_2O$; CrO_3; $FeSO_4 \cdot 7H_2O$, $FeCl_2 \cdot 4H_2O$, $Fe(BF_4)_2$; ZnO, $Zn(CN)_2$, $ZnSO_4 \cdot 7H_2O$; $Na_2SnO_4 \cdot 3H_2O$, $Sn(BF_4)_2$; CdO, $Cd(CN)_2$, $Cd(BF_4)_2$; $Pb(BF_4)_2$, Pb sulfamate; $KAu(CN)_2$; $KAg(CN)_2$, $AgCN$; $In_2(SO_4)_3$, $InCl_3$, $In(BF_4)_3$; $PdCl_2 \cdot 2H_2O$; $H_2PtCl_6 \cdot 6H_2O$, $Pt(NH_3)_2(NO_2)_2$; $Cu_2(CN)_2 + Zn(CN)_2$, $Na_2SnO_3 \cdot 3HO + Cu_2(CN)_2$. For gilding with gold, silver, copper, zinc and cadmium cyano-complex basins are used advantageously, and the pollution problems due to cyanide ion are sometimes serious.

Minamata Disease: The Minamata disease was caused by consumption of fish and shells from Minamata Bay and the causative agent was methylmercury discharged from a factory which belongs to Chisso Corp. Two different kinds of mercury compounds (mercuric chloride and mercuric sulfate) were used as catalysts in the manufacturing processes of vinyl chloride and acetaldehyde. It was estimated that 500 to 1000 gm of mercury were lost for each ton of acetaldehyde produced, and approx. 200 tons of mercury (methylmercury included) were lost in waste water in the years 1949—1963; this amount was just for manufacturing acetaldehyde. The bottom sediments of Minamata Bay contained mercuric oxide and mercuric sulfide, but no organic mercury compound.

Itai-Itai Disease: Environmental pollution due to cadmium was caused in Japan mainly by zinc mining operations. The so-called Itai-Itai disease occurred in Toyama Pref. The analytical values of cadmium in this area are as follows: well water: 0.001 ppm, mine waste water: 0.005—0.061 ppm (average, 0.017 ppm), riverwater: 0.001—0.009 ppm (0.002), sediment near mine: 4.1—238 ppm (121), bottom sediments of river: 0.16—5.0 ppm (3.27), paddy soils: 1—7.5 ppm (2.27), rice: 0.35—4.17 ppm (1.41), cf. Zn 20.8—32.3 ppm (727.4), rice (nonpolluted area): 0.03—0.11 ppm (0.08), cf. Zn 18.9 —28.2 ppm (22.7). The patients are mostly mothers and they ingested cadmium from food (0.6 mg/day) and water (1—1.4 mg/day). In Japan, 35 areas are polluted with cadmium.

Morinaga Milk Incident: This incident occured in the summer of 1955; 12,000 patients and 130 deaths were reported. In the manufacturing process of baby milk technical disodium phosphate was used as stabilizer and the agent was contaminated with arsenic.

In Japan, until August, 1970, 1,778 illnesses and 152 deaths were reported in separate areas in Japan from intake of air polutants (SO_2), cadmium and methylmercury contaminated foods (Table 8). Figure 1 is a pollution map of Japan·

References

[1] Maekawa, M.: Metal Pollution and Chemical Industries. Kagaku 41: 545–550, 1971

[2] Year Book of Chemical Industries Statistics. Published by Minister's Secretariat, Ministry of International Trade and Industry 1970

[3] White Book of Environmental Pollution. Published by Printing Bureau, Ministry of Finance 1970, 1971

[4] White Book of Environmental Pollution. Published by Osaka Prefectural Government 1970

[5] Environmental Pollution and Tokyo. Published by Tokyo Metropolitan Government 1970

[6] Guidebook to the Protection of Human Environment. Published by National Life Council 1970

[7] Pollution Map of Japan. Published by Japan Broadcast Publishing Association 1971

Pesticide Residues in Food — the Situation Today

H. Egan

Laboratory of the Government Chemist, London, England

It is now more than fifteen years since Walker, Goette, and Batchelor, following earlier reports of the occurrence of *residues of DDT in milk, animal tissues, cereal products* and other items of the diet consequent upon the insecticidal uses of this material, published the results of their *residue study of prepared meals*[1]. Earlier, Pearce, Mattson, and colleagues[2,3] had directed attention to the occurrence of small residues of *DDT* and *DDE* in humans who had no record of direct exposure to these materials. It was thus not entirely surprising for Walker et al. to find in 1954 an average of some 50 micrograms of DDT and 25 micrograms of DDE in typical restaurant and institutional meals, corresponding to average concentrations of the order of 0.07 and 0.03 ppm respectively. No meals contained enough DDT to be considered a toxicological health hazard; but it was concluded that DDT residues on food ultimately intended for human consumption should be measured in such a way that the DDE content as well as the DDT content would be determined. Since this time, the subject of residues in food has been one of the dominant themes of the pesticide scene. Many workers have studied the residue levels of DDE and DDT, and of others of the more persistent organochlorine pesticides, in both meals and human tissues. Methods of residue analysis have developed considerably since 1954. It is difficult to assign meaningful figures for the usage of DDT and to deduce from these trends with respect to the likely residues in food. Malaria control programs and usage in non-edible crops such as cotton, for example, tend to obscure the picture. There is, however, a suggestion from U.S. production figures that the overall usage of DDT may have fallen slightly (perhaps by 10—20%) over the past 15 years[4]. It is of interest at the present time, therefore, to review the situation today in relation to that found some fifteen years ago; and if possible to make some simple and direct comparisons of the positions then and now.

To put the matter into fuller perspective, it must be recognized that agricultural chemicals have contributed *unintended residues* to food since long before the advent of synthetic organic insecticides, *arsenic, lead,* and *copper* being examples of those which are still with us today. Indeed, the residues are not always unintended, since to be effective they must be present and it may be unrealistic to expect them to vanish immediately before the consumption of food. Furthermore, although there is an emphasis on pesticides as environmental intruders, there are a number of other trace chemical aspects of the diet not all of which are novel. There are for example technological additives by way of *chemical preservatives, antioxidants, emulsifiers, stabilizers* and so on, some of which have parallels in food processes which have been traditional for hundreds or even thousands of years. Table 1 illustrates

Table 1 Classification of *Toxic Substances in Food*

Natural Contaminants

a) Naturally occuring in food:
    alkaloids
    cyanogenic glycosides

b) Arising from (natural) contamination by microorganisms;
toxins produced by spoilage microorganisms, mycotoxins, fungal toxins

c) Arising directly from environmental cross-contamination:
trace metals
miscellaneous allergens
polynuclear aromatic hydrocarbons(?)

Technological Contaminants

a) From deliberate treatment during the course of production of food:
agricultural pesticides (and other biocides)
veterinary pesticides
growth-promoting substances used in animal husbandry

b) From incidental treatment during processing and distribution:
solven residues (from solvent extraction)
trace metals from processing equipment
extraction of additives from packaging materials from fumigant or other storage treatment (insecticides, rodenticides) before or after processing
detergent, surfactant residues
sterilant, disinfectant residues

c) From deliberate addition during the course of preparation and processing of food:
Smoke constituents (including polynuclear aromatic hydrocarbons)
other preservatives
antioxidants
emulsifiers
stabilizers: flowing agents, anti-caking agents added colours, flavours

d) Secondary contaminants arising from (c) above nitrosamines (from nitrites)
solvent residues (from solvents used to disperse added colors, flavors)

briefly the whole field of *additives* or contaminants in agricultural produce (of which food is by far the most important item). The classification is not concerned with the quantitative aspects of toxicity and it should be understood that some members of the various classes of substances listed may be non-toxic or virtually so. Substances are divided firstly into those which occur in nature and those which are technological in origin. The first group includes substances arising from natural contamination by microorganisms such as the fungal *mycotoxins* and those arising from direct environmental cross-contamination such as traces of *heavy metals,* as well as "natural" poisons such as alkaloids. The second group is subdivided into substances which occur due to deliberate treatment during the course of food production (these include pesticides), from incidental treatment during processing and distribution, from deliberate addition during the course of preparation and processing (preservatives etc.) and by secondary contamination from deliberate additions as in the case of residues of solvents used to disperse added colors and flavors. All of these contribute to the spectrum of chemical and toxicological aspects of the environment and have a bearing on the quality of the environment insofar as this is reflected in the food consumed by man. Let us recognize clearly at the outset, therefore, that pesticide residues are only one aspect of the environmental quality of food. Let us be clear also that pesticides themselves contribute to what broadly may be called a *"positive environmental quality factor"* with respect to food. The emphasis in recent years on pesticide residues has been concerned with the balance of what by analogy can be called the negative and the positive environmental quality factors. That is to say, between those factors which on balance constitute an advantage, an increase in the health and well-being of man and those which on balance constitute a disadvantage or barrier to the improvement of man's health and well-being. Not all men are the same, of course, nor are all agricultural situations. What may to one seem a positive advantage can to another constitute a hazard with no clear advantages. "One man's meat is another man's poison" as we say in Britain.

The views which have been expressed on the *significance* of measured *pesticide residues in food* have, not unexpectedly, shown a wide divergence. They vary from the bluntly antagonistic school of thinking in which any non-traditional method of husbandry is suspect, to the openly optimistic approach that a health hazard no longer exists for the great majority of pesticide applications. When control of residue levels is desirable, the basis for this can vary between two extremes which, on the one hand, call for a strict *mathematical consideration* of toxicological data and daily intakes of food, and on the other hand look to the requirements of *good agricultural practice;* and with many shades of opinion between the two[22].

It is generally accepted that so far as the ingestion of pesticide residues by man is concerned, the main route is by way of food eaten. Campbell et al. in 1965 estimated that for DDT + DDE the annual intake from air might be of the order of 30 micrograms, from water 10 micrograms, and from food some 45 micrograms, with perhaps a further 5 mg from other unspecified sources[5]. Several kinds of surveys of pesticide residues in food have been recorded in the literature and it is of some importance to have an understanding of the terms used to describe the range of samples examined. These are summarized in Table 2. Firstly, we

Table 2   Types of Surveys for Pesticide Residues in Food

1. *Commodity surveys*
   a) non-selective
   b) selective
2. *Dietary surveys*
   a) meals
   b) total diet

can obviously distinguish between *commodity surveys,* conducted on individual food commodities (or types of commodity) such as fruit, or cereals, or meat products and diet surveys designed to provide a measure of the total intake from the diet as a whole· One important difference between the former and the latter is that the latter should always be concerned with the food prepared in the ready-to-eat state, that is to say washed, trimmed, cooked or otherwise prepared as it would be for the table; whereas the former type of survey will often be concerned with produce in commercial channels. Commodity surveys themselves can be divided into selective and non-selective. In the latter, which has been the main method of government appraisal in Britain, special attention is given to those items of the diet known (or thought) to make significant contributions to the dietary residue intake. These taken together can also be considered an approximation to a *total diet study,* if the selection is carefully made. This kind of survey has been described by Lee[6]. Diet studies also can be divided into two broad classes. Firstly, there are *prepared meal surveys* of the type originally described by Walker et al.[1]. In these, meals which otherwise are about to be served, in restaurants or institutions, are intercepted when they arrive at the dining table and taken for analysis. Secondly, there is the total diet study in which the residue is measured in the whole diet, as eaten. For this, individual food is assembled in proportions in accordance with national or regional statistics on dietary intake. The term *"market basket"* is used in the United States to describe the individual samples obtained from normal retail sources which, when prepared for the table and taken together, represent the total diet. A *"market basket"* is thus a food sample as assembled for a total diet study, but in the raw and unprepared state. It is prepared for the table in the normal way before analysis. This type of study has been fully described by Duggan[7]. The total diet itself is normally assembled from a number of clearly defined *"market baskets"* representative of different types of food, for example meats, fats, cereals, fruit, and vegetables. Some of the meal studies described in the literature are very similar to total diet studies. So far as I am aware, no one has analysed a total diet as such, that is a diet assembled from individual meals. There are good reasons for this since such an approach would complicate analysis in an unnecessary manner and result in a lower order of analytical sensitivity.

For all the approaches, interest has concentrated on the *persistent organochlorine pesticides* and it is in this direction that most attention has been given to the *analytical methodology*. Residues of organophosphorus and of other pesticides have also been studied from time to time and although there have been considerable developments in analytical methodology for many of these, their relatively non-persistent character has meant that there is less cause for study. There have in fact been substantial developments in methods of residue analysis in the past 15 years; these have been occasioned largely by the interest in the organochlorine compounds, but the developments themselves apply to all pesticide residues. In 1954 the methods generally available were based on traditional wet processes of analysis. In the case of *DDT* this was the original colorimetric method of Schechter et al.[8] in which DDT is nitrated to give a tetranitro-derivative and the colour developed when this is reacted with alcoholic sodium methoxide measured either by visual comparison or with a spectrophotometer. Following the work first of Mitchell and then of Goodwin et al. paper chromatographic[9] and then gas liquid chromatographic[10] methods of residue analysis were developed. These have since been supplemented by thin layer chromatographic techniques, particularly useful in clean-up as well as diagnostic and quantitative separation processes[11]. It was possible with the Schechter and Haller method for DDT to measure the residues of both DDT and DDE, each of which responded, separately. This was basically a mathematical process based on the differences in the colors of the nitro-compounds eventually measured and was a somewhat awkward and imprecise operation. Modern methods will separate DDT from DDE completely; they can also resolve each of these into their individual isomers and will do this in an unambiguous fashion. When comparing DDT residue results obtained by the earlier colorimetric methods with those obtained by current chromatographic procedures, account must be taken of differences in specificity of the two methods. The method of calculation used in the colorimetric methods was dependent on the ratio of op' — and pp' isomers; Dale and Quinby, studying residues in human fat, showed that if the op'-DDT to pp'-DDT ratio in fat is greater than that in technical DDT, the colorimetric procedure gave results for DDE which were higher than the true values, whilst DDT-levels were less than the true values[12]. In the same study they examined individual samples by both analytical methods and found that the average total

DDT-derived material found by gas chromatography was only 62% of that found by the colorimetric procedure. The same chromatographic methods have superseded the earlier phenylazide method for dieldrin residues, which was insufficiently sensitive or specific for use in human fat or total diet studies.

Commodity Surveys

Results obtained by each of the principal methods of residue study will now be considered. Most *commodity surveys* have been concerned either with animal produce (including dairy produce) or with fruits and vegetables· Although, as with other surveys, they have concentrated on the *persistent organochlorine compounds,* a greater variety of pesticides has been covered by this kind of approach than by any other. Almost all *commodity surveys* are selective in the sense that it is impracticable to look for residues of every possible pesticide; and the principle residues sought are those for which, from a knowledge of agricultural or veterinary practice, there is some possibility of occurrence. This generalization is itself partially obscured, however, when the residue can also arise naturally or from non-pesticidal uses, as in the case of *lead* or *bromide.* Most of the official residue surveys in Britain have been of the selective type[6]. Pocklington and Tatton for example[14] have surveyed apples imported into Britain for *arsenic* and *lead* contents. They found that if only the flesh of the fruit is considered, no sample exceeded the tolerance levels of 1 ppm arsenic and 3 ppm lead, although these levels were very occasionally exceeded if the whole fruit were considered. Raw fruit and vegetables examined by Rymer and Hamence showed no significant levels of *arsenic, lead,* or *mercury.* Thompson and Hill[15] examined maize, pulses, and nuts arriving in Britain in the period 1963—68 for bro-

mide as a possible indication of previous fumigation treatment. Occasional samples were found in which the inorganic *bromides* level exceeded 200 ppm (100 ppm for haricot beans and nuts). Variations from sample to sample from bulk fumigation treatments were large, however. Smart and Hill have examined rice imported into Britain for residues of *mercury* and found that although often negligible, they may rise to 0.01 ppm and occasionally to 0.015 ppm [16]. Raw fruit and vegetables examined in England and Wales by Rymer and Hamence in 1965 showed no significant levels of *thiourea* or *o-phenyl phenol* and less than 10% of the samples examined for organochlorine compounds gave a positive response. In a similar survey of fruit and vegetables in earlier years, Hamence found only relatively infrequent evidence of residues and these at levels which would not normally be considered hazardous [20, 21]. Dickes and Nicholas subsequently examined home grown and imported fruits in Britain during 1967 for residues of organochlorine and organophosphorus compounds [23, 24]. Little organophosphorus residues were found but about half of the samples contained detectable amounts of DDT and BHC; only in a few cases did the levels exceed 1 ppm. More recently local authorities in England and Wales have completed a joint survey in which *organochlorine, organophosphorus, lead, arsenic, and mercury residues* were studied for the period 1966—1967 [25]. All of these surveys relate to residues on produce as purchased by the retail consumer· There have of course been many papers on residues in field trials; these do not come within the scope of the present survey.

The main interest has concerned *persistent organochlorine pesticide residues* in animal products. Arising from the former use of *dieldrin sheep dips* in Britain, it has been possible to follow directly the reduction in residue levels in mutton fat following restriction of the use of these dips. Egan has described the earlier position[26]; in 1964 dieldrin levels ranging from zero to 12.4 ppm with a mean of 0.84 ppm were found. The mean level rose to over 1 ppm in 1965 but has fallen in subsequent years as shown in Table 3: The withdrawal of government approval of dieldrin for sheep dipping in Britain in 1966. Other animal produce, both home grown and imported, has been examined in Britain. Egan et al. have reported organochlorine levels in milk, butter, corned beef, beef and mutton[27] fat. Apart from the latter, only very low residues were found. Butter samples contained up to 0.1 ppm DDT and 0.4 ppm of total DDT equivalent, while the levels in the fats of grass and store fed animals were indistinguishable. Dried full cream milk and other welfare foods including cod liver oil and concentrated orange juice were studied by Ruzicka, Simmons, and Tatton[28]. Small residue losses (of the order of 15—25%) in milk were noted on drying; refined cod liver oil contained less than 0.1 ppm BHC isomers, less than 0.2 ppm dieldrin and an average of about 1.7 ppm of DDT and related compounds including approximately 0.5 ppm pp'-DDT.

Nishimoto et al.[29] have examined cereals, fruit, and vegetables in Japan for persistent organochlorine pesticide residues: levels seldom exceeded 0.02 ppm BHC or total DDT and only occasionally small dieldrin residues were found. Some unpolished rice samples, however, contained up to 0.07 ppm BHC. Ligeti and colleagues[30], have examined foods of animal origin in Hungary for DDT content. Variable levels were found both in dairy produce and in other fats of animal origin, ranging to as high as 20 ppm for pork and poultry fats, and averaging of the

Table 3 Dieldrin Residues in Mutton Kidney Fat in England and Wales 1964–1968 (ppm)

Year	Range	Mean
1964	0.0–12.4	0.84
1965	0.0– 8.2	1.1
1966	0.0– 5.3	0.44
1967	0.0– 8.0	0.24
1968	0.0–10.4	0.21

order of 1 ppm for butter fat. In their study of Danish produce, Bro-Rasmussen et al.[31] attempt to identify the source of the organochlorine residues found in dairy produce by examining animal feed in addition. They concluded that in general some 50% (possibly more) of the mean levels for butter fat of the order of 0.05 ppm BHC (mainly α-BHC), 0.08 ppm total DDT and 0.04 ppm dieldrin, could be accounted for by the residues found in the feed, most of which were imported into Denmark. Corresponding residue levels in Britain in 1967 vary with the origin of the produce but are largely similar for BHC, normally less than 0.04 ppm dieldrin and up to 0.25 ppm total DDT.

Dietary Surveys

The original work of Walker et al. in the United States[1] was conducted in 1953 and was concerned with residues of DDT and DDE in meals prepared in restaurants and in correctional institutions and designed so far as possible to represent a cross section of foods consumed by the general public. Mean levels of 55 micrograms per meal of DDT and 28 micrograms per meal of DDE were found. For the purposes of comparing these results with those found in other meal surveys where the results are expressed on a daily basis, they can perhaps be raised by a factor of 3, giving estimates of 160 and 90 micrograms daily respectively of DDT and DDE and a total DDT equivalent daily intake of about 0.25 mg. This work was followed by a study in 1956—57 by Hayes et al. of DDT and DDE in people with different degrees of exposure of DDT in which both human fat samples and meals were examined[33]. Both meatless and ordinary meals were sampled, the daily DDT content of which were assessed at 0.041 and 0.184 mg respectively, values which do not differ substantially from those found by Walker et al.[1]. In 1963 Durham et al. made a direct comparison of the DDT and DDE content of *completed prepared meals* with the figures found 10 years earlier by Walker et al. and came to the conclusion that the DDT content of the *total diet* had decreased during this period[34]. The average for all meals (restaurant and household) examined in 1962—64, excluding a single meal for which exceptionally high results were found, was 123 micrograms per day of DDT and 57 micrograms per day of DDE, a total of approximately 0.2 mg per day, (just over 0.3 mg per day if the exceptional meal was included) and showed little change with respect to the earlier results. For restaurant meals alone, however, the average DDT + DDE level appeared to fall substantially, from 0.27 mg per day in 1953 —54 to 0.08 mg per day in 1962—64. Although gas chromatographic methods were available, colorimetric methods of analysis were deliberately used in order to facilitate comparison with earlier results. Durham et al.[34] also quote other meal surveys in the United States during the period 1954—60, including one conducted in an Alaskan hospital in 1959—1960[35] and conclude that the results for all of these are in general agreement with their own. They also found that there was no appreciable difference in residue between food items purchased at an ordinary grocery store and similar items purchased at a health food store. The above results

Table 4   Summary of DDT–DDE Residue Results in Meal Surveys in the United States micrograms per day

		DDT	DDE	DDT+DDE
(1)	1953	160	90	250
(33)	1954–56	120	64	184
(34)	1963	123	57	190
(all colorimetric)				

provide some indication of the intake in the total diet and are summarized in Table 4.

As already indicated, gas chromatography is a much more versatile analytical tool for residue analysis than the earlier colorimetric methods. For the most part the chromatographic methods have been used for total diet rather than individual meal studies. However, restaurant meals in Canada have been examined by Swackhamer in 1964.

Systematic studies of residues in the whole diet, representative of the *high consumption level* of a 16 to 19-year-old male have been conducted in the United States since 1961. The organization of these has been fully described by Cummings[37] and Duggan[38]. The diets are made up from food purchased from ordinary retail stores, spread over three cities in the United States, and are examined for a wide range of pesticide residues including organophosphorus and organochlorine compounds. The levels found in foods when prepared for eating are very low. Residues most commonly found have been DDT and its metabolite DDE, as shown in Table 5, in which the maximum levels quoted relate only to the most prominent residue[39]. Duggan and Weatherwax estimated the total daily intake of organochlorine residues to be 0.0014 mg/kg bodyweight, some two-thirds of which is DDT.

McGill and Robinson examined prepared meals representative of the *total diet* of the general population of South-east

England in 1965 [41]. Average daily intake levels of approximately 0.06 mg of DDT-type compounds and 0.02 mg of dieldrin (HEOD) per person were found, no other organochlorine compounds (aldrin, heptachlor and its epoxide, endrin and lindane) being detected using a method of analysis sensitive to 0.002 ppm. The analysis were done on uncooked as well as prepared foods; no significant losses of the residues occured in the cooking process. In a similar survey of dieldrin residues conducted two years later, McGill, Robinson and Stein [42] extended the sample coverage to be representative of Great Britain as a whole and found a mean level of 12.6 micrograms per day of dieldrin (HEOD) in whole prepared meals in 1967 as compared with 19.9 micrograms per day in 1965.

The organization of a total diet study in England and Wales in 1966—67 is des-

Table 5 Pesticide Residues Most Commonly Occurring in U.S. Total Diet Study[38]

Class in food	Most frequent Residue A	No. of samples in which A found*	Maximum levels ppm
Dairy produce	DDT, DDE, TDE	15	0.2
Meat, fish, etc.	DDT, DDE, TDE	18	0.9
Cereals	DDT, γ-BHC	17	0.03
Potatoes	HE, DDE, endrin	19	0.02
Leafy vegetables	DDT, DDE, TDE	12	0.3
Legumes	DDT	5	0.13
Root vegetables	DDT, DDE, dieldrin	9	0.07
Tomatoes, curcubits	DDT, γ-BHC, DDE, dieldrin	16	0.15
Fruit	DDT, DDE, aldrin	16	0.027
Oils and fats	HE, DDT, DDE, TDE	7	0.05
Sugar products	2,4-D, DDT	9	0.16
Beverages	(DDE)	1	0.003

* Out of 18 composite samples

Table 6 Summary of Organochlorine Pesticide Residue Results in Total Diet Studies micrograms per day

		DDT-type	Dieldrin	γ-BHC	
(9)	1964–66	100*	–	–	US
(41)	1965	60	20	–	UK
(44)	1966–67	44	7	7	UK
(42)	1967		13	–	UK

* Total organochlorine compounds based on 70 kg body weight (all gas chromatographic)

cribed by Harries, Jones and Tatton [43] and results for organochlorine residues by Abbott, Holmes, and Tatton [44]. The most frequently detected residues were DDT, dieldrin, and BHC.

The results of total diet surveys for organochlorine pesticide residues, summarized in Table 6, tend to confirm that total DDT equivalent levels in food as consumed are not rising.

Discussion

Both individual commodity and dietary (meals and total diet) studies have been used to estimate the amount of pesticide residue ingested by man. Total diet studies have been developed principally in the United States and the United Kingdom; the results, which to date relate mainly to the *persistent organochlorine compounds,* do not suggest any progressive increase in the levels of ingestion; indeed, the reverse might be true. The results of general studies on individual food commodities and selective surveys tend to support this view. Earlier meal studies have extended over a greater period with similar results. The average levels of total DDT in meals in the United States measured some 15 years ago suggested a daily intake of about 0.25 mg. Recent values in total diet studies using more modern methods indicate less than one-half of this level, while even lower levels are found in Britain. Average dieldrin residue levels in Britain have also fallen in recent years, as measured by both selective commodity studies (particularly mutton kidney fat) and by meal and total studies.

It is no part of this contribution to discuss in detail the significance of the residue results found. But in considering the results of surveys in the past 15 years, it is worth noting that virtually all the work on DDT has also drawn attention to the levels of DDE also present, as originally suggested by Walker and colleagues. Approximately one-third of the total DDT-derived residue in food as eaten is present as the dihydrochlorination product DDE and while the separate measurement of *DDT and DDE residues* is today a much more accurate process than it was 15 years ago, the significance of the presence of DDE residues in the food of man appears to merit fuller consideration. DDE is also associated with DDT residues where these occur in avian and other wildlife specimens and Risebrough et al.[46] have shown that this can affect *sterol degradation* with possible influence on calcium balance resulting in the production of *thin-shelled eggs*. Barnes has recently proposed that where in future DDT residues are studied, these are considered specifically in relation to the pp'-isomer[47]. The concept of *"total equivalent DDT"*, based on the sum of the residues of any DDT, DDE, or TDE isomers found, appears now to be a matter for further consideration.

References

[1] Walker, K. C., M. B. Goette, G. S. Batchelor: J. Agric. Food Chem. 2: 1034–1037, 1954
[2] Pearce, G. W., A. M. Mattson, W. J. Hayes, jr.: Science 116: 254–256, 1952
[3] Mattson, A., J. T. Spillane, C. Baker, G. W. Pearce: Anal. Chem. 25: 1065–1070, 1953
[4] The Pesticide Review 1967, U.S. Department of Agriculture, Washington, D.C.
[5] Campbell, J. E., L. A. Richardson, M. L. Schafer: Arch. Envir. Health 10: 831–836, 1965
[6] Lee, D. F.: J. Sci. Food Agric. 17: 561–562, 1966
[7] Duggan, R. E.: Pesticides Monitoring J. 2: 2–12, 1968
[8] Schechter, M. S., S. B. Soloway, R. A. Hayes, H. L. Haller: Anal. Chem. 17: 704–709, 1945
[9] Mitchell, L. C.: J. Assoc. Offic. Agric. Chem. 40: 999–1029, 1957
[10] Goodwin, E. S., R. Goulden, A. Richardson, J. G. Reynolds: Chem. & Ind. p. 1220, 1960
[11] Abbott, D. C., J. Thomson: Residue Reviews 11: 1–59, 1965
[12] Dale, W. E., G. E. Quinby: Science 142: 593–595, 1963
[13] Shell Method Series, No. 557/53
[14] Pocklington, W. D., J. O'G. Tatton: J. Sci. Food Agric. 17: 570–572, 1966
[15] Thompson, R. H., E. G. Hill: J. Sci. Food Agric. 20: 287–292, 1969
[16] Smart, N. A., A. R. C. Hill: J. Sci. Food Agric. 19: 315–316, 1968
[17] Hill, E. G., R. H. Thompson: J. Sci. Food Agric. 19: 119–124, 1968
[18] Thompson, R. H., E. G. Hill: J. Sci. Food Agric. 20: 293–295, 1969
[19] Rymer, T. E., J. H. Hamence: J. Assoc. Publ. Anal. 5: 24–26, 1967
[20] Hamence, J. H.: J. Assoc. Publ. Anal. 3: 17–20, 1964
[21] Hamence, J. H.: J. Assoc. Publ. Anal. 3: 130, 1965
[22] Egan, H.: Chem. & Ind. p. 1721, 1967
[23] Dickes, G. J., P. V. Nicholas: J. Assoc. Public Analysts 5: 52–57, 1967
[24] Dickes, G. J., P. V. Nicholas: J. Assoc. Public Analysts 6: 60–66, 1968
[25] "Joint Survey of Pesticide Residues in Foodstuffs Sold in England & Wales", Association of Public Analysts. Heffer, Cambridge, England (1969)
[26] Egan, H.: J. Sci. Food Agric. 16: 485–498, 1965
[27] Egan, H., D. C. Holmes, J. Roburn, J. O'G. Tatton: J. Sci. Food Agric. 17: 563–569, 1966
[28] Ruzicka, J. H. A., J. H. Simmons, J. O'G. Tatton: J. Sci. Food Agric. 18: 579–582, 1967
[29] Nishimoto, T., M. Uyeta, S. Taue: Shokuhin Eiseigaku Zasshi 7: 152–162, 1966
[30] Ligeti, G., B. Csiszar, Mindszenty: Die Nahrung 11: 369–374, 1967
[31] Bro-Rasmussen, F., S. Dalgaard-Mikkelsen, T. Jakobsen, S. O. Koch, F. Rodin, E. Uhl, K. Voldum-Clausen: Residue Reviews 23: 55–69, 1968
[32] Report of the Government Chemist, 1967, p. 97, H. M. Stationary Office, London 1968
[33] Hayes, W. J. Jr., G. E. Quinby, K. L. Walker, J. W. Elliot, W. M. Upholt: Arch. Ind. Health 18: 398, 1958

[34] Durham, W. F., J. F. Armstrong, G. E. Quinby: Arch. Environ. Health 11: 641–647, 1965

[35] Durham, W. F., W. M. Upholt, C. Heller: Science 134: 1880–1881, 1961

[36] Swackhamer, A. B.: Pesticide Progress 3: 108–114, 1965

[37] Cummings, J. G.: Residue Reviews 16: 30–45, 1966

[38] Duggan, R. E.: Pesticides Monitoring J. 2: 2–12, 1968

[39] Duggan, R. E., H. C. Barry, L. Y. Johnson: Science 151: 101–104, 1966

[40] Duggan, R. E., J. R. Weatherwax: Science 157: 1006–1010, 1967

[41] McGill, A. E. J., J. Robinson: Food Cosmet. Toxicol. 6: 45–57, 1968

[42] McGill, A. E. J., J. Robinson, M. Stein: Nature 221: 761–762, 1969

[43] Harries, J. M., C. M. Jones, J. O'G. Tatton: J. Sci. Food Agric. 20: 242–245, 1969

[44] Abbott, D. C., D. C. Holmes, J. O'G. Tatton: J. Sci. Food Agric. 20: 245–249, 1969

[45] Abbott, D. C., S. Crisp, K. R. Tarrant, J. O'G. Tatton: J. Sci. Food Agric. (in press)

[46] Risebrough, R. W., P. Reiche, S. G. Herman, D. B. Peakall, M. N. Kirven: Nature 220: 1092–1102, 1968

[47] Barnes, J. M.: Pesticides Abst. and News, 1969

Chemicals in the Environment
Some Aspects of Agricultural Chemicals

H. Hurtig

Research Coordinator (Environmental Quality) Canada Department of Agriculture
Ottawa, Canada

Summary. The possibilities of the experimental investigation of various factors and how environmental chemicals influence environmental quality are discussed. The limitations and necessary knowledge are shown with the example of 3 case studies in Canada on the mercury problem in agriculture, the transport of pesticides in the atmosphere, and a ring analysis on pesticide residues.

Zusammenfassung. Es werden kritisch die Möglichkeiten der experimentellen Untersuchung der verschiedenen Faktoren abgehandelt, wie Umweltchemikalien die Umweltqualität beeinflussen. Die Limitierungen und notwendigen Kenntnisse werden am Beispiel von 3 Untersuchungsreihen in Kanada über das Quecksilberproblem in der Landwirtschaft, den Transport von Pestiziden in der Atmosphäre und eine Ringanalyse auf Pestizidrückstände gezeigt.

The use of *agricultural chemicals* is in a dynamic state of reappraisal in several countries. Rarely does a month pass without some new concern being caused by the premature publication of the results of a partial investigation which is given more prominence by the communications media than is warranted by the adequacy of the research reported. On the other hand, the slow steady accumulation and evaluation of reliable evidence on *cause-effect relationships* and the new concern in several countries about *environmental quality* in general has drawn attention to *fertilizers* and *pesticides*. This is also in part due to the fact that the technology for monitoring some pesticides, their persistence and distribution is more advanced than for other organic chemical pollutants.

Similar concern is also providing the impetus for attempting to establish guidelines for other aspects of environmental quality in which agriculture is either the pollutor or the injured party. Probably the most important are:

1) *Animal and plant waste disposal* in relation to soil and water quality,
2) *Pesticide residue management*,
3) *Plant nutrients, heavy metals* and *trace elements* in soil,
4) Pollutants of non-agricultural origin which present a hazard to agricultural production or the quality of its products.

Many of us who are associated with providing advice on development of government policies have been caught up in the proliferation of international activities concerned with environmental quality.

While we complain about the necessity for preparatory work, case studies and task forces for intergovernmental working groups and the mountains of paper which will accumulate prior to the UN Conference in Stockholm in 1972, possibly for the first time we are casting a critical eye on the duplication of programs already in progress or being newly proposed by international agencies. For example, it appears now that programs for international collaborative monitoring of chemical pollutants which all include pesticides and heavy metals, either exist or are being proposed in the International Biological Program (IBP), World Meteorological Organization (WMO), UNESCO Man and the Biosphere Program, OECD, and the International Council of Scientific Unions (ICSU) new program under Scientific Committee on Problems of the Environment (SCOPE).

These bring into focus the question of what experience have we gained to date which would allow us to make an evaluation of feasibility or value of such programs and the requirements for research they may pose.

I have selected three topics based on recent Canadian experiences and case studies; the agricultural aspects of the mercury problem, a case study on the current state of knowledge concerning the atmospheric transport of pesticides and some preliminary results of a check sample analysis program for pesticide residues.

Agricultural Aspects of the Mercury Problem

As in Japan, Sweden, and several other countries, Canada has been concerned with the problem of *mercury in the environment*. The major problem has been the serious economic losses caused to commercial fisheries by the strict inspec-

Fig. 1 Distribution of samples collected for analysis in 1968

tion which forbids the sale of fish for human consumption containing mercury in excess of 0.5 ppm. In 1968—69 analysis of birds collected by the Canadian Wildlife Service suggested residues in muscle tissue of from 0.1 to 0.8 ppm (Fig. 1). Mercury in various sectors of the environment and its hazard is currently the subject of intensive investigation in Sweden, Japan, Canada, and the U.S.A.

Mercury has been used in agriculture for several years as a broad spectrum *fungicide* which controls many fungus problems especially in cereals and flax. In Canada aside from seed dressings, there were minor uses as foliar sprays and aerosols in greenhouses. Mercury use is gradually being discontinued for these purposes except for use for turf diseases on golf and lawn bowling greens. Cereal and flax seed treatment has seen a steady decrease from a peak year of 1964 due to economic factors, reduced wheat acreage and greater concern for the environment (Fig. 2).

Gurba (1971) has recently reviewed the situation with particular reference to the problem of *mercury residues in game birds, pheasants*, and *Hungarian partridge*, and summarized the results of investigations carried out in Alberta. Due to concern and inadequate information,

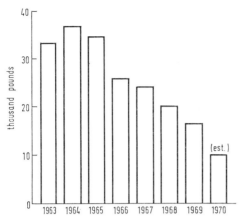

Fig. 2 Canadian use of mercury in seed dressings

the hunting season was closed in 1969, but improvement in management practices for handling treated seed and better investigations of seasonal levels of residues in these birds allowed the hunting season to be re-opened in 1970. Gurba's paper contains a good bibliography of recent pertinent literature.

While this type of management of the problem can alleviate the residue levels in the *seed eating birds,* there is still concern about the effects on reproduction in their predators and other wildlife (Table 1). *The cause-effect relationship* is confused, since eggs of some of the bird predators (e. g. peregrine falcon) contain not only Hg, but also *DDT* and its metabolites and residues of some

cyclodiene insecticides, notably dieldrin and heptachlor epoxide. The agricultural use of mercury in Canada represents only 4 % of the total national use for all purposes, i. e. similar to the U.S.A.

The information assembled by Gurba (1971) leads to the following conclusions:

1) Birds and rodents contained from 0.05 to 3.0 ppm in liver (Table 1). Natural background levels were up to 0.05 ppm.

2) Levels were higher in Alberta than Saskatchewan, on farms using treated seed (Table 2) and where surplus seed was disposed of in a careless manner.

Table 2 Mercury residues from treated and untreated areas (ppm)

		(n)	Treated	(n)	Untreated
Birds					
a	muscle	(4)	1.73	(5)	0.13
b	liver	(5)	2.78	(3)	1.25
c	liver	(6)	1.41	(2)	0.03
Mammals					
a	liver	(3)	2.19	(3)	0.29
b	liver	(2)	0.45	(2)	0.01
c	liver	(20)	2.085	(15)	0.358

n: no of samples; a, b, c: species
(from Gurba, 1971)

3) Initial analyses of residues from four laboratories showed wide variations in results. Confirmatory methods of analysis, the caliber of the analyst involved and experience are essential to produce reliable data (Tables 3, 4).

Table 1 Mercury Residues in Alberta wildlife, 1968 (ppm)

Passerine birds	(8)	Br. muscle	0.49
Passerine birds	(16)	Liver	0.55
Doves and pigeons	(12)	Liver	0.56
Waterfowl	(6)	Eggs	0.05
Upland game birds	(11)	Liver	2.4
Ground squirrels	(5)	Liver	1.5
Large falcons	(12)	Eggs	0.63
Hawks and eagles	(21)	Eggs	0.23
Owls	(3)	Liver	3.0

(from Gurba, 1971)

Table 3 Mercury analyses by four laboratories game birds collected, September 1969 (ppm)

	Cal. Gulf	Univ. Tor.	Prov. Anal.	Pl. Prod.
Range Wet Wt.				
0.001–0.049	40	22	64	0
0.050–0.099	0	1	9	0
0.100–0.499	2	1	24	12
0.500–0.999	1	0	2	0
No. of Analyses	43	24	99	12

Some were split samples, others not.
(from Gurba, 1971)

Table 4 Mercury analyses by three laboratories – 13 game birds – September 1969 split samples (ppm)

No.	Cal. Gulf.	Univ. Tor.	Prov. Anal.
1	0.006	0.050	nil
9	0.035	0.020	nil
15	0.812	0.050	nil
20	0.033	0.026	nil
21	0.012	0.028	nil
23	0.029	0.034	nil
24	0.015	0.052	0.04
25	0.017	0.013	0.25
27	0.014	0.031	0.02
42	0.027	0.028	nil
43	0.106	0.037	0.36
48	0.014	0.021	0.17
89	0.007	0.025	nil

(from Gurba, 1971)

Table 5 Mercury analyses in breast muscle pheasants and H. partridge, 1970 (ppm)

Date*	No. Birds	Pooled	Hg. Av.	Hg. Range
April	130	13	0.02	0.01–0.05
May–June	232	32	0.16	0.02–0.49
July	80	9	0.25	0.14–0.34
September	108	14	0.10	0.04–0.24
Total	550	68		

* Pre-, during, post-seeding; pooled 3 to 11 same species
(from Gurba, 1971)

Table 6 Mercury analyses in breast muscle pheasants and H. partridge, July (ppm)

Year	No. Birds	Pooled	Hg. Av.	Hg. Range
1968	11	11	0.6	0.11–1.48
1969	26	9	0.45	0.24–0.79
1970	80	9	0.25	0.14–0.34

(from Gurba, 1971)

4) In seed-eating game birds, mercury levels increased from an early spring low of 0.02 ppm to a peak of 0.25 ppm in July, dropping below 0.1 ppm in the autumn (Table 5). (None of the pooled averages exceeded 0.5 ppm).

5) Average Hg levels in July decreased from 0.6 ppm in 1968 to 0.45 ppm in 1969 and 0.25 ppm in 1970 (Table 6). Seed treatments decreased in Canada (Fig. 2) also in Alberta by 22% in 1969 and 42% in 1970 (Tables 7 + 8).

A public *education program* on the proper handling of treated seed, the necessity for secure covering of treated seed in transit and disposal of surplus treated seed have had their effect. The plant operators in the 68 Alberta seed treating plants have been trained and licensed. Inspections carried out by 117 provincial and municipal Alberta inspectors reported no surplus treated seed was disposed of in 1970 except at approved sites.

Information reviewed to date indicates that while egg production and hatchability was not affected in domestic chickens fed treated grain, pheasants showed no loss in egg production but a significant

Table 8 Alberta seed-cleaning plants (in million bushels)

	1967	1968	1969	1970
Bus cleaned	18.2	18.7	18.5	19.3
Fungicide	6.0	6.4	5.0	2.9
Insecticide	1.2	1.9	0.7	0.3

(from Gurba, 1971)

Table 7 Annual sale of mercury seed dressing in Canada

	1963	1964	1965	1966	1967	1968	1969
Organic Mercurials							
Powders, lbs.	81,192	376,670	167,844	50,360	55,380	17,579	4,983
Drillbox, lbs.				310,393	170,760	174,018	49,817
Liquids, Imp. gal.	167,620	132,014	147,641	111,050	119,312	104,289	54,441
Organic Mercurials and Insecticides							
Powders, lbs.	131,670	96,505	179,678	28,504	25,394	10,632	
Drillbox, lbs.				107,897	142,109	77,841	
Liquids, Imp. gal.	41,183	25,251	62,251	39,131	20,107	30,172	22,529

Source: DBS (Cat. No. 46–212)
(from Gurba, 1971)

decrease in hatchability after six days of feeding treated grain.

Mercury treated seed is not involved in contamination of water or fish. In soil, treated wheat supplies only 0.011 ounces of Hg per acre. This is equivalent to 0.2 % of the natural occurring Hg in soils and about the same amount which may be supplied by rain each year.

In December 1970 Canada stopped the manufacture or import of further mercury seed treatments but sales of current stocks were permitted for 1971 under revised registration. All registrations are currently under review and the present policy will lead to normal disposal of stocks of formulated material now on hand.

Since the status of aldrin, dieldrin, and heptachlor as cereal seed treatments for wireworm control is also under review it is quite likely that major changes of the total picture of *seed treatment chemicals* will occur. Current research on *lindane* suggests its use for this purpose will present lesser environmental hazards, but a decision cannot really be made until a comprehensive evaluation is made of the hazards associated with the change to combination fungicide-insecticide treatments. Likely replacement mixtures may be combinations of the fungicides 2,3-dihydro-6-methyl-s-phenylcarbamyl-1,4 -oxathiin (the Uniroyal compound *Vitavax*), also known as *"Carboxin"* in the U.K., with *maneb* and *lindane*. The acceptability from the environmental standpoint of *thiram* (bis(dimethylthiocarbamoyl)disulphide) for treatment of corn and pea seeds, as a replacement for mercury also requires careful evaluation. It is doubtful that change to other forms of mercury such as used in Sweden and the U.S.A. will be any advantage, since higher dosages are required and methylation of the other forms of mercury is still a threat.

If nothing else, we have learned a great deal from these experiences and should know what questions to ask concerning the replacement compounds. Similarly, there can be no complacency in programs to educate all concerned in the proper *management of pesticide use.* The experience in Alberta indicated that good management practices in seed-treating plants, transportation and disposal were just as important as proper use in the hands of the farmer.

The most important problem to be resolved is the real *hazard to man* and other important users of our environment *from mercury* currently used for *other than* agricultural purposes and from naturally occurring mercury.

A Case Study on:

Pollution in Relation to Agricultural Pesticides:
Atmospheric Transport of Pesticides*

The Canada Committee on Agrometeorology serves as an advisory committee on national problems in agrometeorology. Its members are drawn from the government and university scientific communities. Each year, in addition to reviewing the state of progress in agrometeorology research and services, the Committee enlists the assistance of additional experts to carry out an in-depth study of a special subject. In 1970—71 Bergsteinsson (Saskatchewan Resarch Council) coordinated the preparation of a series of position papers on "Pollution in Relation to Agricultural Pesticides: *Atmospheric Transport of Pesticides"*. The position papers were reviewed by the CAA and the authors at the 1971 meeting, leading to the development of a set of conclusions and recommendations.

---

* Summarized from: "Report of the 12th Annual Meeting, Canada Committee on Agrometeorology to the Canadian Agricultural Services Coordinating Committee, April 1971."

This study was initiated, since there are well documented instances in which pesticides persist and become pollutants when dispersed or transported beyond the target area. Meteorological conditions directly affect the *release, distribution,* and *deposition of pesticides.* The mechanism of these processes, particularly on a micro-scale, is inadequately understood. Fluctuations of specific meteorological elements such as temperature, humidity, wind and radiation cause a variety of physical and chemical reactions that probably affect the measurable end result — the amount of pesticides detected on plant and soil samples.

Increased use of pesticides in the past 25 years and the growing awareness of the fragility of our environment demand an examination of the role of *pesticides as possible atmospheric pollutants.* Contributions by invited specialists dealt with airborne pesticides in relation to their potential sources, spray dissemination, micro-scale, meso-scale and synoptic-scale transport, analytical techniques and procedures, biological significance, and monitoring by a national network.

The contributors were:
1) Potential Sources of Airborne Pesticides
   J. L. Maybank, Physics Division, Saskatchewan Research Council, Saskatoon, Sask.
2) Pesticide Spray Dissemination,
   J. Maybank, Physics Division, Saskatchewan Research Council, Saskatoon, Sask.
3) Microscale and Mesoscale Transport
   K. D. Hage, Univ. of Alberta, Edmonton, Alta.
4) Synoptic Scale and Long Distance Transport of Pesticides, E. I. Mukammal, Canadian Meteorological Service, Toronto.

5) Deposition of Pesticides,
   J. V. McCallum, Defence Research Board, Suffield, Alta.
6) Analysis of Pesticides,
   H. V. Morley, Analytical Chemistry Research Service, Canada Agriculture, Ottawa, Ont.
7) Biological Significance of Deposited Pesticides, J. R. Hay, Research Station, Canada Agriculture, Regina, Sask.
8) Monitoring of Airborne Pesticides,
   J. L. Bergsteinsson, Physics Division, Saskatchewan Research Council, Saskatoon, Sask.

This complete set of papers is now in press and will be available during 1971 in one specially prepared volume. The conclusions and recommendations resulting from the Committee's review were transmitted to the Canadian Agricultural Services Coordinating Committee in April, 1971.

Conclusions and Recommendations

1) The CCAM found that there is no comprehensive Canadian or international body of statistical data on pesticide types, released amounts and release times in relation to geographic areas. The Canada Committee on Agricultural Meteorology shares the concern of the Canada Committee on Pesticide Use in Agriculture in regard to the need for meaningful *statistics concerning pesticide use.* In order to objectively assess the hazards accruing from *pesticides drift,* it is necessary to have precise information concerning the types, quantities, and times of pesticide releases. It is, therefore, recommended that appropriate action be taken to implement a uniform national system of reporting pesticide use appropriate to quantitative treatment of the pesticide transport problem.

2) The Committee noted that there are no standard procedures outlined for the evaluation of *spraying equipment* and that the relationship between pesticide droplet size and specific toxicity is inadequately understood. In order to minimize pesticide drift hazards, it is necessary to know quantitatively the relationship between pesticide droplet size and *specific droplet toxicity* to target organisms. When this relationship is known, an attack can be made on elimination of smaller, driftprone droplets. A major need for research on this problem is recommended. Furthermore, a need is noted for the development of standardized techniques to evaluate the performance of spraying equipment, both in the areas of laboratory meteorology and field kits for approximate calibration.

3) A variety of post-application sources and *sinks for pesticides* exist in the natural environment. Material entering many sinks can be re-entrained into the horizontal wind field following various retention times. During its dwell time in the free atmosphere or in the various environmental sinks, a pesticide may undergo a matrix of chemical reactions affecting its toxic properties. A lack of knowledge concerning the magnitudes of the various fluxes by which pesticides leave and re-enter the wind field was noted by the Committee. Adequate meteorological and pesticides flux data are not yet available because of lack of suitable instrumentation in the past. The recent development of a flux measuring system, using the eddy correlation method, offers new and promising approaches to the quantitative analyses of *pesticides flux.* It is recommended that research into these problems be intensified.

4) The similarity of large-scale transport of pesticides to that of other particulates such as spores was noted and support for the continuance and expansion of research into this aspect of the total transport problem is recommended.

5) The complexity of numerical modelling the effective *deposition of pesticide droplets* on plant canopies and other types of surface was identified. Research in the field of deposition modelling should be encouraged.

6) The lack of recognized standard techniques and procedures for analysis of *airborne pesticides* was noted. Additionally, there is little knowledge of chemical species and toxicity of pesticide reaction products formed while the present compounds are in transit. This Committee endorses earlier recommendations by the Canada Committee on Pesticide Use in Agriculture that appropriate action be taken to establish *analytical and procedural standards.* Furthermore, it is noted that there is a paucity of work on the interaction between *airborne pesticides* and *other atmospheric pollutants.* Such interactions and the potential of their products as toxicants should be investigated.

7) The *length of exposure to pesticides* is probably more significant than the pesticide concentration. Also the biological effects produced by a pesticide vapor as compared to those produced by droplets of the same compound are not fully appreciated. The encouragement of research in these areas is recommended.

8) While the CCAM does not at present identify a need for a national network for the monitoring of airborne pesticides, it is recommended that one prototype hardware system be developed. Expertise must be gained and a system must be available as a research tool to investigate transport problems and to evaluate standard measurements procedures. Further-

more, it is recommended that a feasibility study be conducted on various strategies for a national network to monitor airborne pesticides.

Check Sample Analysis Program Preliminary Results of Residues in Soil

The Canada Committee on Pesticide Use in Agriculture in 1967 initiated a *check sample program on pesticide residue analysis*. Separate programs were developed for soils, animal feed, and human food. The number of laboratories participating varies on each program annually but the total has averaged approximately 25 federal, provincial, private, and industrial laboratories. The results up to 1970 are currently in the process of being evaluated and prepared for publication under the guidance of the special area coordinators concerned, Dr. H. V. Morley (Soil), Dr. K. A. McCully (Food), and Dr. J. Singh (Feed). The central evaluation of accuracy of analyses is circulated back to the individual laboratories and the summary made available to all without the laboratories being identified except by a code number.

Associated with the program, annual workshops of all interested pesticide chemists are held in eastern and western Canada. Annual attendance is approximately one hundred working chemists.

The object is not to evaluate the performance of any single laboratory but to realistically appraise *the analytical capability for pesticide residue analysis* and provide help and advice when requested. In the early stages only organochlorine pesticides were included. As analytical capability improved, herbicides and organophosphorus compounds have been included in samples sent to the participating laboratories.

The example chosen for presentation here is based on the report (Morley 1968) on six soil samples sent to twelve participating laboratories in 1967. *Performance has improved since that time.*

Previous work (Chiba and Morley 1968, 1968a) had demonstrated that only field treated soils should be used, since fortification is a poor criterion for extraction efficiency. The analysts were to use the method of analysis in vogue at their laboratories and report information on procedures according to an outline provided.

Each laboratory also received solutions of pesticide standards. In 1967, only 68 % of the results were in the ± 20 % range of the quantities in solution. Samples of the same standards were sent out two months later with marked improvement in performance; 90 % of results were in the ± 20 % range and 78 % in the ± 10 % range.

The results of the participating laboratories analysis of field treated samples of three soil types and an extract fortified with known amounts of heptachlor, heptachlor and lindane are summarized in Table 9.

The breakdown of error as determined by reports on individual pesticides and soil samples by each laboratory is illustrated in Fig. 3 — aldrin, Fig. 4 — dieldrin, Fig. 5 — o,p'-DDT, Fig. 6 — p,p'-DDT and Fig. 7 — p,p'-DDE (Morley 1968).

These pilot experiences and the results of the program in the past four years suggest that the final results are only as good as the analyst presenting them.

Table 9

Residue	Total Analyzed	% of Results in Range ±10	±20	>±20
Aldrin	32	40.6	78.0	22.0
Dieldrin	31	35.5	84.0	16.0
p,p'-DDE	23	4.4	26.2	73.8
o,p'-DDT	12	25.0	50.0	50.0
p,p'-DDT	36	11.1	33.3	66.7

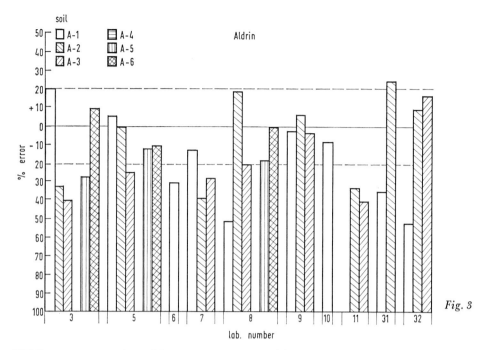

Fig. 3

While some agencies with statutory responsibilities may have a requirement to use routine standardized methods of analysis, at this stage it would be undesirable to do so now. The importance of development of *confirmatory methods of analysis* and *positive identification of residues* has been emphasized.

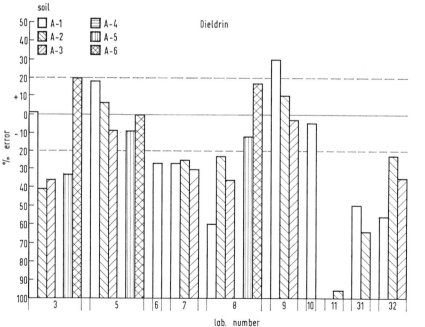

Fig. 4

Conclusions

This program, experience with *check sample analysis for mercury* (Gurba 1971), and the position papers developed by the review of the state of knowledge on *long distance transport of pesticides,* suggest that the skill and experience of the analyst are the most important feature required today. When approaching the subject of *ecological implications of pesticides* or other chemicals in our environment the *lack of methodology* is still a major obstacle, especially in any proposed air, water or soil monitoring program.

and extraction procedures are basic to any proposal to establish sampling stations and procedures.

Similar related experience has been reported in an international cooperative study of organochlorine pesticide residues in terrestial and aquatic wildlife sponsored by OECD in 1967 and 1968 (Holden 1970). In this study another variable was emphasized, i.e. the *range of variation of residues* among individuals of a natural population was much larger than that due to analytical errors or differences between laboratories.

For these reasons, proposals to establish

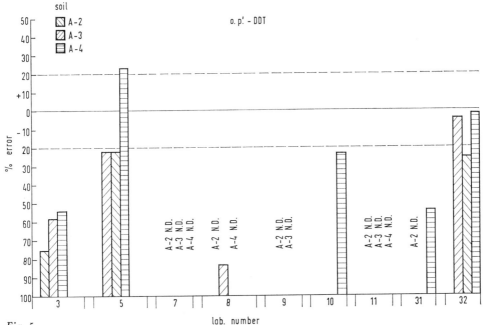

Fig. 5

The sophisticated instrumentation which would be required for global monitoring, its proper operation, maintenance and ultimate evaluation of results would have to be based on *standardized tested techniques* which do not yet exist. Going back deeper into the roots of the conceptual approach, the development of *standardized statistically significant sampling*

international *global monitoring programs* should be examined critically and the research requirements carefully defined and satisfied before international agencies launch such programs. A basic preliminary requirement is to have the scientists concerned with evaluation of cause-effect relationship in establishing criteria for environmental quality to first define

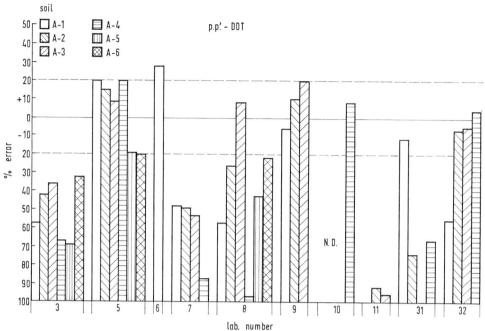

Fig. 6

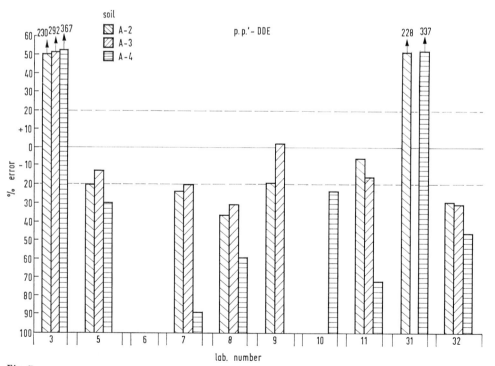

Fig. 7

what chemicals should be monitored and in what substrates. Communication between the disciplines concerned must be achieved first.

References

Chiba, M., H. V. Morley: Studies of Losses of Pesticides During Sample Preparation. J. Assoc. Off. Anal. Chem. 51: 55, 1968

Chiba, M., H. V. Morley: Studies on Factors Influencing the Extraction of Aldrin and Dieldrin Residues from Different Soil Type. J. Agr. Food Chem. 16: Nov./Dec. 1968 a

Gurba, J. B.: Use of Mercury in Canadian Agriculture, Proceedings of Royal Society of Canada Symposium on Mercury in Man's Environment, Feb. 1971 Ottawa, Canada (in press)

Holden, A. V.: International Cooperative Study on Organochlorine Pesticide Residues in Terrestial and Aquatic Wildlife, 1967/ 1968. Pesticide Monitoring Journal 4: 117–135, 1970

Morley, H. V.: The Determination of Organchlorine Pesticide Residues in Soil. Proc. 1st Seminar on Pesticide Residue Analysis, Eastern Canada, 18–19 Nov. 1968. Published by Ontario Department of Agriculture and Food 1968

Food Additives

R. Franck

Max-von-Pettenkofer-Institut des Bundesgesundheitsamtes,
1 Berlin 33, Postfach, West Germany

Summary. In this paper, a survey is given dealing with those substances that are added to food, intentionally or unintentionnally, and that are available at the present time. They are divided according to the kind of activity of these food additives.

Zusammenfassung. Es wird eine Übersicht über die zur Zeit erlaubten und in der Nahrung beabsichtigt sowie zufällig vorkommenden Nahrungsmittel-zusatzstoffe gegeben. Die Aufteilung erfolgt nach der Art der Aktivität dieser Zusatzstoffe.

This paper is to give a survey of those substances that are added to food — intentionally or unintentionally — during food extraction, production, stocking, or preparation. *"Food additives"* are defined as follows by the "Food Additive Committee" of the Joint FAO/WHO Codex Alimentariums Commission:

Food additive means any substance, not normally consumed as a food by itself, whether or not it has nutritive value, the intended use of which results directly or indirectly in it or its by-products becoming a component of, or otherwise affecting the characteristics of a food.

This definition does *not* include contaminants and pesticide residues. Even if these food additives are of topical interest and have been a cause of anxiety to the public for the last two or three decades only, many of them have seen widespread use for some time. Common salt as well as vinegar have been used for seasoning and canning for many centuries. One can read in Homer, too, that the butts of wine had already been sulphurized in the antiquity in order to keep the wine unspoiled, to protect it from

getting brown and from contamination by pathogenous microorganisms. Finally, let us think about the use of *sodium nitrate* for pickling or smoking in order to keep meat, fish, etc., from spoiling.

To get an impression of food additives presently in use, a sectio in Souci* is divided between:

a) *intentional additives*

b) *unintentional additives*.

Group a) are:

Table 1 Substances with chemical activity

1. anti-microbial substances
2. antioxidants
3. complex formers
4. substances against incidental discoloring and decolorants
5. other substances

Table 2 Substances with physical activity

1. dyestuffs
2. thickening and gelling agents
3. interfacially active substances
4. moisture keeping agents
5. anti-foam agents
6. baking agents
7. coating material
8. mould release

* S. W. Souci: Handbuch der Lebensmittelchemie. Bd. I, 1965, p. 1060.

Table 3 Substances with physiological activity

1. Substances for the amelioration of smell and taste
2. dietetically important substances

Intentional additives are also indicators and means of identification, for example, amylum in margarine and hydroxy methylfurfurol in artificial honey.

Substances with Chemical Activity

In this group, we may mention first the substances with *anti-microbial activity*. They are canning material in a more restricted sense, i. e. substances destroying microorganisms during the extraction, preparation, or stocking of food (bactericides) or substances inhibiting the growth of microorganisms (bacteriostatics). These substances are used to avoid or at least to retard the spoiling of food in order to keep it edible for a longer period of time; This is of concern — being of importance in our highly technical world of today — because the growth of our cities, the evolution of supermarkets, etc., forces us to a continually increasing storage of provisions. Others of this group are also substances that have *canning activity* — they have been long known — without being food additives according to the sense of our definition, but rather are used as food by themselves, i. e., sugar, vinegar, common salt, ethanol, acetic acid, lactic acid, tartaric acid, and citric acid.

The following tables illustrate examples of antimicrobial substances (Tables 4 and 5):

Table 4 Customary canning material

1. sorbic acid and compounds
2. benzoic acid
3. p-hydroxybenzoic acid-ethylene ester – propyl ester
4. formic acid and compounds
5. propionic acid and compounds
6. diphenyl
7. orthophenylphenol

8. thiabendazole
9. sulphur dioxide, H_2SO_3, Na_2SO_3, $NaHSO_3$, $Na_2S_2O_3$, $K_2S_2O_3$, $Ca(HSO_3)_2$
10. silver
11. ozone
12. hydrogen peroxide
13. diethylene pyrocarbonate
14. antibiotics (nisine, chlorotetracycline)

Table 5 Possible canning material

p-chlorobenzoic acid
dehydracet acid
monobromacetic acid
monochloracetic acid
formaldehyde, trioxymethylene
hexamethylenetetramine
salicylic acid
chlorine and compounds
chlorine dioxide
hydrofluoric acid
sodium nitrite
hydroxy quinoline
thiocarbamide
sulphonamide
ethylene oxide
methylene bromide

Antioxidants

Antioxidants are especially added to food in order to protect fat from oxidative spoiling. Furthermore, they extend the storage life of food that has been prepared and produced with fat. Antioxidants are used as well for the protection of the lipoid components present in food, e. g. in potatoes and carrots. Autoxidation of fat involves so many partial and chain reactions that numerous conversion products are produced and no uniform interpretation for the use of antioxidants and their synergists is possible. Still today antioxidants are used on an empirical-statistical basis.

It has been observed that fats in the seed of plants and those in animal tissues are more resistent and do not become rancid as long as they remain in their natural protective coverage. This can be explained by the fact that the lipids in their natural cell binding, are associated to natural antioxidants.

Tables 6 and 7 show the *technically used antioxidants.*

Table 6 Antioxidants
_____

a) natural
toxopheroles
L-ascorbic acid
    sodium salt, calcium salt
L-ascorbyl diacetate
L-ascorbyl palmitate
D-ascorbic acid
guaiac resin
NDGA
flavonoide: rutin, quercetin

Table 7 Antioxidants
_____

b) synthetic
gallic acid-alkylene ester
    (propyl-, octyl-, dodecetyl gallate)
butylhydroxy anisole – BHA
butylhydroxytoluol – BHT
hydroquinone
catechol
diphenyl-p-phenylendiamine
thiocarbamide

The group of complex formers comprises, e. g. the ethylenediamine-tetra acetic acid *(EDTA),* which particularly is added to food such as sodium salt or sodium calcium salt, because it forms stable complexes with traces of metal which catalyze the autoxidation of the fat. Thus, EDTA is used, e. g. in potato salad, mayonnaise, baking fat. (In the U.S.A. it is allowed for up to 75—275 mg/kg).
The next group comprises substances used against undesired discoloring and *decolorants.* These are, above all, substances which, upon reductive and oxidative procedures, have a decoloring activity, such as sulphuric acid. Further substances belonging to this group are those having a stabilizing activity on color, such as *nitrites,* by the formation of new compounds. Further examples (Table 8):
The next group consists of *dyestuffs* serving to enhance the appearance of food. For many dyestuffs used in food, the toxicity is determined nationally and internationally, by tests, descriptions, and

Table 8 Color amelioration products and/or decolorants
_____

sulphuric acid
nitrates, nitrites
ozone, hydrogen peroxide
potassium bromate
supersulphate
benzoyl peroxide
chlorine dioxide
sodium-, calcium hypochloride
nitrosyl chloride (NOCl)
nitrogen trichloride (NCl_3)
nitrogen oxides NO, N_2O_4

physical properties required by the WHO. This subject should hardly cause difficulties today.
Instead of a tabulation of the numerous dyestuffs which may possibly be added to foods, I should like to give a compilation of those that have been indicated by the WHO as *undesirable substances in food.* The survey might be of great interest to the analyst because he should pay special attention to these substances when investigating foods (Table 9).

Table 9 Dyestuffs

not to be added to food	color index No.
auramine	41000
methyl yellow	11020
chrysoidine	11270, 11270 B
guinea green B	42085
FD and C green No. 1	
magenta	42510
fuchsine	
oil orange SS	12100
oil orange XO	12140
yellow AB	11380
yellow OB	11390
ponceau 3R	16155
ponceau SX	14700
sudane orange I	12055

(FAO/WHO-Cx 4/50.3-Cx/MAS/70 A 5)

As examples for the *substances with physical activity* I should like to mention the *thickening and geling agents* (Table 10) as well as the interfacially active substances (Table 11). Table 12 shows examples of additives for foods for special, dietary uses.

Table 10 Thickening and gelling agents

gelatine
amylum
  also modified:
  esterified with phosphate or acetate
  decomposed oxidatively
  etherified
pectins
dextrins
cellulose ether
carob core flour
tamarind seed flour
alginate
carragene
agar-agar
gum tragacunth
rubber arabicum (gum arabicum)
caraya gum
guar flour

Table 11 Emulsifiers and Stabilizers

lecithins
aliphatic acid, mono- and -diglyceride
  (monodiacetyl wine acid ester, aceto aliphatics)
sugar aliphatic acid
aliphatic alcohols
polyglycerine ester
oxidated and polymerized vegetable oils
polyethyleneglycols
polyehtylene-sorbitane-aliphatic acid ester
  (Tweens)
sorbitane-aliphatic acid ester (Spans)

Table 12 Additives for foods for special dietary uses

L- and DL-lysine
glutamate
iron compounds
vitamins
trace elements
saccharines:
  saccharin
  dulcine
  cyclamate
  sorbite
  xylite
potassium iodide

The compilations of substances shown here, which are possible food additives, demonstrate once again that the legislator as well as the producer and consumer of food have to consider an extensive and abundant field.

A number of countries have made up legal regulations for the admission of intentional additives. These regulations have been worked out by various experts and they are internationally acknowledged. The most important general points of view are the following:

1) The *use of additives* only seems justified if there is a need in the consumer's interest or if the technical necessity has been proven. If one uses a standard here, the following reasons will justify an admixture of a food additive:

a) keeping of the nutritive value of a food
b) increase of the storage life
c) amelioration of production, working up, preparation, and stocking of a food
d) proven dietetic necessity of a certain food for special diets.

2) The use of additives does not seem justified if the health safety of a substance has not been proven unequivocally. According to general opinion this can only be ascertained by setting up "positive lists" containing all those substances having proven to be not injurious to health.

To examine *health safety*, the following items will be required before admission:

a) sufficient toxicological, biochemical, and physiological nutrition investigations, according to the regulations established by experts of the FAO/WHO;
b) consideration of a sufficient safety factor, in case of permanent intake;
c) examination of possible effects in case of simultaneous intake of several food additives;
d) production of degradation products and conversions in parts of food important for physiologically nutrition.

Furthermore, for the use of food additives, it is absolutely necessary to insure that the specifications have been established by experts. On the one hand, one should pay attention to known possible pollution by inorganic ions such as arsenic, lead, antimony, chromium, tin, cadmium, mercury, selenium, telluride, and thallium. On the other hand, for non-uniform groups of substances, for example, emzymatic preparations, exact specifications will have to be required. Special controls have been developed for dye-stuff, which, in order to eliminate *cancerogenic activities,* must be free from β-naphthylamine, benzidine, policyclic aromatic amines, and others.

Closely related to these requirements would be the obligatory condition to permit admissions only in case where there are recognized analytical methods to detect the food additive in question, quantitatively, and to check its purity.

Furthermore, for the use of food additives, care should be taken that the quantities to be used are chosen, so that a sufficient technological activity is present, which assures health safety. Food additives may only be used with pure foods, and only in cases where the food does not have the appearance of being better than it is, i. e. without deluding the consumer about the real value of the food. It is for this reason that we resist on principle, the marking of additives in quality and quantity (Table 13).

The possible contamination from manufacturing equipment, and later from

packing materials are not indicated in this survey. Furthermore, one should think about the possible *contamination of food* produced technically, e. g. by residues of *detergents* and antiseptics present in the equipment.

The last survey shows those substances reaching our food unintentionally, namely the substances used in agriculture. The permanent increase of the Earth's population requires profitable use of all means offered to increase food production. With the intensification and rationalization of agricultural production, not only the contagious diseases among animals, but also insect pests and plant diseases increase, as a result of better life conditions in large monocultures. In modern intensive breeding this leads to a permanently increasing application of active material. Within plant and provision protection, the use of chemical substances is as unavoidable as the use of chemotherapeutics in medicine (Table 14, Fig. 1).

Table 14 Substances used in agriculture

a) animal breeding
    anti-biotics
    hormones
    anti-parasitics
    drugs
b) soil-treatment material fertilizers
    soil loosening material
    soil desinfection material
c) pesticides
d) provision protectional material

Table 13 Ancillary substances used for production

solvents and extraction material
catalizers
adsorption- and precipitating agents
enzymes
acids (hydrolyses)
leaches (neutralization agents)
buffering agents
propellants (fluor- and chlorine hydrocarbons)

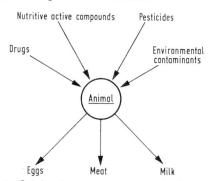

Fig. 1. Origin of various residues in food derived from animals

Toxic Microelements and Therapeutica in Food of Animal Origin

W. Kreuzer

Tierärztliche Fakultät der Universität München, 8 München, Veterinärstraße 13
West Germany

Summary. By means of examples on the occurrence of toxic micro-elements, especially Pb, Hg, Cd, As, and Se, as well as antibiotics in food of animal origin, the necessity for legislation is pointed out, which should follow the rules for the regulation of residues in vegetable foods.

Zusammenfassung. Anhand von Beispielen über das Vorkommen toxischer Mikroelemente, besonders Pb, Hg, Cd, As und Se, sowie Antibiotika in Nahrungsmitteln tierischer Herkunft wird auf die Notwendigkeit einer Gesetzgebung hingewiesen, welche den Richtlinien für die Regulierung der Rückstände in pflanzlichen Nahrungsmitteln folgen sollte.

During 1967 and 1968, approximately 65% of the total protein, and 55% of the fat-content ingested with our food in Germany, were based on animal products. In addtion to this, three quarters of the calcium, 60% of the phosphorus, and 33% of the iron taken in were present in this food. Furthermore, it contributed to the intake of *vitamin B₁* (thiamine) by nearly 71%, of *vitamin B₂* (riboflavin) by about 58%, and of niacin by 51%[1].

These data demonstrate by themselves the importance of food of animal origin in our nourishment. Their impairment by toxic microelements and veterinarian drugs is therefore of topical interest.

As a result of industrial activity, *toxic microelements* in highly industrialized countries and veterinarian drugs in areas with highly developed *intensive-livestock keeping* and breeding have become an urgent food hygiene problem.

The contamination of food of animal origin has become considerably more complex by the utilization of food additives and veterinary drugs than it is in the case of vegetable food; in addition, the contamination situation is further complicated by the link of "animal" within the food chain.

In this process, the animal organism or some of its tissues may act as a selector or accumulator of certain environmental contaminants.

Among the toxic microelements which require a greater interest are those *contaminants of food of animal origin*, the ecological behavior of which favors their concentration in certain ecosystems as well as in plant and animal organisms, and/or those which on account of their high toxicity will be injurious to animals and man, even in the smallest quantities.

At the height of our present knowledge contaminants which are considered especially hazardous to man are Hg, Cd, Pb, As, and Se. Organic Hg and Pb are dangerous mainly because of their damaging effect on the *central nervous system;* inorganic As and Se because of their car-

cinogenicity. The influences of Cd on certain cardiovascular diseases (high-blood pressure) are under discussion [9, 26, 37].

Predominantly chronic intoxications caused by environmental contamination have been reported. In lambs [44], Canadian geese [1], and (Trauer-)pigeons [33] it was caused by Pb; in pigs [11] and birds [5] by Hg; and in lambs, sheep, and cattle by Se [29, 38, 39, 53].

The *Pb-poisoned lambs* [44] had been grazing on the grounds of an old lead mine which had been closed for a long time; the *Hg-poisoned pigs* [11] had been fed for 6 to 8 weeks on refuse of seed grain treated with organo-mercury fungicides. Se-intoxication was especially noticed in domestic animals in certain selenium-rich districts of the USA, Canada, South America, Morocco, and Algeria as well as in Ireland and France [29, 39].

With all intoxications, the feed usually contained increased concentrations of Pb [44], or Hg [11], and/or Se. Small quantities of these toxic microelements have already been found previously in nearly all fodder and food of vegetable and animal origin and, except for just a few ppm of Se, this concentration ranged from ppb to ppm.

The increasing contamination of our environment in the last decades caused a considerable, partly even a multiple, rise of mainly Hg, Pb, and Cd contamination of fodder as well as of food of vegetable and animal origin.

However, in Japan a number of fatal *Hg-poisonings* occurred following the repeated consumption of fish, shell fish, crustaceans, and molluscs which proved to be highly contaminated by methylmercury [49, 51]. The mussels contained up to 102 ppm of Hg [56].

In New Mexico, three children suffered *brain damage* after having eaten pork from pigs fed on refuse of seed grains containing *organic mercury.* In the meat of these animals the content of Hg was found to be 27.5 and 29.4 ppm respectively [11]. Apart from the Se- and As-concentrations in food of animal origin, which generally have always been higher previously, there were no significant differences in the Cd- [12, 30, 31] as well as in the Pb-content of fodder and food of animal and vegetable origin respectively. In general, vegetable products, however, contained somewhat more Pb than products of animal origin [31].

The ability of the animal organism to store toxic microelements to a certain extent supports their accumulation, especially in the case of Pb and Cd. The biological half life time of organic Hg in the human body is 70 days [16, 62].

Especially in the kidneys and liver of slaughtered animals and poultry higher Cd- [30], Se- [36], Pb- [44], and Hg-concentrations are found than in most of the other tissues of the organism. The kidneys contain by far the highest concentrations of Cd [30], Hg [59], and Se [36], whereas the Pb and As concentrations in the liver and kidneys showed only small differences.

In a contaminated environment, considerable concentrations mainly of Se, but also of Hg [60, 61, 63, 64] as well as of Cd and Pb have to be expected in the meat of slaughtered animals and poultry too. Meat from districts rich in Se contained up to 8.0 ppm Se [52].

Pb, Se, Hg, and As were detected in milk in concentrations lower than 1 ppm [15, 26, 34, 65, 66].

Because of their insignificance for the Central European consumer, I will not discuss in detail the contamination of fish, crustaceans, shell fish, and oysters by toxic microelements.

But it seems to be a fact that, under certain circumstances, not only Hg, but also a number of other metals e. g. As, Zn, Co, Cu, and Mn can be present in higher concentrations.

If the present concentrations of toxic microelements in food of animal origin are compared with existing tolerances, it can be seen that certain microelements sometimes exceed these tolerances in individual food. At this point attention should be drawn to the Hg-concentrations in Japanese fish, clams, crustaceans, and oysters, in Japanese and Scandinavian fresh- and sea-water fish; or to the high Se-contents and the occasionally considerable Hg-concentrations in meat. It seems certain that the concentrations of Se and Cd in the livers and kidneys of these animals were still higher.

If in these cases the daily diet consisted exclusively, or for the greater part, of liver and kidney, the tolerable daily amount of one or the other element (Se, Cd) would be exceeded. In the case of As and Se, the consumption approaches the *daily tolerable intake*, but in general not Hg, Cd, and Pb [17, 23, 24, 27, 31, 32, 47]. As regards to Pb there is no generally accepted opinion [27, 32]. The present degree of contamination of our food of animal origin by toxic microelements requires measures of controls and precautions mainly in regard to future developments.

Further problems, especially in the field of *food hygiene* arose recently by the increasing application of veterinarian preparations in nutritive and therapeutic doses. Especially in the latter case, certain *drug residues in slaughtered animals* are often permitted by the producer in order to obtain certain advantages.

In the modern system of effective breeding and keeping of domestic animals in large numbers, agriculture claims the use of such chemical agents and drugs or those which promote growth and improve the utilization of feed, or reduce the risk of keeping animals, to be indispensable. Usually, antibiotics are used, in nutritive doses, on account of their anabolic and/ or microorganism-suppressing qualities.

which have a favorable effect on growth and fodder utilization; hormone products like estrogens, thyroid gland inhibitors (thiouracil derivatives), organo-arsenic compounds, and a number of other chemical substances having a partly bacteriostatic and/or bactericidal effect are also applied. To reduce the risk of keeping and breeding domestic animals in large numbers, *antiparasitic substances*, e.g. coccidiostatics or anthelmintics are added to stocks of poultry and pigs. Moreover, a better rate of breeding is achieved by this method in many cases.

True-bred animals, especially the newest breed of meat pigs, are very sensitive to stress, especially on their transport to the slaughter-house. To reduce the transport risk with calves, illegal doses of drugs are applied. Pigs and young cattle often are very excited when they are transported. Sometimes tranquilizers and neuroplegics are administered to these animals, by the personnel conducting the transport, just prior to the beginning of the transport, in order to keep the animals in a sedated condition. Pigs which are especially susceptible to stress, are given glucocorticosteroids.

This outlines the most essential therapeutica frequently used in slaughtered animals.

Residues in the meat are scarcely to be worried about if the feeding of nutritive doses of antibiotics and other food-additives has been done in a correct manner [3, 6, 7, 21, 25, 35, 41, 50] and if the application has been discontinued three to seven days prior to slaughtering.

Only in the milk and eggs could small residues be demonstrated occasionally.

It is generally considered to be unlikely that residues of antibiotics are directly injurious to the health of the consumer. The increasing dissemination of antibiotic-resistent bacteria caused by selective and transmissible resistance [8, 48] is under discussion as an indirect detriment, even when small quantities of nutritive antibiotics are added to the feed.

On the other hand, if therapeutic doses are parenterally administered, larger residues in meat and the organs will be present.

Residues of tranquilizers, neuroplegics, and glucocorticosteroids are rather unlikely since they are either easily excreted and/or are normally quickly metabolized.

Following parenteral application of antibiotics, especially those in oil suspension, or following application of extremely high doses of *estrogens*, however, residues in the meat have been repeatedly found [10, 28, 45, 46], especially in and around the site of application [4, 18, 19, 20, 22, 24]. In the case of estrogens they were nearly always below the doses of 1 mg/day [35] which is considered as therapeutically effective.

The extent of illegally administered therapeutic doses of antibiotics, given to calves just prior to the transport for slaughtering, may be guessed through results of control analyses conducted at slaughter houses. Out of 91 slaughtered calves proved to be clinically healthy, and of 1508 animals, 30 % and 67,3 % respectively, were found to contain antibiotics. In addition to primary health aspects, for instance, the occurrence of symptoms of sensitization and allergy after the consumption of food of animal origin, health and technological aspects are also important. Residues of antibiotics for example, will mask the true status of the samples in microbiological analysis tests for establishing the healthiness and/or for the qualitative determination of food and will be the cause of incorrect determination. In this respect the *bacteriological examination* of meat may serve as an example. Residues of antibiotics in meat may also endanger the production, for instance, of fermented meat, that is, of pickled meat.

Further, the detrimental effects of antibiotic residues in milk used for the production of cheese and curdled milk are known.

The necessity for legal regulation of the residue problems is evident.

References

[1] Baglay, G. E., L. N. Locke, G. T. Nightingale: Avian Diseases 11: 601, 1967
[2] Bartels, K. H.: Schlacht- u. Viehhof-Ztg. 8: 309, 1970
[3] Bird, H. R.: Amer. J. Clin. Nutrit. 9: 260, 1961
[4] Bird, S. L., I. Pugsley, M. O. Klotz: Endocrinology 41: 1947
[5] Borg, K., H. Waunторp, K. Erne, E. Hanko: J. Appl. Ecol. 3: 171, 1966
[6] Boyd, J., H. H. Weiser, A. R. Winter: Poultry Sci. 39: 1067, 1960
[7] Brüggemann, J., M. Merkenschlager: Arch. Lebensmittelhyg. 9: 197, 1958
[8] Bulling, E., R. Stephan: Bundesgesundheitsblatt 33: Nr. 3, 1970
[9] Carroll, R. E.: J. Amer. Med. Ass. 198: 267, 1966
[10] Clegg, M. T., H. H. Cole: J. Anim. Sci. 13: 108, 1954
[11] Curley, A., V. A. Sedlak, E. F. Girling, R. E. Hawk, W. F. Barthel, P. E. Pierce, W. H. Likosky: Science 172: 65, 1971
[12] Essing, H. G., K. H. Schaller, D. Szadkowski, G. Lehnert: Arch. Hyg. 153: 490, 1969
[13] Gißke, W., S. Wenzel, J. Pichnarcik, J. Schaper, E. Ennen: Arch. Lebensmittelhyg. 21: 25, 1970
[14] Goto, M.: Inorganic chemicals in the environment with special references to the pollution in Japan. Report to the 2nd Internationalen Symposium: Aspekte der chemischen und toxikologischen Beschaffenheit unserer Umwelt, München 27 and 28 May 1971
[15] Hadjimarkos, D. M., C. W. Bonhorst: J. Prediat. 59: 256, 1961
[16] Hammond, A. L.: Science 171: 788, 1971
[17] Horiuchi, K.: Osaka City Med. J. 11: 225, 1965
[18] Huis in't Veld, L. G., E. B. Jonkman van den Broek, W. C. de Groot, C. Laan: Diergeneesk. 92: 1272, 1967
[19] Huis in't Veld, L. G., E. B. Jonkman van den Broek, W. C. de Groot, H. O. Krusel: Diergeneesk. 94: 169, 1969
[20] Huis in't Veld, L. G., E. B. Jonkman van den Broek, W. C. de Groot: Neth. J. Vet. Sci. 2: 85, 1969
[21] Jepsen, A.: Proc. Holmenkollen Symposium, Oslo, 3–5 March, 1967
[22] Karg, H., H. Schaarschmidt, W. Waldschmidt: Arch. Lebensmittelhyg. 21: 1, 1970

23 Kehoe, R. A., F. Thamann, J. Cholak: J. Amer. Med. Ass. 134: 90, 1935
24 Kehoe, R. A.: J. Royal Inst. Publ. Hlth. Hyg. 24: 81, 1961a, 24: 101, 129, 1961b, 24: 177, 1961c
25 Kelch, F.: Fleischwirtschaft 12: 439, 1960
26 Kirchgessner, M.: Ztschr. Tierphysiol. Tierernährung und Futtermittelk. 12: 156, 1957
27 Kloke, A., K. Riebartsch, H. O. Leh: Landw. Forschg. Sonderheft 20: 119, 1966
28 Koch, W., G. Heim: Endokrinologie 30: 336, 1953
29 Kolthoff, J. H., P. J. Elving, E. B. Sandell: Treatise on Analytical Chem. 7: 141, 142, 214, 1961
30 Kropf, R., M. Geldmacher-v. Mallinckrodt: Arch. Hyg. 152/153: 218, 1968
31 Lehnert, G., G. Stadelmann, K. H. Schaller, D. Szadkowski: Arch. Hyg. 153: 403, 1969
32 Lehnert, G.: Arch. Hyg. 152: 559, 1968
33 Locke, L. N., G. E. Baglay: J. of Wildlife Management 31: 515, 1967
34 Marshall, S. P., F. W. Hayward, W. R. Meagher: J. Dairy Sci. 46: 581, 1963
35 Mitchell, G. E., A. L. Heumann, H. K. Draper: ref. J. Anim. Sci. 15: 1285, 1960
36 Oelschläger, W., K. H. Menke: Ztschr. f. Ernährungswiss. 9: 216, 1969
37 Perry, H. M., H. A. Schroeder: J. Lab. Clin. Med. 46: 936, 1955
38 Rosenberger, G.: Staub 23: 15, 1963
39 Rosenfeld, J., O. A. Beath: Selenium, Geobotany, Biochemistry, Toxicity, and Nutrition, Academic Press, New York, London 1964
40 Schaarschmidt, H.: Morphologische und biologische Nachweismethoden der Östrogenbehandlung von Masttieren. Vet. Med. Inaug. Diss. München 1969
41 Scheibner, G.: Mh. Vet. Med. 24: 739, 1969
42 Schroeder, H. A., J. J. Balassa: J. Chron. Dis. 14: 236, 1961
43 Schroeder, H. A.: Circulation 35: 570, 1967
44 Stewart, L. W., R. Allcroft: Vet. Rec. 68: 723, 1956
45 Stob, M., T. W. Perry, F. N. Andrews, W. M. Beeson: J. Anim. Sci. 15: 997, 1956
46 Stob, M., F. N. Andrews, M. X. Zarrow, W. M. Beeson: J. Anim. Sci. 13: 138, 1964
47 Stöfen, D.: Arch. Hyg. 153: 284, 1969
48 Swann Report London, Her Majesty's Stationary Office 1969
49 Takizawza, Y., Y. Kosaka: Acta Med. Biol. 14: 153, 1966
50 Tiews, J.: Fortschritte der Medizin 87: 1387, 1969
51 Uchida, M., K. Hirakawa, T. Inoue: Kumamoto Med. J. 14: 181, 1961
52 Underwood, E. J.: Trace Elements in Human and Animal Nutrition Acad. Press, New York 1956
53 Underwood, E. J.: The Mineral Nutrition of Livestock. Aberdeen 1966
54 Ernährungsbericht 1969 der Deutschen Gesellschaft für Ernährung e.V. Frankfurt a. M.
55 World Health Organization Technical Report Series Nr. 373. Specifications for the identity and purity of food additives and their toxicological evelution: Some emulsifiers and stabilizers and certain other substances.
56 Evaluations of some pesticide residues in food. Food and Agriculture Organization of the United Nations World Health Organization, Rome 1968
57 Westöö, G.: Acta Chem. Scand. 21: 1790, 1967
58 Westöö, G.: Acta Chem. Scand. 22: 2280, 1968
59 Stock, A., F. Cucuel: Naturwiss. 22: 390, 1934
60 Smart, N., M. Lloyd: J. Sci. Food Agr. 14: 734, 1963
61 Lagervall, M., G. Westöö: Var Föda Nr. 2: 9, 1969
62 Ekman, L.: Nord. Hyg. Tidskr. 50: 116, 1969
63 Loosmore, R. M., J. D. J. Harding, G. Lewis: Vet. Rec. 81: 269, 1967
64 Harvey, F. T.: Vet. Rec. 12: 328, 1932
65 Furutani, S., Y. Osajima: Kyushu Daigaku Nogakubu Gakugai Zasshi 22: 45, (1)
66 Reigo, J.: Sv. Mejeritian 62: 312, 1970

Power Generation Chemicals

J. J. Went

N. V. KEMA, Arnhem, Utrechteweg 310, Netherlands

Summary. In this article a few of the most important pollution effects of power generation are reviewed. It is outlined that for a large reduction of pollution effects separation of production and consumption of energy are required; therefore electricity should be an intermediate step for supplying energy to densily populated areas. For fossil fuels the harmful combustion products are discussed and the importance of the atmospheric chemistry for the removal of SO_2 and NOx is outlined. For the future nuclear energy can solve the problem of air pollution due to the small amount of "combustion products" (fossile products). Requirements and the environmental consequences of cooling capacity of electricity production are mentioned briefly.

Zusammenfassung. Dieser Beitrag behandelt die wichtigsten Auswirkungen der Umweltverschmutzung durch Energieerzeugung. Es wird betont, daß zur Einschränkung der Verunreinigung eine Trennung zwischen Energieproduktion und Verbrauch erforderlich ist, deshalb sollte Elektrizität ein Zwischenglied zur Energieversorgung dicht bevölkerter Gebiete sein. Die schädlichen Verbrennungsprodukte fossiler Brennstoffe und die Bedeutung der Chemie in der Atmosphäre zur Beseitigung von SO_2 und NOx werden betont. In Zukunft wird die Kernenergie das Problem der Luftverschmutzung durch geringe Mengen von „Verbrennungsprodukten" lösen können. Forderungen und Konsequenzen, welche für die Umwelt durch die Kühlkapazität zur Elektrizitätserzeugung entstehen, werden kurz erwähnt.

Bulk power generation is used for different purposes: as process heat, mechanical energy, but also for generation of electrical energy. In this paper, mainly the latter — electric power — in relation to environmental problems will be discussed.

Introduction

A few fundamental facts should be pointed out first [1].
Energy production, essential for the well-being of mankind is, except in the case of the production of hydroelectric power, always accompanied by pollution effects; this is also true if this energy is transferred into another form of energy such as electrical energy. But by application of electrical energy to lighting, heating, mechanical power or otherwise, no additional pollution effects are created; *electric energy as such is completely clean.* Therefore, it is important that it is possible to produce electric power outside populated areas but still use it in populated areas.

However, transport of electric energy is

then necessary. It is regretable that such transport is very expensive. This can be demonstrated by a rough split up of cost for an integrated electric system as it exists, e. g. in the Netherlands.

Present-day extra high voltage transmission cables are 10—20 times more expensive than overhead lines. In Table 1 the transmission system considered is an overhead system. Therefore it is evident that present-day cables, perhaps desirable for avoiding landscape pollution, would increase the cost of electricity excessively. A second point of interest is that large-scale energy production, e. g. by burning coal, oil, or gas, can be equipped with high stacks, so that gaseous pollution products can be diluted and even partially removed before reaching ground level, which is not so easy with pollution products produced at the ground level from small energy units such as cars or house heating installations. Therefore it might be desirable in the future "to go more electrical".

Table 1   Division of the cost of electrical energy in arbitrary units

Production		$1/3$ unit
Transmission	>100 kV	$1/6$ unit
Distribution	> 10 kV	$1/4$ unit
	< 10 kV	$1/4$ unit
Total cost		1 unit

Table 2   Average electric power density in a few highly developed areas in the world (1968)

Country	GWh th/km² year	kWh/inha- bitants year	inhabi- tants/km²
Netherlands	0.95	2,470	380
Belgium	0.81	2,400	320
Germany	0.69	3,270	242
Italy	0.18	1,920	175
France	0.12	2,400	91
EEC	0.30	2,550	159
United Kingdom	0.74	3,880	227
United States	0.13	7,140	21
Japan	0.57	1,900	340

This trend has already been observed for a long time. The *total energy consumption* in the last few decades has been increasing on the average of 3—5%/year; electric energy at 6—10%/year. Currently, about 20% of all primary energy is transformed into electric energy. Using growth rates of 4 and 8% respectively, this would mean at the end of this century that 65% of all primary energy would be transformed into electric energy.

There is another point of interest relating to *electrical energy and pollution*. In highly developed countries much more electricity is used per inhabitant than in underdeveloped countries. A rather straightforward relationship exists between energy consumption and the gross national product per inhabitant. However, for pollution it is more important how much electric energy is produced per km² than per inhabitant in each area. As *hydro-electric energy* production neither pollutes the air nor needs cooling water, hydro-electric energy is excluded from this consideration. In Table 2 a survey is given of such data for certain highly developed countries but with considerable differences in population densities.

As already mentioned, non-polluting hydro-electric power production will not be considered in this contribution. However, it should be realized that the importance of hydro-electricity, although of great value in the first half of this century, is decreasing fast. In Western Europe, e. g. practically all hydro-electric possibilities have already been used and therefore this method of electricity production cannot grow anymore in an era of rapidly growing electricity production. For non-hydro-electric power production, at the moment, more than 95% is produced with fossil fuels — coal, oil, or natural gas. Nuclear energy is just starting but not yet of great importance. It is generally believed however, that at the end of this century the ratio between fossil and nuclear might be about 1 : 1. This will result in very different gaseous pollution products.

Pollution Products from Fossil Fueled Power Stations

By the application of fossil fuels, the following gaseous combustion products are formed:

CO_2

The mass of CO_2 formed is about three times the mass of the fossil fuel itself. Therefore it is not practical to store this combustion product. Besides, it is one of the most important natural components of the atmosphere, useful to produce, through photosynthesis, new organic material in plants. The concentration in the atmosphere of CO_2 however has increased during the last few years at about 0.2%/year. This means that so much CO_2 is produced that neither photosynthesis nor chemical take-up in the oceans is sufficient to keep the CO_2 concentration constant. Hypotheses about the *meteorological consequences* of this effect are controversial. If, however, this increase in the CO_2 concentration continues, in the next hundred years a doubling of the CO_2 concentration will result, which could cause perhaps some respiratory problems, while the growth rate of plants would increase. However, a reduction of the oxygen content of the lower atmosphere would not occur[2].

H_2O

The hydrogen content of the hydrocarbons is burned to H_2O. This combustion product is naturally acceptable; only a very large local water production might cause some local meteorological difficulties. It should be realized that if so-called

wet cooling towers are used these effects are even aggravated.

SO_2

SO_2 production depends on the sulphur content of the fuel. This sulphur, in general, amounts to a few percent of the fuel. However, *natural gas* is often sulphur free. Due to the switch over from coal or oil to sulphur free natural gas in the Netherlands, the SO_2 emission from the power works is diminishing considerably; (Fig. 1) in addition the overall SO_2 concentration in the country is slightly decreasing. It is interesting that from measurements in Greenland[3] it can be demonstrated that the sulphur content is about twice in new ice the sulphur content in old ice since the beginning of this century. The old ice sulphur content must have a partly vulcanic origin and is partly produced by oxidation of organic matter. The new ice therefore might indicate the general increase of SO_2 content in the atmosphere due to burning fossil fuels.

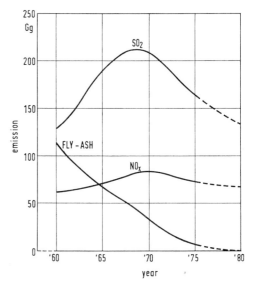

Fig. 1 *Total emission of air pollutants from Dutch power stations (1 Gg = 10^3 t)*

It is, however, questionable how harmful SO_2 is or can be in combination with other contaminants. Elaborate studies over several years, under supervision of the Edison Electric Institute in the U.S., have demonstrated that even SO_2 concentrations much higher than the normally accepted limits are harmless for guinea pigs and monkeys. These animals have been exposed for years to clean air, air with different SO_2 contents and air contaminated with a combination of SO_2 and fly ash. No biological or health changes have been observed[4]. On the contrary, several *plants* are much more sensitive to SO_2 exposures specifically at high relative humidities.

Although it is probable that SO_2 as such is not the most harmful contaminant in normal air, the present-day large scale control of SO_2 concentrations in industrial or urban areas might still be useful. A certain figure of merit for the *quality of the air* in general and for the meteorological conditions which might be favorable for concentrating or diluting SO_2 pollution can be given by these measurements.

In the case of SO_2, it is important to disperse and dilute the SO_2 before it reaches ground level. High stacks and favorable meteorological conditions can stimulate this.

However, equally important might be the *chemical clean-up of SO_2 in the air.* If SO_2 is oxidized to the more toxic SO_3 and reacts, e.g. with NH_3 in the air, it is transformed into a sulfate that is no longer a serious health hazard for animals and plants.

Quantitatively, little is known about this *meteorological chemistry.* Nevertheless, measurements of SO_2 concentrations in the Netherlands under such meteorological conditions that mainly SO_2 pollution from Rhineland/Westphalia is observed, indicated a half-life of SO_2 between 5 and 25 hours. More should be known about this chemical clean-up. Research in this field is also progressing in the Netherlands.

For reducing the effects of large potential SO_2 sources several technical methods can be used.

First of all substitution of low sulphur content fuel for high sulphur content fuel is possible. When such fuel is not available, it is expensive but possible to *desulfurize the fuel* partially before combustion. This switch over to low sulphur content fuel can be done, e.g. under unfavorable meteorological conditions.

Secondly, it is also possible to remove the SO_2 from the flue gas in the stack. Much work in this latter direction is underway. It is believed that it might be possible in the future to extract about 80% of the SO_2 content from the flue gases but with an increase of production costs by 10—15%.

NO_x

The production of *nitrogen oxides* NO and NO_2 is not so easy to predict as the SO_2 production. In all combustion processes, including natural gas, oxygen, and nitrogen from the air are present. This means that at high flame temperatures NO is formed. This NO can be further oxidized even in the atmosphere to the much more toxic NO_2, one of the essential components for the formation of the dangerous *Los Angeles smog.* How much NOx is formed depends on the excess amount of oxygen in the combustion chamber and on many constructional details of the boilers determining the flame temperatures and the time the gas is at high temperatures. In addition to these circumstances, it has been found that, if organically bound nitrogen is present in the fuel, this nitrogen reacts much easier and faster with oxygen than molecular N_2 from the air. Therefore, if the flame conditions are the same, fuel containing *organic nitrogen* will produce more NOx than fuel without organic nitrogen as, e.g. natural gas. In the Netherlands, a survey covering 13 power generating units has given upper limits of NOx for coal, oil and natural gas of 3.0, 1.5, and 0.75 kg/Gcal. respectively. Due to the switch-over from oil to natural gas in the Netherlands, the NOx contents of the flue gases are decreasing, notwithstanding the increasing electricity production (Fig. 1). It has been, however, difficult up until now to reduce the NOx by a factor of two or more.

Finally, it should be realized that just as the clean-up of SO_2 in the air by conversion to a sulfate, NOx can be transformed as in a fertilizer to a nitrate. It seems to me very desirable to investigate more thoroughly these chemical meteorological reactions as well.

Fly Ash

The importance of and the difficulties with *fly ash* have decreased very fast in the last few years. *Electrostatic precipitators* with efficiencies between 92 and more than 99% are cleaning up the flue gas sufficiently (cf. Fig. 1).

Pollution Products from Nuclear Fueled Power Stations

By the application of *nuclear fueled power stations*, the production of gaseous and non-gaseous radioactive materials mainly inside fuel elements is unavoidable. Although a very large amount of radioactivity is produced, the mass of this radioactive material is very small compared with the mass of fossil gaseous combustion products. This mass ratio is about 1 to 10 million. Therefore, in principle, storage of these *radioactive waste products* is possible.

A very small part of the radioactive products is discharged in the nuclear power station itself; so that under normal operational conditions and even more under the present-day developed so-called "zero release" conditions, no health problems for the surroundings exist.

For this introduction it is not necessary to discuss accident conditions, but it should be realized that the consequences of accidents should be handled by the reactor containments to prevent a serious dissemination of radioactivity in the surroundings after an accident.

Therefore, nuclear power stations are in principle not polluting to the air in the surroundings. As a consequence air pollution can be better controlled with nuclear power stations than with fossil fuel power stations.

The amount of radioactivity produced inside the fuel elements is released, not in the nuclear power plant, but in the reprocessing plants and should consequently be controlled very thoroughly in those installations. Therefore, the handling of gaseous and non-gaseous radioactive products, mainly the fission products but, of course, also other materials activated by neutrons, needs to be discussed separately.

All present-day methods for reprocessing fuel elements start by dissolving the elements in a nitric acid solution. After taking out the valuable nuclear materials U and PU, the solution still containing the radioactive fission products is further concentrated by evaporation. For storing this highly active waste, the most favored present-day method is to solidify the waste in the form of a glass or other ceramic substance. Thereafter, it should be stored in a geologically safe place.

For the *gaseous radioactive materials*, the long lived isotopes are the most difficult ones. Krypton 85 with a half-life of 10.8 years and tritium with a half-life of 12.3 years should be stored for about ten times the half-life (100 years). That would result in a radioactive decay of a factor of about 1,000. Without a safe storage for such a long time the build-up of radioactivity in the air might become unacceptable after the year 2000. It is therefore of interest that Oak Ridge National Laboratory in the U.S. has developed a rather simple method to fixate the rare gases krypton and xenon. Both can be adsorbed easily and preferably in cooled freon. The krypton removal efficiency should be even better than 99.9%. For tritium still more research is required to find an acceptable method to concentrate and fixate it.

Cooling water for the Condensor

Not all pollution problems can be discussed in detail in a short review article. Therefore, the *cooling water problems* will be mentioned only briefly. This is the more so as the biological problems related to the thermal pollution (calefaction) of the water are completely dependent on local circumstances. However, I hope to convince you that the cooling water problems are probably the most difficult problems to solve for future large scale electricity production, notwithstanding the fact that thermal effects of cooling water discharge may be beneficial as well.

First of all, it should be realized that about 80% of all the cooling water needed is used for electricity production. The second point is that electricity production with a steam cycle will have a higher thermodynamic efficiency for a high temperature system, e.g. 540° C (fossil fuel plants) than for a low temperature system, e.g. 300° C (present-day nuclear fuel plants) but that in both cases 60% or 70% respectively of the heat has to be rejected. Between present-day nuclear reactors and conventional power

stations the ratio of calories rejected in the cooling water might even be a factor of 1.5 but with growing electricity production with a doubling time of 10 years, this factor 1.5 only means postponing the problem for a few years.

As a rule of thumb, a cooling water demand of 40—50 m³/sec for 1,000 MW of generated power (the minimum flow of the Rhine river is about 1,000 m³/sec) can be used. Before using this water again it must be cooled through the water surface to the air. For such at cooling water surface a value estimated at 10 km² for 1,000 MW (the fresh water surface in the Netherlands is about 2,000 km²) is sometimes used. A large river, such as the Rhine, has a small surface, this means that only after several hundred kilometers is the water sufficiently cooled again for reuse.

These values are related to the fact that for the time being *maximum temperatures* not higher than 30° C are believed to be acceptable. As the natural temperatures in summertime in our countries are generally not higher than 24° C, a temperature rise of about 6° C seems acceptable.

It can be expected that too large a temperature increase in the water will stimulate plankton growth considerably if the water contains sufficient minerals. This increased growth rate can result in an *increased oxygen consumption*. If the oxygen content of the water becomes too low the biological life can be destroyed, resulting in dangerous water pollution.

This temperature limitation is related to the ecology — the biological life in the water. Often the fish content of the water is discussed. However, it is the opinion of this writer that the *plankton ecology* might be in fact of greater importance. The difficult question, however, is how to define and to determine an acceptable ecology. If a sample of the water is taken it has been impossible to investigate the stability of the sample. In a pure culture it is possible to cultivate one type of plankton and therefore to investigate perhaps the influence of a short thermal shock of 6° C in a condensor of a power station. But up till now it has not been possible to cultivate the sample even without a thermal shock. Research in this direction is underway, but it will be difficult to make it possible to investigate this in different places in the world. A microscopic survey has indicated that practically no mechanical damage is observed, except perhaps for very large zooplanktons, by pumping cooling water through a condensor circuit of a power station.

Conclusion

The following conclusions are, according to my ideas, essential for the future:

1) Electrical energy should be the form of energy to be used in densely populated areas.

2) Nuclear energy is, apart from the raw material supply, also desirable to control and reduce air pollution.

3) A great effort is needed in the biological effects of water calefaction.

References

[1] More details can be found in: Electrical energy needs and environmental problems, now and in the future. Future shape of technology publications. Number 7. Koninklijk Instituut van Ingenieurs, The Hague, Netherlands

[2] van Valen, L.: The History and Stability of Atmospheric Oxygen. Science 171: 3970, 439, 1971

[3] Weiss, H. V., M. Koide, E. D. Goldberg: Selenium and sulfur in a Greenland ice sheet: Relation to fossil fuel combustion. Science 172: 3980, 261, 1971

[4] Chronic exposure of cynamolgus monkeys to sulfar dioxide. Submitted to Electric Research Council, Project RP-74. Hazleton Laboratories Mc., Virginia U.S.A. (1969)
Chronic exposure of guinea pigs to sulfuric acid mist. Submitted to Electric Research Council, Project RP-74. Hazleton Laboratories Mc., Virginia U.S.A. (1970)

Toxicological Evaluation of Special Organochlorinated Compounds

R. Roll

Max-von-Pettenkofer-Institut des Bundesgesundheitsamtes, 1 Berlin 33, Postfach
West Germany

Summary. A summarizing account of the problems of polychlorinated biphenyls (PCBs), edematous disease in chickens, and of 2,4,5-trichloro-phenoxy-acetic acid (2,4,5-T) is given.
A survey is given of the relevant toxicological studies made so far and the fact is stated that a full picture of the toxicology of PCBs is not yet available. considering the ubiquitous distribution of these substances and also with a view to the evaluation of health hazards it will be necessary to devote increased attention to the PCB problem in the future.
On the subject of 2,4,5-T, the course of events that resulted in the well-known restricting measures adopted in the U.S. is described and a brief survey of other toxicological studies is given. There is no hazard for pregnant women after ingestion of admissible 2,4,5-T residues in food despite evidence of its teratogenic effect. Considering the dose/effect relationships and the teratogenic no-effect level which should be rated at approx. 20 mg/kg/day at present, a general ban on the use of herbicides containing 2,4,5-T is not believed to be necessary.

Zusammenfassung. Es wird eine zusammenfassende Darstellung über die Probleme der polychlorierten Biphenyle (PCB), der Ödemerkrankungen bei Küken sowie der 2,4,5-Trichlorphenoxyessigsäure (2,4,5-T) gegeben.
Es wird über die bisher durchgeführten toxikologischen Untersuchungen berichtet und festgestellt, daß noch kein abgerundetes Bild über die Toxikologie der PCB vorliegt. In Anbetracht der ubiquitären Verbreitung dieser Substanzen sowie im Hinblick auf die gesundheitliche Beurteilung ist es erforderlich, dem PCB-Problem in der Zukunft verstärkte Aufmerksamkeit zu widmen.
Zum „Fall 2,4,5-T" wird neben der Entwicklung, die zu den bekannten Restriktionsmaßnahmen in den USA führte, eine kurze Übersicht über weitere toxikologische Untersuchungen gegeben. Trotz der nachgewiesenen teratogenen Wirksamkeit besteht keine Gefährdung von schwangeren Frauen nach Aufnahme von zulässigen 2,4,5-T-Rückständen in Lebensmitteln. Unter Berücksichtigung von Dosis-Wirkungs-Relationen sowie des „teratogenen no effect levels", der gegenwärtig mit ca. 20 mg/kg/Tag anzusetzen ist, wird ein generelles Anwendungsverbot für 2,4,5-T-haltige Herbizide nicht für erforderlich gehalten.

This report covers the toxocological evaluation of organochlorinated compounds.

Polychlorinated Biphenyls (PCB)

After the introduction of *PCB* in 1929, it took nearly 40 years until the importance of such substances for the environment was recognized. Today you can say that PCBs are approximately as common as DDT. These substances are used in 3 industrial branches:

1. in electrotechnics as uncombustible lubricants, insulating material, and refrigerants (transformers and condensers),
2. in the dye-industry as additive substances, and
3. in plastic fabrication as softeners.

In 1966, Jensen (Institute of Analytical Chemistry, Stockholm) showed that most of the unknown compounds which were found during the analysis of DDT residues, were PCBs. PCBs were found in fish and birds in the Swedish regions. Moreover there have been indications that these compounds appear to be ubiquitous, like DDT. Further investigations showed that they can also be found in organisms in Great Britain, the Netherlands, the USA, and Finland. In addition to this, Jensen demonstrated the *presence of PCBs* in a dead eagle from the Stockholm Archipel, in his own *hair,* in his wife's, and his 5-month-old daughter's hair. Jensen concluded from this, that the PCBs found in his daughter's hair are due to PCBs in *mother's milk.*

Further investigations carried out during the last years proved the *presence of PCBs* in different sea animals like salmon, herrings, plaice, mussels, seals, etc. The decrease of the *sea-eagle population* in Sweden is remarkable. All eagles which where examined contained DDT, PCBs, and Hg-compounds. 3.4 ppm DDT and 3.9 ppm PCBs were found in the muscular tissues of 2 eagles from North Sweden, 290—400 ppm DDT and 150—240 ppm PCBs in the muscular tissues of eagles from the Stockholm Archipel. Although most of the adult birds do not die from the consequences of PCBs and DDT content, the important decrease of the *rate of reproduction* is obvious. According to environmental protection, it should not be accepted without contradiction. This decrease in the population density can be established on many facts, e. g. delayed egg-laying, failure in egg production, thinning of egg shells and high mortality rates of eggs and young birds. One can assume that these symptoms are due to disturbances in *hormone equilibrium*[1,2,3]. You can find the reasons for the accumulation of PCBs in living organisms in their chemical and physical properties. They are stable (even more than DDT) and are soluble in fat like DDT. When entering the organism, they do not change, but will rather be stored in adipose tissues. Their stability enables them to raise their concentration in increasing quantities.

A knowledge of the *toxic properties* of these substances is necessary, if you want to judge the environmental situation connected with PCBs. Although there is no complete picture about the toxicity of PCBs, a number of individual investigations are available.

Working with animals, Miller[4] has already described some damages in rats, guinea pigs, and rabbits. In the first place came *liver injury* and *skin illnesses,* where the degree of toxic effect increased proportionally with the chlorine content of the examined biphenyls. It could also be shown that the PCBs *stimulated liver microsomes* and enhanced the metabolism of *sex hormones* (progesterone, testosterone, and estradiol). In this respect it was found that the PCBs are 5 times more active than DDT, DDE, or dieldrin. Exactly like DDT, the PCBs stimulate the liver to produce enzymes which, e. g. catabolyze barbiturates faster than usual. That is why it is impossible to predict the influence of these drugs in every individual case.

To estimate the toxicological risks with regard to the *PCBs*, it is further important to achieve knowledge of their *distribution in the body*, their metabolism, and the elimination and toxicity of the metabolites. Concerning the distribution, it is known that these compounds can accumulate in adipose tissues, and they can raise their concentration in the lipid-containing tissues of the brain. Up to now, knowledge of metabolism is rather small. It is only known that compounds with lower chlorine content are metabolized easier than those with higher chlorine content. Except for the fact that liver cells are interfered with such PCBs concentrations to which the human being is exposed through industrial contamination, the *toxicity on different cell systems* is practically unknown. For further hygienic evaluation of the PCBs in the environment of man, it is very necessary to carry out studies on the metabolism, to elucidate their mechanism of action.

In connection with the *intoxication caused by biphenyls*, it is interesting to mention the fatal event which happened in Finland. A 31-year-old industrial worker was exposed to a concentration of 100 mg/m³ (corresponds to 10 folds of MAK-value) of chlorinated biphenyl for 11 years. Beside a histological manifestation of liver insufficiency, a general atrophy of brain cortex was detected. The medical examination of 120 workers from this industrial branch discovered 3 more cases of liver injury and 2 cases of pathological EEG [5].

Also in animal experiments, different effects of CNS were observed with DDT and PCBs (Aroclor). Following oral administration of several doses of DDT

(0—25 mg/kg) and aroclor (0—10 mg/kg) you could discover obvious influences on the nervous system of mice [6].

After the manifestation of the *estrogenic activity of o,p'-DDT*, it was obvious to study such effects with compounds of similar structure. In a recent work, Bitman and Cecil [7] said that the PCBs are effective on estrogen, where the low glycogen content in the uterus was taken as a basis. These studies have shown that only PCBs with chlorine content up to 48% were active, whereas those with more than 54% did not cause any effect on estrogens. Apparently this effect could only be observed, if the distance between the active sites amounts to around 9—11 Å.

Concerning the *teratogenic effect* which was described in literature, it is noticed that such observed effects with chicken eggs (deformity of the peak) after injection of aroclor 1242, are not of special value for hygienic toxicological evaluation [8]. In recent years, a point of view was established, that the chicken embryo cannot be used as a relevant test object, because contrary to mammals, the metabolism and *placenta barriers* remain unconsidered. In mice it was proved, that a passage of chlorinated biphenyl-compounds through the placenta as well as concentration in the fetus is possible. These compounds were found as well in the skin of a stillborn child whose mother had suffered from chlorinated biphenyl poisoning, which indicates that the presence of these substances in human beings can occur through placental barriers [9].

PCBs have been detected not only in animal but also *in human adipose tissues*. With the help of gas chromatographic and spectrometric methods, considerable amounts of PCBs were found in 2 samples of human adipose tissues. They extended from penta- to decachloro-biphenyl and included 14 isomers and homologues. Although the origin of the PCBs in adipose tissues is not known in detail, the indications show, that at least the possibility of identifying such compounds in human adipose tissues is given[10].

While it has been detected that out of the 3 examined PCB-preparations only 2 had a strongly toxic effect, recent investigations have shown the presence of polar compounds in these 2 PCBs, whereas the 3rd non-toxic dissection did not contain any of these compounds[11]. Additional analyses elucidated the identity of these polar compounds which are *tetra- and pentachlorodibenzofuran*. It can therefore be concluded that the evaluation of animal experiments with PCBs may be extremely difficult due to the presence of chlorinated dibenzofurans in some commercial PCB-preparations.

If we consider the PCBs from the hygienic-toxicological point of view it can be concluded that these compounds, although they are not used as *biocides*, exhibit a *mammalian toxicity* which has been detected recently in different species including the human being. At present it is not clear how these quantities got into the human organism. Furthermore, if we consider that the content of *PCBs in mother's milk* and human adipose tissues has already reached that of DDT and sometimes exceeded it, we will be justified in drawing more attention to the PCB-problem in future.

Acker and Schulte[12] detected 3.5 ppm PCB and 3.8 ppm DDT and DDE in human milk (relative to milk fat) as well as 5.7 ppm PCB and 3.3 ppm DDT and DDE in human fat. Their physical properties and their special stability lead us to expect that these increase in their application and will lead to an increasing distribution in the environment.

Edema Disease

After the appearance of *edema disease* in 1957 in chickens in different parts of the USA, which led partly to great losses, the first studies were carried out which showed that the causes were definite adipose samples, whereas animal medicaments or heavy metals (Pb) as well as infections by bacteria, virus, or parasites were found not to be responsible for illness or death. It has been proven that toxicological symptoms (e. g. hydropericardin, ascites, liver and kidney injury) are attributed to the unsaponified fraction in definite adipose samples. In industrially produced *triolane*, a toxic chlorinated product was isolated in unsaponified fraction which was not only toxic to chickens in concentrations of 0.1 ppm when given in fodder, but also led to toxic symptoms in monkeys. In this connection it is of interest to mention that *chicken edema-like* symptoms were observed in chickens which used to pick in sand containing an epoxy dye that was placed in a floor. These symptoms, however, were not identical with the real chicken edema factor. In this case, chlorinated diphenyl derivates in the dye were ascertained as the toxic factor[13].

In 1961, Wootton succeeded in isolating 4 mg of the toxic substance from 100 pounds of contaminated fat. After preliminary analyses had been carried out, one assumed the active substance might be a hexachlorinated hexahydrophenanthrene. This hypothesis was later rejected and replaced by anthracene constitution. Improved methods (crystallography) led to a correction of the formula. It was noted that the toxic compound is a *1,2,3,-7,8,9-hexachlorodibenzo-p-dioxin*[14]. As it is improbable that this dioxin represents a natural constituent of the fat from which it was isolated, speculations were clearly developed about the origin of these compounds.

According to an observation by Tomita, after heating, the chlorophenols and their salts give rise to condensation reactions and chlorinated derivatives of dibenzo-p-dioxin, different chlorophenols (2,4-dichlorophenol, 2,4,5-trichlorophenol, 2,4,6-trichlorophenol, 2,3,4,6-tetrachlorophenol, pentachlorophenol) were subjected to pyrolysis. The chlorinated derivatives of *dibenzo-p-dioxin* produced were similar to the toxic material found in fat during the chicken-embryo test. The pyrolysate of 2,3,4,6-tetrachlorophenol was the only product which showed the presence of chicken edema factor on chromatogram. In the standard chicken-embryo test, 0.1 ppm already caused serious illness.

The results of these studies led to the assumption that *chlorophenols* will be regarded as precursors of *hydropericardic factors,* if fat contaminated with chlorophenols is exposed to heat treatment. In this connection, it should be mentioned that chlorophenols are widely used for the preparation of contact herbicides and defoliants as well as in termite control [15]. In the meantime, up to 60 compounds of the *dibenzo-p-dioxin family* were found in *chlorophenols* (including *fungicides* and *herbicides).* It is therefore assumed that dioxin could reach the atmosphere as a result of burning papers and products containing chlorophenols. Considering the persistence of dioxins, the worries grow more and more that the probability of accumulation and transport in the food chain exists [16].

2,4,5-T: As you certainly know, the herbicide *2,4,5-trichloroacetic acid* (2,4,5-T) which is prevalent in corn, forestry, and meadow, has been in the public eye for more than one year. Lately, many articles have appeared in the newspapers, in which the "2,4,5-T case" was not always relevant and objectively considered.

The case was released in the USA by Bionetics Research Laboratories through studies in which an examination of a large number of pest control substances and their carcinogenic effects was carried out. During these studies, however, no carcinogenic effects were proven, but there have been *teratogenic effects in mice and rats* [17]. Based on these findings and in view of its use in Vietnam as a defoliant, a world wide discussion arose about the evaluation of these studies.

After the declaration of this information, it was proven that a 2,4,5-T sample stored for years, containing about 30 ppm of the highly toxic *2,3,7,8-tetrachlorodibenzo-p-dioxin,* is responsible for teratogenic activity and produces *chlorine acne* in human beings. These studies, especially the teratogenic experiments in chicken eggs, which, as already mentioned in connection with PCBs, were relevantly evaluated or not transmissable to human beings, led to the decision of the American Government to suspend certain applications of 2,4,5-T. The members of the Committee have apparently been under public pressure, particularly during the hearing, since the television cameras demonstrated the abnormal chickens in considerable enlargement.

As is well known, the use of liquid formula of 2,4,5-T in the vicinity of populated areas, in the sea, ponds, and brook shores, has been prohibited ("suspension") since 15 April, 1970. As further safety measurements, the use of a granulated formula in house gardens as well as in treating of food stuffs was also banned ("cancellation"). Excluded from these measurements was its use for weed control in parks, pastures, or roads, against underwood in forestry, and use in various non-agricultural areas.

By reason of the topicality of this problem, further investigations were carried out on different animal species in different places, including the FRG. The availability of the result through publications and other information sources gives a better picture than before. Additionally, the responsible authorities obtained so many discrepancies from the first spectacular results, that the problem nowadays is regarded more objectively. In the USA, the situation was also critically considered, and one expects a corresponding attitude to develop, shortly.

The present results of the teratogenic experiments from the USA show the following: The subcutaneous application of the so-called pure ($<$ 0.05 ppm dioxin) and technical (0.5 ppm dioxin) 2,4,5-T led to teratogenic effects in 2 breeds of mice (Charles River and DBA) after a dosage of 100 and 150 mg/kg/day. However, in another breed (C57 BL/6), following subcutaneous administration of 100 mg/kg/day of technical 2,4,5-T, no teratogenic effects were observed. Oral administration of pure and technical 2,4,5-T in rats (Charles River) did not show any teratogenic effects with dosages of 150 and 80 mg/kg/day respectively. Moreover, the oral administration of 2,4,5-T ($<$ 1 ppm dioxin) did lead to some teratogenic damage in rats and rabbits which received dosages of 24 and 40 mg/kg/day, respectively [18, 19].

As the preliminary results indicated that dioxin was responsible for the *toxic side effects*, it was obvious to work with dioxin itself. The results show that, following oral administration of dioxin in rats with doses up to 8 μg/kg, no teratogenic effects but only dose-dependant *fetal toxicity* was detected [20].

In the meantime, data from the FRG are available, the investigations were carried out in C. H. Boehringer Sohn/Ingelheim as well as by the Bundesgesundheitsamt, and the results will be published soon.

In private investigations, in which a 2,4,5-T sample containing 0.5 ppm dioxin was used, a significant increase in *cleft-palate* was found to be dose-dependant after oral administration of 35—130 mg/kg/day in NMRI-mice. The "*teratogenic no effect level*" was given at 20 mg/kg[21,22]. In view of the fact that the 2,4,5-T sample used contained only 0.05 ppm dioxin, our opinion was, that it is highly probable that dioxin alone was not responsible for the teratogenic activity. For this reason, a further experiment has been carried out, in which a warranted *dioxin-free* sample of 2,4,5-T was used for the same mice breed. This 2,4,5-T, which did not contain any traceable amounts of dioxin (within the analytical limit of 0.02 ppm) was prepared by Boehringer Sohn on laboratory scale using a synthetic reaction in which theoretical consideration would not allow any dioxin formation. As we already suspected, a dose-dependant increase in the rate of deformity was achieved with this 2,4,5-T concentration. The concentrations used were practically similar to those in the experiment with 0.05 ppm dioxin.

Contrary to the mouse experiments, no teratogenic effects were detected in FW 49-rats following oral administration of 25-150 mg/kg/day of dioxin-free 2,4,5-T. Also the administration of a dose of the running production ($<$ 0.1 ppm dioxin) did not induce any deformity in FW-rats [23, 24, 25].

In order to obtain more accurate knowledge about the *excretion and metabolism of 2,4,5-T in animals*, the renal excretion of rats has been studied in the Toxicology section of the Bundesgesundheitsamt[26]. The up-to-date valid assumption that complete excretion in the unchanged form takes place in urine, could not be substantiated. The results show that after one administration of 50 mg/kg in rats, excretion proceeded slower than mentioned in the literature. Moreover it was found that beside the unchanged 2,4,5-T, considerable amounts of derivatives were also excreted. One of these derivatives was identified as *N-(2,4,5-trichlorophenoxy-acetyl)-glycine*.

If one considers the up-to-date teratogenic experiments, it can be shown that contradictory results were obtained. Consequently some experiments have to be critically considered from the view of transmission to humans. In particular, those investigations in which teratogenic effects were detected after subcutaneous injections, using dimethyl-sulfoxide as a solvent, give only very little relevant information. Moreover, the *experiments with chicken embryos* cannot be used in health-toxicological evaluation, because the experimental methods are to a great extent artificial. It is also known that, e. g. the injections of glass-dust and other inert substances can lead to deformity. Nevertheless, with these experimental assignments, as performed, the *placental barrier* was absent and the metabolic process remained unconsidered. In the light of the present data it can be concluded that, with regard to the *teratogenic sensitivity*, considerable species and strain differences exist in mice. As a whole, the mouse seems to be sensitive, while the rat shows lower sensitivity.

Considering the necessity for performing *mutagenic studies*, more than ever, in the toxicological program and in order to establish better ideas about the *genetic risks* in the human being, after absorption of foreign substances, a mutagenic test of 2,4,5-T in mice has been carried out in the Bundesgesundheitsamt in addition to the teratological experiments and with the help of *dominant-letal-method*. This experiment, however, did not show an increase in induction of letal mutations. Whether after one or seven treatments of males with 50 or 150 mg/kg (= ca 7 and 20% of LD_{50} respectively), the coupling with untreated females for four series (4 weeks) did not reveal any differences from the control, relative to the dead implants which were taken in this case as a measure of the mutagenic effect [27].

In connection with these investigations, it should be emphatically noted that it is urgently necessary to perform a step-by-step study of the *possible mutagenic effects* of all substances which are present in the environment and are taken up by the general population. Although a large number of relevant investigations has been carried out in the field of drugs, it looks, on the whole, disconsolate in the field of *pesticides, food additives*, etc. To put weight on mutation research, a society for environmental mutation research was established by famous geneticists in the FRG. The mission of this society is to supply the necessary information to governmental and administrative authorities as well as to the industry.

To consider once more the problem of *teratogenic effects* in correlation with 2,4,5-T, I would like to conclude that in spite of the teratogenic effect demonstrated in animal experiments, there is no danger at present to the fertility of pregnant women. If one relates the *"teratogenic no-effect level"* which at present is estimated to be approx. 20 mg/kg, and can be taken as a basis for safety considerations, with that in which the highest amount — VO of tolerance value is fixed at 0.01 ppm, and the actual residues (0.04 ppm in corn), it would be obvious that a very high safety factor can be established. Judged by human standards, this factor would not create any danger after absorption of 2,4,5-T residues with food stuffs. On basic considerations it must be expressly indicated, that only the fact that the observed teratogenic activity of a pesticide in animal experiments is detected, would not justify general banning. Also with plant protection chemicals, it is necessary to establish a reasonable relation between toxic concentrations, tolerant maximum amounts, and residue values with regard to an appropriate safety factor. A classification of teratogenic substances in the so-called Delaney-clause, which recommends a zero-tolerance for *food additives with carcinogenic activity* in the USA, is rejected by almost all specialists. On the contrary, one trends to support the view of *dose-activity relations,* which clearly stand, in the present case, against a general banning of 2,4,5-T.

References

1 Jensen, S.: Report of a new chemical hazard. New Scient. 32: 612, 1966
2 Jensen, S., A. G. Johnels, M. Olsson, G. Otterlind: DDT and PCB in marine animals from Swedish waters. Nature 224: 247–250, 1969
3 Risebrough, R. W., P. Rieche, D. B. Peakall, S. G. Hermann, M. N. Kirven: Polychlorinated Biphenyls in the Global Ecosystem. Nature 220: 1098–1102, 1968
4 Miller, J. W.: Pathologic changes in animals exposed to a commercial chlorinated diphenyl. Publ. Hlth. Rep. 59: 1085–1093, 1944
5 Berlin, M.: PCB-effects on mammals. PCB Conference, Stockholm, September 29, 1970
6 Sobotka, T. J.: Differential effects of DDT and polychlorinated biphenyls on the central nervous system. Inf. Bull. BIBRA 9: 363, 1970
7 Bitman, J., H. C. Cecil: Estrogenic activity of DDT analogs and polychlorinated biphenyls. J. Agr. Food Chem. 18: 1108–1112, 1970
8 McLaughlin, J., J. P. Marliac, M. J. Verret, M. K. Mutchler, O. G. Fitzhugh: The injection of chemicals into the yolk sac of fertile eggs prior to incubation as a toxicity test. Toxicol. appl. Pharmacol. 5: 760, 1963
9 Kojima, T., H. Fukumoto, S. Makisumi: Chlorobiphenyl poisoning. Gas chromatographic detection of chlorobiphenyls in rice oil and biological materials. Nippon Hoigaku Zasshi 23: 415–420, 1969
10 Biros, F. J., A. C. Walker, A. Medberry: Polychlorinated biphenyls in human adipose tissue. Bull. Environm. Contam. Toxicol. 5: 317, 1970
11 Vos, J. G., J. H. Koeman, H. L. van der Maas, M. C. ten Noever de Brauw, R. H. de Vos: Identification and toxicological evaluation of chlorinated dibenzofuran and chlorinated naphthalene in two commercial polychlorinated biphenyls. Fd Cosmet. 8: 625–633, 1970

12 Acker, L., E. Schulte: Über das Vorkommen chlorierter Kohlenwasserstoffe im menschlichen Fettgewebe und in Humanmilch. Deutsche Lebensmittel-Rundschau 66: 385–390, 1970
13 Flick, D. F., R. G. O'Dell, V. A. Childs: Development of chick oedema disease-like symptoms with chlorinated biphenyl. Proc. Fed. Am. Soc. exp. Biol. 23: 406, 1964
14 Anonym: Search for chick oedema factor. Chem. Engin. News 45: 10, 1967
15 Higginbotham, G. R., A. Huang, D. Firestone, J. Verrett, J. Ress, A. D. Campbell: Chemical and toxicological evaluations of isolated and synthetic chloro-derivates of dibenzo-p-dioxin. Nature 220: 702–703, 1968
16 Inf. Bull. BIBRA 9: 265, 1970
17 Courtney, K. D., D. W. Gaylor, M. D. Hogan, H. L. Falk, R. R. Bates, J. Mitchell: Teratogenic evaluation of 2,4,5-T. Science 168: 864–866, 1970
18 DOW-Report of 4 Feb. 1971 to the Environmental Protection Agency
19 Emerson, J. L., D. J. Thompson, J. Strebing, C. G. Gerbig, V. B. Robinson: Teratogenic studies on 2,4,5-Trichlorophenoxyacetic acid in the rat and rabbit. Fd Cosmet. Toxicol. 9: 395–404, 1971
20 Sparschu, G. L., F. L. Dunn, V. K. Rowe: Study of the teratogenicity of 2,3,7,8-Tetrachloro-dibenzo-p-dioxin in the rat. Fd Cosmet. Toxicol. 9: 405–412, 1971
21 Roll, R.: Untersuchungen über die teratogene Wirkung von 2,4,5-T bei Mäusen. Fd Cosmet. Toxicol. 9: 671–676, 1971a
22 Roll, R.: Zur hygienisch-toxikologischen Bewertung des 2,4,5-T. Bundesgesundheitsblatt 14: 342–345, 1971b
23 C. H. Boehringer Sohn: Prüfung der Substanz 2,4,5-T auf Teratogenität an Mäusen. 25 November 1970a
24 C. H. Boehringer Sohn: Prüfung der Substanz 2,4,5-T auf Teratogenität in Ratten. 10 December 1970b
25 C. H. Boehringer Sohn: Prüfung der Substanz 2,4,5-T auf Teratogenität an Ratten. (Technische Charge 70 200/371.) 18 June 1971
26 Grunow, W., Chr. Böhme, B. Budczies: Renale Ausscheidung von 2,4,5-T bei Ratten. Fd Cosmet. Toxicol. 9: 667–670, 1971
27 Roll, R.: In press

Evaluating Toxicological Data as Regards Environmental Significance

F. Coulston

Institute of Experimental Pathology and Toxicology, Albany Medical College, Albany
New York 12208, USA

Summary. The methodology used by the toxicologist to assess the safety of environmental chemicals is discussed using as examples the results of studies with an organophosphate insecticide and airborne lead particulates. Animal experiments, using dosage levels sufficiently high to produce toxic effects, are done to disclose target organs and systems, to establish the patterns of absorption, metabolism, excretion and/or storage and to develop sensitive methods for monitoring these effects in man. When widespread human exposure is likely, well-controlled studies in man are warranted to check on the relevance of the animal data to man.

Animal experiments with the organophosphate insecticide indicated that the red-blood cell cholinesterase level was the most sensitive indicator of exposure and possible toxic effect. Liver toxicity and elevation of serum SGPT levels in dogs, but not in monkeys or rats, was a noteworthy species difference. Subsequent studies in man, using levels higher than those to which the general population would likely be exposed to, demonstrated the safety of this compound.

In animals exposed to airborne particulate lead, blood levels increased for a time and then reached an equilibrium in spite of continual exposure. Man responded in a similar manner. Measurements of the activity of erythrocyte δ-aminolevulinic acid dehydrase and blood levels of lead provide sensitive indicators for monitoring exposure to airborne lead.

Zusammenfassung. Die toxikologische Methodologie zur Bestimmung der Sicherheit von Umweltchemikalien wird am Beispiel von Versuchsergebnissen mit einem Organophosphat-Insektizid und staubförmigem Blei in der Luft diskutiert.

Zur Erkennung von beeinflußbaren Organen und Systemen, zur Festsetzung der Muster für Absorption, Metabolismus, Ausscheidung und/oder Speicherung und zur Entwicklung von Methoden zur Überwachung dieser Auswirkungen im Menschen werden Tierversuche gemacht, bei denen Dosen angewandt werden, die hoch genug sind, um toxische Wirkungen hervorzurufen. Wenn es wahrscheinlich ist, daß weitverbreitet der Mensch diesen Substanzen ausgesetzt ist, sollen Untersuchungen im Menschen unter sehr gut kontrollierten Bedingungen durchgeführt werden, um die Relevanz von Daten aus Tierversuchen für den Menschen zu überprüfen.

Tierversuche mit dem Organophosphat-Insektizid zeigten, daß die Cholinesterase der roten Blutkörperchen der empfindlichste Indikator für die Belastung möglicher toxischer Effekte ist. Lebertoxizität und Erhöhung des Serum SGPT-Spiegels in Hunden, aber nicht in Affen oder Ratten, war ein nennenswerter Unterschied innerhalb der untersuchten Spezies. Anschließende Untersuchungen in Menschen mit höheren Dosen, als der Belastung der normalen Bevölkerung entspricht, zeigten die Sicherheit dieser Verbindung. Bei Tieren, die staubförmigem Blei der Luft ausgesetzt waren, stieg der Blutspiegel für einige Zeit an, erreichte aber später, trotz kontinuierlicher Belastung, ein Gleichgewicht. Der Mensch reagierte entsprechend.

Messungen der Aktivität von Erythrozyten-δ-Aminolävulinsäuredehydrase und des Bleiblutspiegels sind empfindliche Anzeiger zur Überwachung von der Belastung aus der Luft.

The role of the toxicologist has gone full circle from the responsibility of identifying the natural environmental poisons to which primitive man was exposed, through the evaluation of the safety of natural and synthetic substances used as drugs, and back again to a renewed concern with the hazards of environmental substances. The early toxicologists were concerned with relatively few substances and most of these were of *bacterial or plant origin* with a few poisonous compounds produced by animals and a few inorganic materials such as *arsenic* or *phosphorus*. With the coming of the industrial revolution his modern counterpart has become concerned with a great variety of substances introduced into the environment by man himself such as insecticides, *automobile exhausts,* industrial waste products, and various radioactive substances. In short, the task of the toxicologist in protecting man's environment has become immensely complex not only because of the large number of environmental contaminants but also because of the possible *interactions* between them and drugs, food additives and natural materials vital to man's existence.

Along with the increase in number of substances to be studied, there has been an increase in the complexity of the methodology for the evaluation of environmental contaminants. The evaluation of safety, which is the final goal of toxicology, has truly evolved into a *multidisciplinary* science. All of the basic disciplines including pharmacology, pathology, physiology, biochemistry, immunology, genetics, and analytical chemistry are now represented in the typical toxicology study. The multidisciplinary approach has made it possible for the toxicologist to discover, quite rapidly, the extent to which a chemical is absorbed, how it is metabolized and excreted, what its distribution and storage in the body is, and what organs are likely to be injured. It has become possible in some instances to determine the specific enzyme or subcellular structural component that is attacked by the chemical. In spite of our increasing sophistication in the methods of studying a chemical in the intact animal and in various in vitro systems, in our search for effects on the *molecular level,* species differences soon return us to the question of the relevance of these findings to *man.*

One of the major problems of toxicology is that of relating data derived from animal experiments to man. Even more difficult is the problem of determining the relevance of results derived from *in vitro* systems. In the case of a new drug, the chemical is cautiously administered to man relatively early in its development to determine the accuracy of animal data and to learn something of its route of *metabolism* so that the most appropriate species can be selected for the long-term animal studies. Observations, made in a species that metabolizes the chemical in a manner similar to that of man, are more likely to have *predictive value* for effects in man than those made in species in which the chemical is metabolized quite differently. These considerations make it imperative that the principal environmental chemicals be studied in man in a manner similar to that used for drugs, for man is sometimes unique in his responses and this information can be obtained at present in no other way.

For substances of environmental significance, a pattern of research has evolved whereby the areas of concern are first identified by in depth studies in several animal species. With this information, studies in man can be designed to search for these specific effects, using techniques which are extremely sensitive, thereby minimizing the risks involved in experiments in man. Two examples of how this can be done are presented in the following studies on Supracide, an *organophosphate insecticide* and with *lead:*

Supracide*

Studies conducted by the Institute of Experimental Pathology and Toxicology of Albany Medical College with Supracide®, an organophosphate insecticide, exemplify this approach to the problem of evaluating the toxicologic significance of a potential environmental contaminant.

* We wish to thank Ciba-Geigy for their support of these studies.

Acute studies in rats, dogs, and monkeys revealed effects commonly associated with cholinesterase inhibition. These effects included salivation, lacrimation, miosis, muscle fasciculation, tremors, prostration, and death by apparent respiratory paralysis. When determinations of *erythrocyte,* plasma, and *brain cholinesterase* were made, marked inhibition of *erythrocyte* and *brain* cholinesterase were always found to accompany the occurrence of the severe signs of intoxication. Inhibition of the pseudocholinesterase of the plasma, on the other hand, was not a consistent finding. For example, when a regimen of increasing dosage was given to a monkey, beginning at a dose of 1.2 mg/kg and doubling the dose each subsequent day until 38.4 mg/kg per day was reached, signs characteristic of cholinesterase inhibition occurred after a few days at this high level. Red blood cell cholinesterase activity fell from an initial value of 321 to 12 in terms of ml CO_2/30 min. Similar results were observed in a second monkey given dosages increasing from 1.2 to 19.2 mg/kg and maintained at the 19.2 mg/kg dose for seven days. The monkey became prostrate and red blood cell cholinesterase fell from an initial value of 366 to 14. Both monkeys survived and six weeks were required for the erythrocyte cholinesterase activity to return to normal. In both cases, only slight decreases in the plasma cholinesterase activity were observed.

Even after long term administration of doses sufficiently high to kill some of the monkeys, *plasma cholinesterase* activity proved to be the least reliable indicator of toxic effect (Fig. 1). At a daily dose of 5 mg/kg, two of four monkeys died within a few weeks. Yet at 16 months, the plasma cholinesterase activity in the two survivors was the same as that in the control monkeys. At a dose of 10 mg/kg, which also killed two of four monkeys, a mode

CHOLINESTERASE ACTIVITY

In Rhesus monkeys given Supracide for 16 months

Dosage Level	RBC ChE	Plasma ChE	Brain ChE
Control	308	231	179
5 mg/kg	46	229	127
10 mg/kg	28	153	69

Results in μl CO_2/30 min.

3 monkeys in control group, 2 each in other groups

Fig. 1

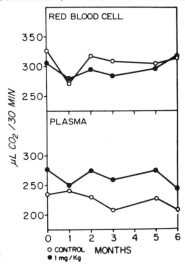

CHOLINESTERASE ACTIVITY
IN MONKEYS GIVEN SINGLE DAILY DOSES
OF SUPRACIDE (MANOMETRIC PROCEDURE)

Fig. 2   o CONTROL
            ● 1 mg/Kg

rate depression of plasma cholinesterase activity was found. At the same time marked inhibition of red blood cell cholinesterase was observed in both groups. When the monkeys were sacrificed, determinations of brain cholinesterase revealed a marked inhibition of activity in the 10 mg/kg group and a moderate decrease in the 5 mg/kg group. These studies demonstrated that the erythrocyte cholinesterase activity was the best indicator of toxic effect for this particular *organophosphate*, while the plasma level was the least reliable. Depression of brain cholinesterase levels also correlated well with toxic effects, but this measurement is of no value of course for the estimation of toxic effects in man.

The studies discussed so far utilized dosage levels which produced overt signs of intoxication. Would the level of red blood cell cholinesterase activity provide an index of exposure at levels that did not produce any observable signs? In the first group of four monkeys given 1.0 mg/kg for six months, there was no indication of effect upon erythrocyte or plasma cholinesterase activity (Fig. 2). Brain cholinesterase was also normal.

In subsequent groups of 12 monkeys each given 0.25 and 1.0 mg/kg over a two-year period there was a slight indication of a dose-related effect upon the red blood

cell cholinesterase activity (Fig. 3). At each of the determinations made at 6, 12, 18 and 22 months, the values in the groups given these low doses of Supracide were somewhat lower than those of the control group, but the dose of 0.25 was never statistically lower and only occasionally was the difference between the 1.0 mg/kg dose and the control valid. Since no overt signs of *cholinesterase inhibition* were seen at any time during the study, these low levels of inhibition were perhaps best interpreted as an index of expo-

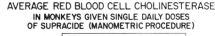

AVERAGE RED BLOOD CELL CHOLINESTERASE
IN MONKEYS GIVEN SINGLE DAILY DOSES
OF SUPRACIDE (MANOMETRIC PROCEDURE)

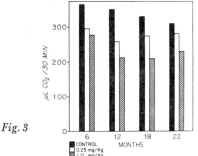

Fig. 3

CONTROL
0.25 mg/Kg
1.0 mg/Kg

sure no necessarily reflecting a toxic effect and they could be considered to be the no-effect level.

With these monkey studies, and similar eperiments in dogs and rats as a guide, a six-week time-dose response study at dose levels of 0.04 and 0.11 mg/kg was conducted in 18 male volunteers divided into 3 groups of 6 each at one of the New York State prisons. Supracide was administered orally in capsules via a staggered schedule (Fig. 4), so that four men took the low dose for two weeks before four other men started receiving the high dose. After three additional weeks the remaining four men to receive each of the doses of *Supracide* started taking their capsules. All volunteers ingested capsules for six weeks, except those in the placebo group, who took theirs throughout the duration of the study. This type of administration schedule is designed to

minimize the risks involved in a human study. Administration of the low dose for two weeks prior to starting the high dose provided an opportunity to detect adverse effects prior to initiation of the higher dose level. Continuation of the low dose aftcr the high-dose started provided the time for accumulative effects to occur and the higher dosage could be stopped before serious effects might occur on a time-dose basis.

The volunteers received their appropriate dosage by capsule each morning under rigidly controlled conditions and close medical supervision. Recordings of the *EEG* were made before, during and at the termination of ingestion of capsules containing Supracide or placebo. The volunteers were questioned daily about symptoms they had experienced that might be attributed to the administration of Supracide. No significant signs or symptoms were reported during the course of the study. The EEGs showed none of the effects that are regarded as being typical of those induced by inhibitors of *cholinesterase*, such as irregularities in rhythm, variable increase in potential, and intermittent bursts of abnormally slow waves of comparatively high voltage.

Red blood cell and plasma cholinesterase levels were closely monitored (Fig. 4). At no time during the six-week study was there any indication of an effect upon either of these parameters. The erythrocyte cholinesterase levels in the high-dose group declined in the 4th and 5th weeks, but a similar decline was observed in the placebo group, and the levels in the high dose group did not fall below the levels in the placebo group. There was no discernible pattern for the changes in plasma cholinesterase observed in the study to suggest an effect related to Supracide.

In contrast to the findings in this study, decreases in plasma cholinesterase were

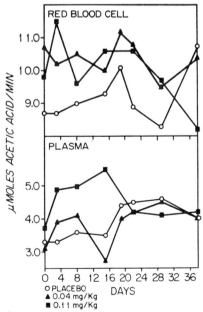

CHOLINESTERASE ACTIVITY IN MAN GIVEN SINGLE DAILY DOSES OF SUPRACIDE (pH STAT METHOD)

RED BLOOD CELL

PLASMA

μMOLES ACETIC ACID/MIN

O PLACEBO
▲ 0.04 mg/Kg
■ 0.11 mg/Kg

DAYS

Fig. 4

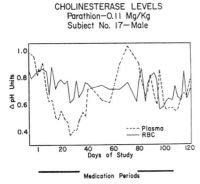

CHOLINESTERASE LEVELS
Parathion—0.11 Mg/Kg
Subject No. 17—Male

Fig. 5

observed in a previous study in which *Parathion* was given to human volunteers. For example, in one subject given 0.11 mg/kg of Parathion (Fig. 5) a sharp decline in plasma cholinesterase was observed during a four-week administration period. When the Parathion was withheld the plasma cholinesterase gradually returned to the pre-administration level. Decline in the plasma cholinesterase activity occurred a second time when administration was begun again. In a second subject (Fig. 6), a sharp decline in plasma cholinesterase occurred during the second administration period, which also returned to normal when administration

was stopped. In the Parathion study symptoms were observed which may have been related to the intake of *Parathion,* such as headache, nausea, occasional vomiting, nervousness, sweating, giddiness, and muscle cramps. However, correlation between occurrence of symptoms and decrease in plasma cholinesterase levels was poor. Symptoms often occurred with no detectable changes in cholinesterase levels and sharp declines in cholinesterase were not always accompanied by symptoms. Nevertheless, determinations of red blood cell and *plasma cholinesterase* activity do provide an index of exposure and can in

AVERAGE SERUM GLUTAMIC PYRUVIC TRANSAMINASE VALUES (R.F.U.)

In Beagle Dogs Given Single Daily Doses of Supracide

GROUP*	Initial	1 wk.	2 wk.	3 wk	5 wk.	7 wk.
Control	60	70	81	80	57	84
Batch 99	109	109	211	253	180	315
Batch 101	77	84	172	200	225	343
Batch 102	137	153	257	357	325	338
Batch 103	96	123	230	333	283	328

* 3 dogs per group; given 2 mg/kg daily by stomach tube.
   Control given gum tragacanth vehicle.

Fig. 7

some instances give warning of too high an exposure before clinical signs and symptoms occur.

Preliminary studies in animals with Supracide had disclosed one additional effect which was sought in the human study. When *Supracide* was administered to beagle dogs (even at the relatively low level of 2 mg/kg) an increase in the serum glutamic pyruvic transaminase values was observed (Fig. 7). With four different batches of the compound, slight elevations of the *transaminase* values were observed at about two weeks. The average values gradually increased and values over 300 were uniformly found at 7 weeks. When the dogs were sacrificed mild changes consisting of slight hydropic swelling and

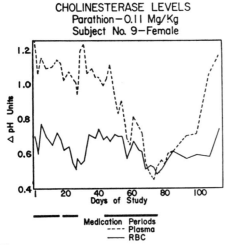

CHOLINESTERASE LEVELS
Parathion—0.11 Mg/Kg
Subject No. 9—Female

Fig. 6

focal fatty vacuolation were found in the liver. Similar elevations of SGPT values were not observed in monkeys or rats.

Frequent determinations of SGPT values were made in the human study (Fig. 8). No elevations of this transaminase ocurred in man. In fact, average values in the groups given Supracide remained lower than those of the placebo group throughout the six-week period.

Lead*

The value of using human volunteers, as well as various animal species in evaluating the effects of environmental chemicals, is demonstrated by another study involving inhalation of *lead sesquioxide* particulates. These studies were undertaken to ascertain whether certain selected levels of continuous exposure to lead particulates would produce any detectable effects using the most sensitive criteria of exposure. Work of a similar nature, by Kehoe and his colleagues at the Kettering Laboratory in Cincinnati, had indicated that continued exposure to lead would result in a constantly increasing body burden of lead and that this would be reflected by a continually increasing level of blood lead. It should be noted that the inhalation studies of Kehoe were not continuous; but were intermittant and similar to the types of exposure encountered in *industrial situations*. Another aim of our study was to determine whether a continually increasing blood lead would be encountered under conditions of continuous daily exposure to lead sesquioxide particulates.

Preliminary studies were carried out in two animal species, the rat and the rhesus monkey, at an average daily level of

* We wish to thank the American Petroleum Institute, the International Lead and Zinc Organization and the Environmental Protection Agency for their support of these studies.

AVERAGE SERUM GLUTAMIC TRANSAMINASE VALUES

In Man Given Single Daily Doses of Supracide

GROUP *	Day of Study				
	3	8	19	29	38
Placebo	23 ± 3	23 ± 3	29 ± 3	20 ± 2	38 ± 4
0.04 mg/kg	19 ± 2	16 ± 2	21 ± 2	13 ± 1	27 ± 2
0.11 mg/kg	17 ± 2	17 ± 1	14 ± 7	10 ± 3	12 ± 5

* 4 men in placebo group; 8 men in each of other two groups.

Fig. 8

21.5 µg lead per cubic meter of air. These exposures were for 22 hours/day, 7 days per week for one year. Blood lead levels were monitored at monthly intervals. Changes occurring in the *blood lead* concentration in the animals during the exposure phase of the experiment are shown in Fig. 9. In both species there was a significant rise in the blood concentration of lead. In the rats, there was no further increase after the third month of exposure and in the monkeys there was no apparent increase after four months. A

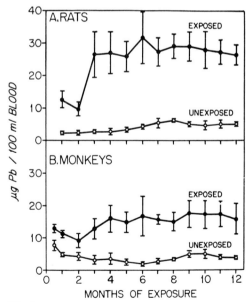

Fig. 9

species difference is evident since the blood lead level in the exposed rats increased to almost twice the level found in exposed monkeys despite the fact that they were exposed to identical atmospheric concentrations of lead.

The enzyme δ-*aminolevulinic acid dehydrase* (ALAD), which catalyzes the formation of porphobilinogen during the synthesis of heme, is inhibited by lead and has been demonstrated to be a sensitive indicator of lead absorption when measured in the blood of some animal species and in man. We found that after twelve months exposure, the rats in the lead chamber had a mean ALAD activity of 3.7 Bonsignore units, while the controls had a mean of 12.5 units. It is not possible to assess this effect in the *rhesus monkey erythrocytes,* because the normal levels are too low to measure, by the usual procedures.

The results obtained in the animals suggested that blood lead in humans might reasonably be expected to rise to an apparent stable level under similar conditions. To accomplish such an exposure in human volunteers, a ward in the hospital at the Clinton Correctional Facility at Dannemora, N. Y., was converted into an appropriate exposure chamber. The ward was equipped with its own separate air conditioning system which was also designed to provide a constant input of lead sesquioxide and distribute it throughout the chamber. The chamber had room for fourteen beds, some recreational space and integral toilet and shower facilities. The volunteers who participated in the study lived in the room and were out only about one hour/day for meals. Exposure was therefore continuous for about 23 hours/day, 7 days/week for up to four months. The chamber was operated to provide 10 μg Pb/cubic meter of air. Results of daily air sampling and analysis showed the average level

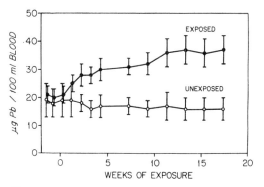

Fig. 10

throughout the study was 10.9 μg/cubic meter.

The concentration of blood lead in the men are shown graphically in Slide 10 for the eight men who remained on the study during the four months exposure phase. The controls shown in the figure were men residing outside the chamber, but who otherwise lived under similar circumstances. The blood lead concentration in the exposed group increased to nearly twice their pre-exposure levels while the level remained relatively unchanged in the controls. During the final six weeks of the exposure phase of the experiment, there was no apparent further increase in the blood lead of the exposed men.

The activity of *ALAD* was also estimated in the blood samples from the volunteers. After approximately six weeks of exposure, the enzyme activity decreased to about $^{1}/_{2}$ the pre-exposure value in the men.

The results of these studies not only demonstrate the predictive value of animal experiments, but also point out some of the limitations of extrapolating the findings from animal studies to man. Both species, the rat and the monkey, predicted that when man was exposed to a constant level of lead in the atmosphere a rising blood level would be observed initially and that an equilibrium would be reached after some interval of exposure. This prediction was verified by the results in the volunteers. However, neither the results in the rat nor those in the monkey could be used to predict the magnitude of the blood lead increase nor the rate at which an equilibrium would be established in man. An example of the problems which result from species differences also was observed in these studies. The rat predicted that measurable reductions in *ALAD* might occur in man exposed to lead, while the monkey proved to be useless in the assessment of this criterion of exposure. These results point out the necessity of using two or more species in the preliminary work designed to identify the most sensitive parameters of exposure and toxicity.

Conclusion

The two examples presented illustrate the most effective methods, available at present, for the evaluation of toxicologic data concerning environmental contaminants. Preliminary studies in animals point out the areas of concern, the target organs or enzyme systems which are affected, and provide the opportunity to develop sensitive methods that can be used for monitoring exposure and as indices of toxic effect. Because of species differences the results of animal studies often leave us with the question: which set of results is applicable to man? Many times this question can be answered only by cautious trial in man using the sensitive criteria developed in the animal studies.

In the first example, animal studies with *Supracide* established that toxic effects resulted from the inhibition of *cholinesterase* as would be expected with an *organophosphate*. These studies further indicated that the erythrocyte cholineste-

rase level was the best indicator of toxic effect for this particular compound, while the plasma activity was less consistently associated with adverse effects. Results from the dog suggested that liver toxicity, as indicated by a rising level of SGPT, was a possibility. This latter result was not observed in the monkey or rat. Studies in human volunteers with levels up to 0.11 mg/kg did not result in any significant decreases in plasma or *erythrocyte cholinesterase* levels. Other criteria of cholinesterase inhibition such as subjective symptoms and *EEG* patterns were also negative. No indications of liver toxicity, using *SGPT* as a criterion of deleterious effect, were observed. Information such as this provides the basis for more intelligent establishment of ADI's. A comparison of the results of these studies with those conducted previously with Parathion suggest that Supracide has considerably less potential for acute intoxication than has Parathion.

In the second example, animal studies predicted that, under a constant level of exposure, blood levels of lead would reach an equilibrium after some interval of time. This was verified in man. Studies in the rat indicated that the level of δ-*aminolevulinic acid dehydrase* would provide a sensitive indicator of exposure to lead. This was also verified in man. These studies provide the bases for the establishment of acceptable exposure levels of atmospheric lead and demonstrate that determinations of blood levels of lead and the activity of *ALAD* provide a practical method for monitoring exposure.

Man has exposed himself to many kinds of environmental contaminants. More knowledge regarding the possible long-term effects of these chemicals is necessary. Research has provided some answers, but we cannot afford to be complacent with our achievements. Ways to improve our ability to extrapolate the animal data to man must be sought, so that man himself will not continue to be the species of choice for the safety evaluation of chemicals and drugs in his own environment.

The Problem of Permissible Limits of Potentially Toxic Substances in the Working and General Environment of Modern Man

R. Truhaut

Centre de Recherches Toxicologiques, Faculté des Sciences Pharmaceutiques et Biologiques de l'Université René Descartes 4, Avenue de l'Observatoire Paris VI° (France)

Summary* An old — and correct — saying by Paracelsus "The dose makes the poison" finds its modern application for toxicologists and pharmacologists in the *dose-effect* relationship and in the *toxicity threshold*. Fixation of the toxicity threshold is the basis for any attempt to establish *tolerable or admissible limits* — not to mention effective control and prevention measures — for potentially toxic air pollutants in urban and industrial environments.

The Nature of the Risks

There are three principle categories of risks:

1) *Acute toxicity risks* occur most often in occupational environments: relatively high, short-term concentrations of the pollutant (sulphur oxides, nitrogen oxides, resulting immediately or after a period of latency, from industrial processes, spillage, etc.).

2) *Chronic toxicity risks* refer to more or less long-term exposure to substances with *cumulative properties*, substances likely to be retained by the organism for greater or lesser periods of time (e.g. fluoride ions).

3) Risks of ingestion of cancerogenic substances. Much recent research has shown that for some substances (e. g. diethylnitrosamine) the activity seems to be determined by a sum of absolutely irreversible effects. This assumption is based on the fact that it is not so much the stored fraction of the absorbed amount, but the total absorbed quantity independent of metabolic breakdown and excretion. In terms of the dose-effect relation for cancerogenic substances, this means the percentage of tumors induced or the mean time necessary for provoking the tumors. With the present state-of-the-art, it is nearly impossible to establish *dose thresholds* for these substances.

Remarks on the Implementation of a Methodology for Evaluating Risks

Evaluation of risks involves two fundamental objectives:

1) fixing the morphological and biochemical sites of attack — the qualitative aspects, and

---

* Editors' note: In keeping with the international approach of EQS and because of the high literary quality of this paper, it is being published in the original language. However, in order to make Monsieur Truhaut's very important comments available to *all* readers, we have compiled an extensive summary in English of the whole paper.

2) establishment of the relationships between doses (in this case the concentrations) and effects for the purpose of fixing toxicity thresholds — the quantitative aspects.

As regards the methodology of evaluation, it is important to emphasize that it cannot be rigorously codified because it depends on:

1) the nature of the toxic substance *(physicochemical viewpoint)* and of possible changes of it due to various factors,

2) *the condition of exposure* changes, among others, with the geographic site (climatic factors, meteorological conditions, etc.),

3) the nature of the toxic potential as revealed by preliminary tests *(toxicological viewpoint)*,

4) the results obtained in biochemical tests, particularly those concerning comparative *metabolic studies* among man and laboratory animals which determines the choice of species to be subjected to ultimate and elaborate tests *(the biochemical viewpoint)*.

Experiments with Laboratory Animals

Animal experiments consist primarily of general classical toxicity tests, acute and subacute, *short-term* or *semi-chronic toxicity*, and *long-term toxicity*. The latter tests are indispensable for the evaluation of toxic air pollutants as well as for food additives and contaminants because these are things man is most often exposed to during his life.

Experiments in Humans

The ideal method would be experiments on humans, but for obvious reasons, it cannot be systematically applied. Experiments with human volunteers using a small number of toxicants (e. g. of pesticides or solvents or substances likely to be encountered in urban air: lead or tetraethyl lead vapors) are justified and valuable. In any case, it must not be forgotten that in the case of air pollution, *experimentation on human beings* is taking place automatically and this is why epidemological inquiries into the morbidity and mortality from air pollution are of special importance. The results of these experiments show the importance of such concepts as "synergic toxicant", "co-cancerogen" or "promoter", "pharmocodynamic antagonist" in establishing toxicity thresholds.

The Concept of Tolerable Admissible Limits for Toxic Substances in Occupational Environments

Various types of limits — *threshold limit values, MAC values, ceiling values, emergency exposure limits*, etc. — are under study in this regard. The adjectives "tolerable" or "admissible" are usually taken to mean that, in the case of occupational exposure, neither short nor prolonged exposure of workers to the concentration may cause toxic effects or hazards in the place of work. However, it must not be forgotten that factors other than the concentration in the environment are important for determining toxic phenomena, namely the *conditions of absorption and elimination* and *individual variations of sensitivity* to the toxic products.

The *conditions of absorption and elimination* are a function of various parameters:

1) Increased respiratory rhythm due to progressive work or sports training with the resulting increased taking up of possibly toxic substances or disturbance of renal function resulting in decrease of excretions.

2) the state of the toxicant, particularly in case of dust or particles where dimensions (granulometry) determine the possibility of stronger or weaker penetration into the respiratory system.

3) the presence of associated substances which enhance absorption of toxic substances (solvents, surface active agents . . .).

Individual sensitivity depends on

1) physiological condition (age, sex, race, temperament, hormonal balance, state of nutrition, fatigue, emotional shock . . .),

2) higher or lower degrees of functional activity in the organs in the detoxification or elimination chain,

3) alcohol intake which is particularly important because of its sensitizing effect for a number of toxic substances.

Even with these aspects considered, the limits established as being tolerable or admissible serve only as guidelines, not strong demarcations between harmless or dangerous concentrations.

Various national governments (the USA, USSR, etc.) and international organizations (World Health Organization, International Labor Organization, International Commission of Occupational Medicine, etc.) are continuing their efforts to establish admissible limits (see tables).

Pollutants may be classified into *three categories* as regards the nature of their toxic effects and the rapidity of their manifestation:

1) Substances whose primary effects are signs of *irritation, sensitization* or *acute intoxication* caused by short-term, immediate, or delayed exposure to concentrations of the type encountered in practice.

2) Substances involving *cumulative effects* caused by repeated exposure to concentrations usually encountered in practice (working 6 to 8 hours daily, for 5 to 6 days per week). Suggestions here include the concepts of *"time-weighted averages"* and *"threshold limit values"* (heavy metals, arsenic, insecticides).

3) Substances which are cancerogenic. With the present state of knowledge, it is not possible to establish admissible limits for these substances with any sort of certainty.

Among the dificulties to be faced are the complexity of urban air pollution — it is not a question of discrete substances which can be individually treated but rather a "soup" of pollutants — and the frequent chemical modifications of the pollutants under the influence of a number of environmental factors and social and economic imperatives.

Present Possibilities for the Application of a Tolerable Limit for Chemical Air Pollutants in Urban and Industrial Environments

Here it is important to note that *limits* established for *occupational environments* cannot be applied to pollutants encountered by the general population. There are two main reasons for this:

1) Workers are exposed to pollutants on a discontinuous basis (6—8 hours daily for 5—6 days per week) whereas the general population is subject to relatively continuous exposure, and

2) in nearly all cases, workers are adults subject to medical checks and controls; the general population, on the other hand, includes not only these adults but also children, elderly people, pregnant women, ill persons and persons with non-apparent deficiencies or afflictions who may be very sensitive.

With these principles in mind, limits for some substances have been proposed (lead, beryllium, aldehydes. USSR see Table 6). In any case, active international cooperation is a task in studies aimed at establishing toxicity thresholds and admissible limits to serve as guides for industry and sanitary policy makers.

Zusammenfassung. Dieser Beitrag behandelt Gefahren für den Menschen, die durch in der Luft enthaltene Chemikalien entstehen. Dabei werden langfristige Einwirkungen hervorgehoben, zu denen auch Karzinogene gehören. Die Methodik zur Bewertung dieser Gefahren wird kritisch geprüft, und die Notwendigkeit der Auswahl angemessener Kriterien zur Bestimmung toxischer Werte bei Labortieren und bei betroffenen Menschen wird betont. Weiterhin werden die tolerierbaren Grenzwerte für toxische Substanzen innerhalb der beruflichen Umwelt (arbeitenden Bevölkerung) und ihre Anwendung auf die allgemeine Luftverschmutzung angegeben. Ergebnisse aus verschiedenen Ländern werden erwähnt. Unter anderem wird auf die Schwierigkeiten hingewiesen, die durch den Umfang der Luftverschmutzung in bewohnten Gebieten entstehen, und die chemischen Veränderungen der Verunreinigungssubstanzen unter dem Einfluß einer bestimmten Anzahl von Umweltfaktoren sowie die sozialen und ökonomischen Gegebenheiten werden diskutiert. Die angemessenen analytischen Methoden zur Kontrolle der Luftverschmutzung erfordern eine enge Zusammenarbeit zwischen Toxikologen und Analytikern. Schließlich wird die Notwendigkeit einer aktiven, internationalen Kooperation für Untersuchungen zur Festsetzung toxischer Grenzwerte und höchstzulässiger Werte betont, die der Industrie und den Gesundheitsbehörden als Richtlinien dienen sollen.

«Dosis fecit venenum»; c'est la dose qui fait le poison, a dit, il y a bien longtemps, le grand Paracelse. De cette vérité découle la règle d'or que constitue, en Pharmacodynamie et en Toxicologie, l'établissement des *relations doses-effets,* ainsi que le principe fondamental de prévention des intoxications, notamment de celles pouvant résulter, pour l'homme, de la pollution de son environnement au sens large de ce terme: diminuer les doses susceptibles d'être absorbées par des sujets exposés jusqu'à des valeurs situées en dessous de celles représentant *les seuils de toxicité.*

Ce principe est particulièrement valable pour la surveillance de la pollution chimique de l'environnement, et c'est pourquoi la fixation de *limites admissibles ou tolérables* pour les polluants potentiellement toxiques de l'air des villes et des environnements industriels revêt une grande importance.

J'ai choisi de traiter ce sujet, car j'estime que les données, qualitatives et quantitatives, fournies par les toxicoloques sont primordiales pour guider et orienter le travail à effectuer par les analystes en ce qui concerne les qualités, spécialement la *spécificité* et la *sensibilité,* que doivent posséder les *méthodes* qu'ils ont à mettre au point pour le contrôle de *la pollution de l'air,* ainsi que pour les *enquêtes épidémiologiques* dont le but ultime est l'établissement de relations entre les doses et les effets. L'objectif commun primordial de ces deux groupes de spécialistes, la protection des populations contre les risques

pour la santé que peut comporter l'inhalation répétée, jour après jour, d'un air pollué, impose, à mon avis, une liaison étroite entre eux.

C'est un lieu commun de rappeler que l'air est l'élément le plus indispensable à la vie de l'homme. Ce dernier ne peut survivre plus de cinq minutes en son absence et, de sa naissance jusqu'à sa mort, il en absorbe journellement environ 12 m³, soit un peu plus de 15 kgs. Toute altération de *la pureté de l'air* est, par suite, d'une importance primordiale, car les agents qui peuvent s'y trouver incorporés, au moins les gaz et les vapeurs, ainsi que les vésicules et les particules de taille suffisamment fine pour ne pas être arrêtés mécaniquement au niveau des voies aériennes supérieures, pénètrent jusque dans les profondeurs des alvéoles pulmonaires. Si leur concentration est suffisante, ils peuvent y exercer des effets nocifs locaux ou même, très souvent, passer dans la circulation générale et provoquer alors des symptômes de toxicité au niveau des récepteurs sensibles.

Or, depuis une cinquantaine d'années, le développement de l'industrie et notamment de l'industrie chimique, la de grandes *aglomérations urbaines* et la motorisation des transports ont eu comme conséquence un accroissement de la teneur de l'air en impuretés. Dans certaines cités ou aux alentours de certains grands complexes industriels, la *pollution de l'air* a pu atteindre un degré suffisant pour provoquer des maladies graves et même des décès dans la population. Il suffit de mentionner, à cet égard, entre autres exemples classiques d'accidents dramatiques, ceux de la Vallée de la Meuse, en décembre 1930, ceux de Donora, petite ville industrielle de Pennsylvanie, en 1948 et ceux de Londres, en Novembre-Décembre 1952 et en Décembre 1962.

Il s'agit là *d'épidodes aigus* qui créent une situation d'alarme par leur caractère brusque et leur gravité spectaculaire. Mais les toxicologues et les hygiénistes sont, à juste titre, encore bien plus préoccupés par les *risques de nocivité à long terme,* beaucoup plus insidieux, qui peuvent éventuellement résulter de *l'exposition longtemps répétée* à un air renfermant des concentrations, même très minimes, d'agents de pollution. Le bon sens populaire ne s'y trompe pas d'ailleurs et redoute fort justement les effets, pour la santé, de la respiration, par les habitants de villes, d'un air chargé de vapeurs, de poussières et de fumées, au point de diminuer parfois notablement la visibilité, cependant qu'il fait dépérir les arbres des boulevards et qu'il attaque la pierre des maisons et le zinc de toitures.

On conçoit, dans ces conditions, l'intérêt qui s'attache aux recherches sur les risques de nocivité pouvant résulter pour la santé publique, de la pollution de l'air. Comme dans les autres domaines de la toxicologie, le problème le plus important à résoudre dans ces recherches est la fixation des *seuils de nocivité* dont la connaissance est primordiale et conditionne la mise en œuvre de mesures de prévention vraiment efficaces. C'est l'étude, forcément sommaire de ce problème qui constitue l'objet du présent article.

J'envisagerai successivement, d'une façon très générale:

1) *La nature des risques,* en me limitant à l'étude sommaire de ceux liés aux agents chimiques de pollution, avec exclusion, toutefois, des agents radioactifs dont l'étude me paraît devoir être réservée à des experts hautement spécialisés.

2) *Des aperçus sur la méthodologie* à mettre en œuvre pour évaluer ces risques et fixer des euils de nocivité.

3) *Le concept des limites tolérables* pour les substances toxiques dans les ambiances professionnelles.

4) *Possibilités actuelles* d'application de ce concept aux agents de pollution chimique de l'air des cités ou des environnements industriels pour l'établissement de normes de pureté de l'air, ainsi que les perspectives d'actions concertées sur le plan international dans le domaine de la recherche toxicologique.

La Nature des Risques

Comme dans les cas des agents chimiques de *pollution des ambiances professionnelles,* il convient de distinguer trois principales catégories de risques [36].

Risques de toxicité immédiate

Risques de toxicité immédiate ou après une phase de latence, pouvant résulter de l'exposition, pendant un temps court, à une concentration relativement élevée de polluant.

Ces risques sont beaucoup moins fréquemment rencontrés que dans les ambiances professionnelles et ne s'observent, à vrai dire, que dans des circonstances exceptionnelles, notamment lors d'émissions accidentelles ou de situations météorologiques spéciales (brouillard, inversion de température...), telles que celles réalisées au cours d'épisodes aigus du type de ceux que j'ai mentionnés à titre d'exemples. Les principaux agents de pollution étaient alors l'anhydride sulfureux et l'acide sulfurique, formés par grillage de minerais soufrés ou lors de la combustion des charbons ou huiles minérales (mazout) qui renferment toujours une certaine quantité de soufre. Il faut également mentionner, parmi les risques du même type, les effets d'irritation des muqueuses oculaire et trachéobronchique, si fréquemment observés dans des villes, comme Los-Angelès, à type de pollution photochimique oxydante (oxydes d'azote, ozone, aldéhydes, cétones, acides, ozonides, peroxydes organiques, nitrates de peracyle . . .) comportant des transformations complexes des polluants, sous *l'influence de la lumière solaire,* après leur émission.

Risques insidieux de toxicité

Risques insidieux de toxicité à plus ou moins long terme, pouvant résulter de l'exposition à des substances possédant des *propriétés cumulatives,* c'est-à-dire susceptibles d'êtres retenues plus ou moins longtemps dans l'organisme [35]. L'absorption répétée de petites doses de ces substances, qui, si leur élimination était suffisamment rapide, serait sans conséquence discernables, provoque, au bout d'un certain temps, dépendant de la grandeur des doses et de la vitesse d'élimination, d'atteindre les seuils de concentration toxique au niveau des récepteurs sensibles.

Il en est ainsi, par exemple, avec les *dérivés fluorés* pouvant être présents sous forme de produits gazeux (HF, SiF_4, H_2SiF_6) ou de poussières (fluorure de calcium, CaF_2, et cryolithe ou fluorure double d'aluminium et de sodium, Al_2F_6, 6Na F, employés comme fondants) dans les fumées rejetées, entre autres, par les usines d'aluminium ou de superphosphates. *L'anion F^- s'accumule en effet,* en raison de son affinité pour les phosphates de calcium avec formation de fluoapatites insolubles, dans les tissus calcifiés (os, dents). Lorsqu'une certaine concen-

tration est atteinte, il provoque ainsi, au niveau de ces tissus, des lésions génératrices de fragilité qui, avec la cachexie et certains troubles endocriniens, liés à une accumulation au niveau d'autres tissus, caractérisent les intoxications chroniques connues sous le nom de fluoroses.

Il en est ainsi, également, avec des éléments comme l'arsenic ou les métaux lourds (plomb, mercure, cadmium . . .) présents dans les déchets rejetés par certaines fonderies et qui sont fixés, entre autres, par les groupements thiols des protéines.

En tenant compte des *vitesses d'élimination* et des concentrations au niveau des récepteurs sensibles nécessaires à la manifestation des effets toxiques, il est possible de fixer des concentrations pour lesquelles et en deçà desquelles il n'y a pratiquement pas de danger.

Parmi les polluants à l'exposition prolongée desquels sont liés des risques de nocivité à long terme, il faut également mentionner certains produits irritants, et en particulier l'anhydride sulfureux et l'acide sulfurique sous forme de fines vésicules, qui, par répétition des agressions, paraissent, à la lumière de diverses enquêtes, susceptibles de contribuer à l'apparition de bronchites chroniques, particulièrement dans des villes comme Londres à type de pollution dite acide.

Risques d'induction de proliférations malignes

Ils peuvent résulter de l'exposition à des substances dites *cancérogènes* ou *carcinogènes.*

Les travaux de Druckrey et collaborateurs concernant, entre autres, le *p-diméthylaminoazobenzène,* le *p-diméthylaminostilbène* et la *diéthylnitrosamine,* tendent à montrer que la manifestation de l'activité de ces composés est une fonction, non pas de la fraction des doses absorbées retenue dans l'organisme comme dans le cas des poisons typiquement cumulatifs, mais de la somme totale des doses absorbées, quel que soit leur fractionnement dans le temps et le jeu des éliminations et des destructions métaboliques. Tout se passe *comme s'il y avait sommation totale d'effets absolument irréversibles,* comme le sont les impressions successives d'une plaque ou d'un film photographique, par exemple. Cette notion de l'irréversibilité absolue des effets est actuellement discutée à la suite de résultats obtenus dans les domaines de la Biologie moléculaire (possibilité de réfractions des lésions provoquées au niveau des macromolécules nuclériques par certains agents physiques) et de l'Immunologie. Cette question dont l'importance, théorique et pratique, est considérable, sera traitée par l'auteur dans un article ultérieur.

En conséquence, bien qu'il existe indiscutablement, dans le cas des substances cancérogènes comme dans celui des substances toxiques en général, une relation entre les *doses* absorbées et les *réponses* obtenues (pourcentage de tumeurs induites ou temps moyen nécessaire pour les provoquer), il est pratiquement impossible, dans l'état actuel de nos connaissances, de fixer, pour des expositions répétées pendant une grande partie de la vie, des *doses seuils* en deçà desquelles il n'y a plus de danger. En effet, si l'on admet la persistance de l'effet après l'élimination de la substance qui en est responsable (cessante causa, non cessat effectus), même des doses infimes peuvent être dangereuses, si leur absorption se répète pendant une période suffisamment longue ou si un temps suffisant s'écoule leur permettant de manifester leur activité, ce qui les rend particulièrement redoutables lorsque l'exposition commence dès les premiers stades de la vie.

Je me permets, pour des développements plus élaborés, de reporter les lecteurs à des articles antérieurs [31, 35, 37, 38], dans lesquels ils trouveront, d'autre part les références des travaux sur lesquels sont basées ces conclusions.

Le danger est d'autant plus grand que, bien que des phénomènes d'antagonisme aient pu être démontrés entre certaines substances cancérogènes ou potentiellement cancérogènes[*], on ne peut, à priori, exclure la possibilité d'une sommation des effets d'autres substances cancérogènes auxquelles l'homme peut se trouver exposé dans les conditions de la vie moderne, non seulement du fait de la pollution de l'air, mais encore du fait de la consommation du tabac, de l'absorption de certains produit médicamentaux, ou de l'ingestion d'aliments dans lesquels ont été incorporés, intentionnellement ou non, certains produits chimiques. Les résultats obtenus par Nakahara et Fukuoka [13] dans l'application successive sur la peau de la Souris de *méthylcholanthrène* et de *N-oxyde de nitro-4-quinoléine,* plaident en faveur d'une telle éventualité et le concept de la syncarcinogénèse, mis en avant par Bauer [4] se trouve, dans ce cas, matérialisé. Il faut objectivement souligner que beaucoup d'autres résultats avec divers cancérogènes aboutissent à des conclusions opposées.

Parmi les agents chimiques de pollution de l'air des villes présentant une activité cancérogène, il faut surtout citer certains *hydrocarbures aromatiques polycycliques,* notamment le *benzo(a)pyrène,* qui, présent dans les imbrûlés dits lourds (suies), rejetés en permanence dans l'atmosphère des villes par les foyers domestiques, les

---

[*] On entend par *"cancérogène potentiel"* toute substance ayant été démontrée, de façon certaine, douée d'action cancérogène chez l'animal et pouvant être suspectée d'exercer une telle action chez l'homme, bien que la preuve n'en ait pas été faite. Un sens tout à fait différent s'attache à "substance suspecte d'action cancérogène", cette dénomination impliquant que les essais effectués sont considérés comme insuffisants.

foyers industriels et les *gaz d'échappement des moteurs de véhicules* automobiles et notamment des moteurs Diesel mal réglés, se retrouvent constamment en petites quantités dans l'air des cités industrielles où leur association avec des solvants tels que les essences de pétrole favorise, par des processus d'élution à partir des particules de suies, la manifestation de leur activité au niveau du parenchyme pulmonaire.

Comme autres composés cancérogènes ou potentiellement cancérogènes pouvant être éventuellement présents dans l'atmosphère des villes, il faut encore mentionner les *époxydes aliphatiques* formés dans l'oxydation photochimique des hydrocarbures aliphatiques non saturés provenant des essences de pétrole[7] et certains dérivés minéraux, notamment des dérivés de *l'arsenic* et, plus rarement, des dérivés du *cobalt*, du *nickel*, du *zinc*, du *plomb*, du *sélénium*, ainsi que les *chromates* [29, 18].

Il faut mentionner également le caractère pour le moins extrèmement suspect des radicaux libres, caractérisés, entre autres, dans l'atmosphère à pollution de type oxydant de *Los Angelès,* par Stephens et collaborateurs [21, 20].

L'existence d'agents chimiques de pollution cancérogènes ou potentiellement cancérogènes dans l'air des villes est à prendre en particulière considération, si l'on rappelle que les données statistiques rassemblées dans différents pays ont révélé, de façon indiscutable, une augmentation spectaculaire de la fréquence des morts dues au cancer du poumon, au cours des dernières décennies.

Bien que cette augmentation puisse s'expliquer en partie par une amélioration des méthodes de diagnostic, il est aujourd'hui admis qu'elle traduit l'intervention accrue de facteurs de causalité.

Parmi ceux-ci, il faut souligner l'importance reconnue de la *fumée de tabac,* ce qui rend difficile l'évaluation exacte de la part réelle qui revient à la pollution de l'air comme facteur de causalité dans l'étiologie du *cancer pulmonaire* qui, d'après Lawther et Waller[9], tendrait à être exagérée par certains. Bien que reconaissant la valeur des observations de Lawther en ce qui concerne l'absence d'une augmentation significative de la fréquence du *cancer bronchique* chez les ouvriers des garages, pourtant relativement exposés aux *hydrocarbures polycycliques cancérogènes,* je pense personnellement que les deux facteurs (consommation du tabac et pollution de l'air) peuvent intervenir en association. En faveur du rôle de la pollution de l'air s'inscrit le fait que la fréquence du cancer pulmonaire est, ainsi que l'ont montré les résultats de nombreuses enquêtes dans différents pays, plus élevée dans les zones urbaines que dans les zones rurales.

Naturellement, de telles données statistiques doivent être interprétées en tenant compte de toute une série de facteurs, en particulier des variations dans des conditions de diagnostic et dans les méthodes d'enregistrement, de l'habitat des malades, de leurs occupations professionnelles et enfin des différences dans l'habitude de fumer.

Apercus sur la Methodologie a mettre en Oeuvre pour l'Evaluation des Risques

L'évaluation des risques comporte deux objectifs fondamentaux:

a) Réveler *la nature des risques,* c'est à dire fixer les sites, morphologiques ou biochimiques, d'agression: c'est l'aspect qualitatif.

b) Etablir une *relation* entre les *doses* (en l'espèce les concentrations) et les *effets,* de manière à fixer les *seuils de toxicité:* c'est l'aspect quantitatif.

Il est bien évident que, lorsque l'on considère des polluants bien déterminés, tels que, par exemple, l'anhydride sulfureux et l'acide sulfurique, l'oxyde de carbone, l'ozone, les oxydes de l'azote, l'acide fluorhydrique et les fluorures, les éléments toxiques . . ., l'évaluation de leurs potentialités toxiques doit faire appel à la méthodologie classiquement adoptée dans d'autres domaines, en gardant cependant dans l'esprit que la voie de pénétration des polluants de l'air dans l'organisme est la voie pulmonaire. Il en découle la nécessité primordiale de recourir à des épreuves par inhalation. C'est là une analogie avec les toxiques présents dans *l'atmosphère des lieux de travail*. Il faut toutefois bien souligner une importante différence: les ouvriers, contrairement à la population (en général, sont non seulement exposés à l'inhalation des) toxiques, mais encore, du fait que, le plus souvent, ils les manipulent à la *pénétration par la voie cutanée* de ceux d'entre eux qui sont liposolubles.

Néanmoins, il est clair que les informations toxicologiques rassemblées dans les domaines de *l'hygiène industrielle* et de la Médecine du travail sur des polluants pouvant se rencontrer également dans l'air des villes et des environnements industriels ont un intérêt considérable en ce qui concerne l'évaluation des risques pouvant résulter de la pollution de l'air pour la santé de la population en général. En ce qui concerne la *méthodologie d'évaluation* de ces risques, il faut bien souligner qu'elle ne saurait être codifiée de façon rigoureuse, car elle dépend:

a) de la nature de la substance toxique *(panorama physicochimique)* et de ses éventuelles transformations sous l'influence de divers facteurs.

b) des *conditions d'exposition*, variables, entre autres, avec les localisations géographiques (facteurs climatiques, conditions météorologiques etc.).

c) de la nature des potentialités toxiques révélées par les épreuves préliminaires *(panorama toxicologique)*.

d) des résultats obtenus dans les épreuves biochimiques, notamment celles concernant *l'étude du métabolisme* comparé chez l'homme et les animaux de laboratoire, qui conditionnent le choix des espèces à soumettre à des épreuves ultérieures plus approfondies *(panorama biochimique)*.

Le protocole d'étude doit donc, tout en respectant certains principes généraux, être laissé à l'initiative des experts toxicologues, dans un contexte de perpétuelle évolution, à la lumière des progrès de nos connaissances toxicologiques et de l'apparition de nouvelles approches méthodologiques.

Expérimentation sur les animaux de laboratoire

En ce qui concerne *l'experimentation sur l'animal,* elle comporte en premier lieu des épreuves générales classiques de toxicité aiguë ou subaiguë, de *toxicité à court terme* ou *semichronique* et de *toxicité à long terme*. Ces dernières épreuves revêtant, à mon avis, pour l'évaluation toxicologique des polluants de l'air un caractère aussi indispensable que pour les additifs aux aliments ou les contaminants alimentaires, puisque c'est le plus souvent pendant toute sa vie que, comme dans le cas de ces derniers, l'homme risque d'y être exposé. En ce qui concerne les modalités de ces épreuves et l'interprétation des résultats, je me permettrai de renvoyer à des revues générales antérieures [30, 32, 33, 35].

Mais je tiens à souligner que ces épreuves générales, qui sont, dans une certaine mesure, des épreuves de routine, doivent être complétées par des investigations orientées vers des explorations fonctionnelles en profondeur de tel ou tel organe, tissu ou système, dépendant de la nature et du type d'action toxique des agents chimiques à examiner.

A cet égard, il convient de souligner tout d'abord l'intérêt croissant porté, sous l'impulsion des toxicologues et des hygiénistes soviétiques, à l'exploration des fonctions du système nerveux et notamment du système *nerveux* central, considéré à juste titre comme l'un des récepteurs les plus sensibles aux agressions toxiques. Qu'il s'agisse de l'étude des réflexes, conditionnés ou non, ou de la mesure des constantes électrophysiologiques (chronaxie, tracés électroencéphalographiques...), les travaux dans cette direction ont, dans le cas des nombreux toxiques industriels, fourni des informations d'un grand intérêt pour la fixation des *limites tolérables* dans les ambiances professionnelles, fixation qui pose, dans l'ensemble, des problèmes de même nature que l'établissement des *seuils de nocivité* pour les polluants de l'air respiré par la population générale. Les deux symposiums internationaux, tenus à Prague d'abord, en Avril 1959, puis, à Paris, en Avril 1963, ont, surtout le dernier, attiré l'attention sur l'intérêt de l'étude de *l'action des substances toxiques sur le système nerveux,* volontaire au autonome, pour la fixation des seuils de nocivité et des limites tolérables [39, 40].

Mais il ne faut pas, pour autant, négliger les épreuves d'exploration d'autres systèmes, parmi lesquels je mentionnerai, entre autres: le *système cardio-vasculaire,* le *système hémopoiétique,* la *fonction pulmonaire,* la *fonction rénale,* les *fonctions endocriniennes* et, notamment, à côté des fonctions de reproduction dont l'étude nécessite la prolongation des essais sur plusieurs générations, la fonction thyroïdienne que sont susceptibles de perturber, à partir d'une certaine concentration, de nombreux polluants de l'air, en particulier *l'ozone,* les *vapeurs nitreuses* et *l'oxyde de carbone.*

Les *approches biochimiques* présentent également une très grande importance; en premier lieu pour l'étude des modalités d'absorption, de la répartition dans l'organisme et du rythme de défixation; ensuite, qu'il s'agisse de l'examen des effets sur les *système enzymatiques fondamentaux* (cholinestérases, transaminases, glucose - 6 - phosphate - deshydrogénases, phosphatases, cytochromoxydase etc....) qui permet souvent de révéler les mécanismes intimes de toxicité, ou de l'étude des modifications de concentrations, tissulaires, sanguines ou urinaires, de certains éléments, tels que les alcalins, le cuivre, le zinc, etc....ou de certains composés intervenant comme produits intermédiaires dans les métabolismes, elles peuvent conduire à des tests de dépistage précoce des agressions toxiques et contribuer ainsi, de façon primordiale, à la fixation des *seuils de nocivité,* particulièrement dans les cas *d'expositions prolongées* qui intéressent tout spécialement les problèmes toxicologiques posés par la pollution de l'air [25]. Toujours dans le domaine biochimique, je soulignerai une fois encore, l'importance des données concernant les transformations métaboliques des produits soumis à l'évaluation toxicologique, car la nature de ces transformations conditionne très souvent la toxicité, les différences de réceptivité entre les espèces dépendant, dans beaucoup de cas, des différences de métabolisme.

D'autres directions de recherche spécialisée peuvent se révéler extrêmement fécondes et pour certains toxiques, diverses disciplines, telles que, par exemple, l'histochimie, l'immunochimie, la génétique biochimique et la biologie dite moléculaire, peuvent apporter des renseignements d'un grand intérêt pour la fixation des *seuils de nocivité,* tout en contribuant à révéler les *mécanismes d'action toxique.* Il apparaît donc, de plus en plus, que la révélation des effets toxiques, notamment des effets les plus subtils, exige la mise en œuvre dynamique de multiples disciplines. Les sciences de l'analyse y ont une place importante, car leurs progrès spectaculaires en ces dernières années permettent de résoudre maintenant des problèmes tels que la détermination des localisations fines au niveau des fractions infra-cellulaires et même de certaines molécules, la révélation de l'éventail des transformations métaboliques et la détermination d'activités enzymatiques très spécialisées, grâce, par exemple, au dosage de co-enzymes.

Observations sur les sujets exposés

On retrouve les mêmes impératifs en ce qui concerne les *observations chez l'homme*. Elles sont de toute évidence fondamentales, en raison des différences de réceptivité aux substances toxiques en fonction de l'espèce et de l'incertitude que comporte toujours, de ce fait, l'extrapolation à l'homme des données obtenues par expérimentation sur l'animal.

La méthode idéale serait d'expérimenter sur l'homme, mais, pour des raisons évidentes, elle est inapplicable de façon systématique. A cet égard, les expérimentations sur sujets volontaires humains, réalisées avec un petit nombre de toxiques, tels que certains pesticides et divers solvants, ou, pour prendre des exemples de produits pouvant éventuellement se rencontrer dans l'air des villes, les poussières de plomb minéral ou les vapeurs de plomb tétraéthyle, ont donné lieu à de vives critiques de la part de certains, pour lesquels l'intérêt qui s'attache à des observations sur des groupes limités, même dans le but louable de baser sur elles l'établissement de mesures de prévention pour la population en général, ne constitue pas, sur le plan moral, une justification suffisante.

Il ne faut cependant pas oublier que, dans le cas de la pollution de l'air, *l'expérimentation sur l'homme* se trouve automatiquement réalisée et c'est pourquoi les enquêtes épidémiologiques sur la morbidité et la mortalité en rapport avec la pollution de l'air présentent un intérêt primordial. Ces enquêtes sont très délicates, en raison de la multiplicité des facteurs à prendre en considération. Pour conduire à des résultats significatifs, elles doivent, comme dans le cas des ambiances professionnelles, être conduites en respectant, de façon rigoureuse, certains critères et, en premier lieu, ceux d'ordre statistique, avec constitution de groupes témoins, choisis de façon à permettre des comparaisons valables. Comme il est souligné dans le rapport d'un Comité OMS d'Experts des polluants atmosphériques, réuni à Genève du 15 au 21 Octobre 1963 [41], rapport à la rédaction duquel j'ai eu l'honneur de participer, pour l'étude de la morbidité, les enquêtes sur la fréquence de symptômes ou de syndrômes soigneusement définis sont nettement préférables aux comparaisons de maladies dont la définition est fonction de critères de diagnostic qui varient parfois beaucoup d'un pays à l'autre. De toutes façons, l'emploi de questionnaires standards, minutieusement étudiés, comme ceux que recommandent Lawther et collaborateurs [11], s'impose pour éviter les erreurs dues à l'ambiguïté des questions, ainsi qu'aux idées préconçues des enquêteurs ou des sujets soumis à l'enquête.

En ce qui concerne les *études de mortalité*, on devra s'assurer, par tous les moyens, que les causes de décès déclarées ont des chances raisonables de correspondre à la réalité. Il est bien connu, en effet, que les certificats de décès sont souvent sujets à caution et il est, par suite, hautement désirable d'obtenir des renseignements précis par autopsie suivie d'examens histologiques adéquats. Les données publiées dans la littérature toxicologique sont malheureusement, pour la grande majorité, loin de satisfaire ces exigences. Naturellement, les informations *pathologiques,* même convenablement rassemblées, sont absolument ininterprétables, si elles ne sont pas accompagnées de données adéquates concernant la composition, qualitative et quantitative, de l'air pollué et ses fluctuations dans l'espace et dans le temps. Comme la valeur de ces dernières données dépend des méthodes analytiques mises en œuvre et des conditions des prélèvements, notamment de leur durée, il s'impose de fournir, à cet égard, des indications précises et détaillées. L'Union Internationale de Chimie Pure et Appliquée (Section de Toxicologie et d'Hygiène industrielle de la Division de Chimie Appliquée) a, dans cette direction, un rôle très important à jouer en ce qui concerne la recommandation de techniques d'analyse adéquates par leur sensibilité, leur spécificité et leur précision, en ce qui concerne également l'établissement de méthodes normalisées rendant plus valide la comparaison des résultats. Le dossier, rassemblant les informations d'ordre pathologique et analytique, concernant une enquête épidémiologique relative à la pollution de l'air, ne peut d'ailleurs être interprété valablement que si les pathologistes et les analystes collaborent étroitement à cette interprétation. C'est seulement en respectant cette règle que des conclusions valables peuvent être obtenues concernant les seuils de nocivité de tel ou tel polluant pour l'homme.

Comme exemples d'enquêtes ayant conduit à des résultats exploitables, il faut surtout citer celles concernant les rapports entre la fréquence de la *bronchite chronique* et le degré de pollution de l'air par *l'anhydride sulfureux* [15, 16, 10]. Des résultats intéressants, mais pas aussi démonstratifs, ont également été obtenus dans les enquêtes concernant les rapports entre la fréquence du *cancer du poumon* et le degré de pollution par les *hydrocarbures polycycliques cancérogènes*. La grosse difficulté, dans ce cas, est, ainsi que je l'ai déjà souligné, l'intervention du facteur tabac.

Certaines observations chez l'homme peuvent avoir un grand intérêt pour l'établissement du protocole des expérimentations sur les animaux de laboratoire. Il en est surtout ainsi des données concernant les transformations métaboliques des polluants de nature organique. Il est en effet recommandable, comme nous l'avons déjà indiqué de retenir, pour les investigations expérimentales, les espèces qui *métabolisent le polluant* soumis à l'étude d'une manière analogue à l'homme. Cette remarque m'amène à souligner l'intérêt qui s'attache à effectuer les expérimentations sur l'animal et les enquêtes concernant l'homme en étroite liaison. Là encore, une collaboration entre les spécialistes des diverses disciplines intéressées s'impose.

J'ai, jusqu'ici discuté l'évaluation toxicologique de tel ou tel polluant considéré isolément. Or, dans la pratique, ce qui caractérise la pollution de l'air des villes, c'est sa complexité.

On a, en réalité, affaire à une «soupe» de polluants chimiques de nature très diverse et ce qu'il convient d'évaluer, ce sont les risques que comporte l'exposition de l'homme à de telles associations. Dans certains cas, il peut en résulter, pour un polluant déterminé, une *diminution d'agressivité,* par suite de phénomènes physicochimiques, de réactions chimiques neutralisantes ou *d'antagonismes pharmacodynamiques*. De ce dernier point de vue, on peut citer, à titre d'exemple, l'action antagoniste vis-à-vis des effets de toxicité aiguë de l'ozone et des vapeurs nitreuses de divers composés soufrés pouvant contribuer à la pollution de l'air, tels que l'hydrogène sulfuré, les

mercaptans et certains disulfures organiques[24]. Mais, dans d'autres cas, il
peut y avoir *addition des effets* de polluants ayant une potentialité toxique
orientée vers les mêmes récepteurs ou,
ce qui est plus grave encore pour la santé,
potentialisation des effets. De nombreux
exemples de telles *synergies toxiques* peuvent être fournis. Je citerai, entre autres,
la synergie entre l'anhydride sulfureux et
l'acide sulfurique, mise en évidence par
Amdur et collaborateurs (1952)[1]. Alors
que les sujets respirant de l'air renfermant 0,62 ml/m³ de SO_2 ou un *aérosol
d'acide sulfurique* à la concentration de
0,12 mg/m³ ne présentent aucune modification sensible du rythme respiratoire ou
du nombre de pulsations, l'inhalation
d'air renfermant les deux agents de pollution associés aux mêmes concentrations
entraîne une accélération nette des deux
rythmes. Expérimentant sur le Cobaye,
Miss Amdur (1959)[2] a observé une
synergie entre SO_2 et des aérosols d'acide
sulfurique, lorsque les gouttelettes de ce
dernier ont une taille suffisamment petite
(0,8 µ); elle a en outre observé une *potentialisation des effets toxiques de SO_2* par
association avec des aérosols contenant
des particules submicroniques (0.04 µ) de
chlorure de sodium à la concentration totale de 10 mg/m³. Ces constatations confir-
ment les observations antérieures de La
Belle et collaborateurs[8] concernant le
rôle potentialisant joué vis-à-vis des
effets de gaz toxiques par certaines particules présentes sous forme d'aérosols.
Un autre exemple de synergie est, d'après
Stokinger et collaborateurs[21], l'augmentation de toxicité du *sulfate de béryllium*, lorsqu'il est associé, dans l'atmosphère à laquelle sont soumis les animaux d'expérience, à de l'acide fluorhydrique.
D'autres exemples de synergie concernant des polluants existent sous le même
état physique. Je rappellerai, à cet égard,

les observations de von Oettingen[14]
sur la potentialisation des effets toxiques
de l'oxyde de carbone par l'acide cyanhydrique, les vapeurs nitreuses ou l'hydrogène sulfuré. Un bel exemple de synergie toxique a enfin été apporté par les
travaux de Svirbely, Stokinger et collaborateurs[27]. Ils ont démontré que l'eau
oxygénée, H_2O_2, associée à l'ozone dans
l'air d'une chambre d'inhalation. aug-
mentait notablement la nocivité de ce
gaz, au point que, chez la souris, l'exposition pendant quatre heures à une
concentration de 1 ml/m³ d'ozone, en
présence de 3 ml/m³ d'H_2O_2, entraîne
la mort, alors que, pour atteindre le même
résultat avec chacun des composés pris
isolément, il faut atteindre les concentrations respectives de 5 à 6 ml/m³ et de
113 à 226 ml/m³. Les auteurs ont, au
cours de leur étude, observé qu'une exposition préalable à l'ozone permettait
aux animaux de supporter des concentrations beaucoup plus fortes des deux
oxydants. Ces constatations s'ajoutent à
celles faites antérieurement chez le rat,
la souris et le lapin, concernant l'acquisition d'une tolérance à l'ozone à la suite
d'expositions successives, même de courte
durée et à de faibles concentrations[22,
12, 19]. Il est intéressant de noter que la
tolérance acquise vaut, au moins dans
une certaine mesure, pour d'autres *irritants pulmonaires* de caractère oxydant
(H_2O_2, NO_2 ...). Mais il faut bien souligner qu'elle s'applique aux effets de toxicité aiguë (oedème pulmonaire) et non
aux effets de toxicité chronique (fibrose
pulmonaire)[23]. On trouvera dans un article de Stokinger et coll.[5] une discussion
sur les mécanismes pouvant être invoqués
pour expliquer les faits constatés.
Il faut enfin ne pas oublier l'existence
dans l'environnement de nombreux facteurs «*co-cancérogènes*» ou «*promoteurs*»,
c'est à dire de facteurs qui ne sont pas
cancérogènes par eux mêmes, mais dont

l'intervention, en même temps que ou à la suite d'un agent cancérogène, permet à ce dernier d'exercer son activité à des doses qui, en leur absence, seraient sans effet (cf. à cet égard: 35). Je ne puis m'étendre, mais je crois utile de souligner que les *polluants* à caractère *irritant* pour l'épithélium pulmonaire sont, d'après Kotin, aptes à manifester des effets co-cancérogènes. Il peut en être de même de certaines bactéries ou virus à tropisme pulmonaire.

Cette éventualité de *synergies toxiques* entre divers polluants, qu'illustrent les quelques exemples que je viens de donner, doit retenir la plus grande attention, car elle a très certainement une grande importance pratique en ce qui concerne les effets nocifs pouvant résulter de la pollution de l'air. Rappelons, à cet égard, que les enquêtes effectuées à l'occasion des épisodes graves de pollution que j'ai mentionnés (Meuse, Donora, Londres) ont révélé que les concentrations de tel ou tel polluant considéré séparément n'étaient pas suffisantes pour expliquer les effets nocifs constatés chez les groupes humains exposés. Ces constatations ne sont pas de nature à faciliter la fixation de seuils de nocivité.

Il faut également prendre en grande considération *l'état physico-chimique des polluants dans l'atmosphère.*

Il convient de souligner tout d'abord que, particulièrement dans les atmosphères à type de pollution oxydante d'origine photochimique, certains polluants peuvent apparaître sous forme de radicaux libres à vie courte, dont les capacités réactionnelles particulièrement marquées peuvent conditionner une agressivité élevée.

Plus fréquent est le cas des vésicules liquides ou des particules dont la taille est suffisamment réduite pour permettre la constitution d'aérosols ou de micro-brouillards, particulièrement aptes à pénétrer dans la profondeur des voies respiratoires.

Dans la constitution de ces aérosols et micro-brouillards, les *facteurs météorologiques* ont, à côté de la taille des particules, un rôle important. On connaît trop à cet égard l'influence du phénomène *d'inversion de température* sur la constitution des «smogs» pour qu'il me soit nécessaire de m'étendre.

Quoi qu'il en soit, pour toutes ces raisons, un grand intérêt s'attache à l'étude expérimentale des effets des atmosphères polluées considérées dans leur complexité, c'est-à-dire dans l'état même où l'homme s'y trouve exposé, avec des fluctuations souvent considérables dans les concentrations de divers agents de pollution et des variations parfois très amples des conditions météorologiques. C'est dans cet esprit que, en collaboration avec Bourbon, nous poursuivons, depuis plusieurs années, une expérimentation à long terme sur des animaux de laboratoire dans la région française du complexe industriel de Lacq [37].

Le Concept de Limite Tolerable ou Admissible pour les Substances Toxiques dans les Ambiances Professionnelles

Le but essentiel poursuivi dans l'évaluation des risques par les diverses approches méthodologiques que j'ai, très sommairement, examinées, évaluation qui doit toujours faire intervenir une large marge de sécurité, est *l'établissement de seuils de concentrations toxiques* et, en conséquence la possibilité de fixer des *limites tolérables* ou *admissibles.* L'adjectif «tolérables» signifie, dans le cas des *expositions professionnelles,* que l'exposition, courte ou prolongée, des ouvriers à de telles concentrations ne doit pas être susceptible de provoquer l'apparition d'effets nocifs ou de gêne dans le travail. La concentration d'un agent chimique dans l'air des lieux de travail est, de toute évidence, un facteur prépondérant à cet égard, car, d'après la notion bien établie des relations entre les doses et les effets, cette concentration conditionne les risques d'imprégnation toxique. En découle logiquement le principe fondamental de la prévention: diminuer le plus possible les concentrations des substances potentiellement nocives dans l'air et, en tout cas, jusqu'à des valeurs inférieures à celles constituant les seuils de toxicité, en tenant compte des conditions d'exposition.

Il ne faut cependant pas oublier que d'autres facteurs que la concentration dans le milieu ambiant sont importants à considérer pour le déterminisme des phénomènes toxiques, notamment les *conditions d'absorption et d'élimination* et, toutes choses égales d'ailleurs, en ce qui concerne la concentration dans l'air et les conditions d'absorption, les *variations individuelles de sensibilité* aux produits toxiques absorbés.

Les conditions *d'absorption et d'élimination* peuvent varier dans une large mesure en fonction de divers paramètres, tenant:

1) soit aux sujets, dont, par exemple, le rythme respiratoire peut être accru par un entraînement progressif au travail ou aux épreuves sportives, avec, comme conséquence, une captation accrue des produits éventuellement toxiques, ou dont le fonctionnement rénal est imparfait, avec, comme conséquence, une diminution du pouvoir d'excrétion;

2) soit à l'état du toxique, surtout dans le cas des poussières ou des particules dont les dimensions (granulométrie) conditionnent les possibilités de pénétration plus ou moins profonde dans les ramifications de l'arbre pulmonaire;

3) soit à la présence de produits associés pouvant favoriser l'absorption (solvants, agents surfactifs...).

Pour toutes ces raisons, de même que dans le cas des expositions professionnelles, un grand intérêt s'attache à la détermination quantitative des polluants étudiés dans l'organisme même des sujets exposés [28, 30], avec, comme objectif, la fixation de *limites tolérables dans les milieux biologiques*, complétant celle de limites tolérables dans l'air (I). Je ne puis, malheuresement, m'étendre sur ce point très important dans cet article.

Mais, même en mettant à part les réactions *d'intolérance congénitale* ou *acquise* et les *hypersensibilités* liées à certains états physiologiques (grossesses, par exemple) ou pathologiques (déficiences rénale, hépatique, endocriniennes; perturbations du système nerveux volontaire ou autonome...), il faut également tenir compte des variations de *sensibilité individuelle liées*, par exemple:

a) soit à des conditions physiologiques (âge, sexe, race, tempérament, état hormonal, nature de l'alimentation, fatigue, chocs émotionnels...);

b) soit au plus ou moins haut degré d'activité fonctionnelle des organes intervenant dans les détoxifications ou l'élimination;

c) soit à l'Imprégnation éthylique, très importante à considérer, en raison de son influence sensibilisante aux effets de nombreux toxiques.

Il faut enfin se souvenir que, ainsi que je l'ai déjà signalé, certains produits associés peuvent être à l'origine de phénomènes *d'antagonisme* ou de *synergie* [3]. Ces remarques sont destinées à attirer l'attention sur le fait que les chiffres pouvant être recommandés comme limites de *concentrations tolérables* ou *admissibles*, ne doivent pas être considérés comme des lignes de démarcation tranchées entre les concentrations inoffensives et les concentrations dangéreuses. Ils ne sauraient, par suite, avoir une valeur absolue et sont seulement destinés à servir de *guides,* pour la protection de la santé et le maintien du bien être et du confort des travailleurs. Ces guides sont destinés:

1) aux technologistes, aux ingénieurs de sécurité et aux chimistes, chargés de la mise au point des procédés de fabrication et d'application, ainsi que de la réalisation des installations ou dispositifs conçus pour assurer le contrôle des émissions et la prévention technique (captation des vapeurs et poussières, ventilation etc. . . .).

2) aux analystes, qui ont pour tâche d'établir des méthodes adéquates, quant à leur spécificité, leur sensibilité et leur précision, pour la surveillance des ambiances de travail.

3) aux médecins et aux pathologistes, engagés dans les *enquêtes épidémiologiques,* qui ne sauraient effectuer des interprétations valables sans confronter les données cliniques ou biologiques avec celles relatives aux conditions d'exposition, comportant la connaissance précise des concentrations dans l'air des locaux de travail.

Ce problème des limites tolérables ou admissibles intéresse au premier chef les spécialistes de la *Médecine du travail* et de *l'Hygiène industrielle* et c'est pourquoi il a été, depuis une vingtaine d'années, très étudié dans le cas des ambiances de travail, aussi bien par des organismes nationaux, comme, par exemple, l'Association des Hygiénistes industriels gouvernementaux des Etats-Unis, que par des organismes internationaux: Organisation Mondiale de la Santé (OMS) et Organisation Internationale du Travail (OIT) et, surtout, Commission Internationale permanente de Médecine du Travail.

Cette dernière organisation a constitué, en 1957, une Sous-Commission chargée de l'étude de ce problème, dont j'ai eu l'honneur d'assumer la présidence et qui a travaillé en liaison avec l'Organisation Mondiale de la Santé et l'Organisation Internationale du Travail, ainsi qu'avec l'Union Internationale de Chimie Pure et Appliquée sur le plan analytique.

Je voudrais tenter de présenter sommairement les résultats actuellement obtenus en y incluant des commentaires concernant des recommandations adoptées dans un certain nombre de pays.

J'essaierai ensuite d'esquisser les perspectives qui s'offrent aux toxicologues pour appliquer à l'air respiré par la population en général cette approche de prévention.

Je crois opportun de bien souligner, tout d'abord, que l'une des principales raisons pour une coopération internationale dans ce domaine a été l'existence de divergences parfois considérables entre les valeurs de limites tolérables adoptées par les organismes de certains pays, notamment des Etats-Unis et de l'Union Soviétique.

Le *Tableau 1* montre clairement l'ampleur des *divergences* dans le cas d'un certain nombre d'éléments ou de composés pris, à titre d'exemples, parmi les toxiques industriels ou agricoles pouvant être à l'origine d'expositions professionnelles. L'ampleur de ces divergences est encore accentuée par le fait que les chiffres soviétiques sont des valeurs plafonds, alors que les chiffres américains sont, le plus souvent, des valeurs moyennes.

En présence de telles divergences, on ne peut qu'être conduit à penser que les valeurs ont été proposées, soit sur la base de données obtenues par une méthodologie complètement différente, soit par suite d'une interprétation complètement différente d'informations toxicologiques identiques ou analogues.

Tableau 1   Exemples de divergences entre les *limites admissibles* proposées ou adoptées aux Etats-Unis d'Amérique et en Union Soviétique

Hygiénistes industriels gouvernementaux des Etats-Unis	ppm (parties par million vol/vol)	mg/m³	Experts de l'Union Soviétique (derniers chiffres parvenus à notre connaissance) mg/m³	Rapport
Oxyde d'éthylène	50	90	1	90
Acide cyanhydrique	10	11	0,3	37
Acétone	1000	2400	200	12
Formaldéhyde	5 (valeur plafond)	6	1	6
Chlorure de méthylène (Dichlorométhane)	500	1740	50	35
Trichloréthylène	100	535	20	27
Dichloro 1-1 éthane	100	400	10	40
Toluène	200	750	50	15
Plomb		0,2	0,01	20
Mercure		5	0,3	17
Manganèse		0,05 (valeur plafond)	0,01	5
Chlordane		0,5	0,01	50
Lindane		0,5	0,05	10
Heptachlore		0,5	0,01	50
Malathion		15	0,5	30

Quoi qu'il en soit, la conclusion qui s'imposait à la lecture d'un tel tableau est qu'un grand intérêt s'attachait à une *confrontation des méthodes* mises en œuvre et des résultats obtenus entre les experts des Etats-Unis et ceux d'Union Soviétique.

C'est pourquoi le premier Symposium international sur les limites tolérables pour les substances toxiques dans les ambiances professionnelles a été organisé à Prague au début d'avril 1959.

Dans ses résolutions terminales, ce Symposium tenu sous le patronage conjoint de la Commission Internationale de Médecine du Travail et de l'Union Internationale de Chimie pure et appliquée, a recommandé que:

«par *concentration maximale tolérable*, on entend, pour une substance déterminée, la concentration moyenne dans l'air ne provoquant, sauf cas d'hypersensibilité, chez aucun des ouvriers ex-

posés de façon continue en raison de leur travail journalier, aucun signe ou aucun symptôme de maladie ou de mauvaise condition physique, pouvant être mis en évidence par les tests les plus sensibles acceptés internationalement.»

Cette définition me paraît appeler les remarques suivantes:

1) les chiffres proposés comme limites tolérables représentent des concentrations moyennes, c'est-à-dire intégrées par rapport aux temps, auxquelles les ouvriers, industriels ou agricoles, peuvent être exposés pendant les 7 à 8 heures de leur journée de travail, cinq à six jours par semaine. Ceci correspond à l'ancienne définition adoptée pour les maximums de concentration tolérables, «maximum allowable concentrations» des auteurs anglo-

saxons, désignées classiquement par l'abréviation MAC.

2) les cas d'hypersensibilité sont exclus.

3) l'induction, en dehors de signes ou de symptômes d'intoxication, d'un état de mauvaise condition physique, doit être prise en considération, ce qui souligne l'importance accordée au maintien du bien être ou du confort dans le travail, ceci en conformité avec la définition de la santé dans la charte de l'Organisation Mondiale de la Santé.

4) pour l'évaluation des limites tolérables, doivent être mis en œuvre les tests les plus sensibles acceptés internationalement.

Le concept des limites tolérables a fait l'objet, à nouveau, de discussions approfondies à l'occasion d'un 2ème Symposium international, tenu à Paris, en avril 1963, sous le patronage des deux mêmes organisations et, en plus, de l'Organisation Internationale du travail. Il a été reconnu qu'une définition univoque ne pouvait être valable pour toutes les substances et qu'il fallait tenir compte de leur type de potentialités toxiques.

Les *polluants* ont été *classés en 3 catégories* en fonction de la nature de leurs effets toxiques et de la vitesse de leur manifestation:

1. Substances dont les effets principaux consistent en phénomènes *d'irritation,* de *sensibilisation* ou *d'intoxication aiguë,* provoqués, immédiatement ou après une phase de latence, par une exposition de courte durée à des concentrations de l'ordre de celles pouvant être recontrées dans la pratique.

Pour les substances de cette catégorie, les limites de concentration tolérables doivent être considérées comme des valeurs ne devant jamais être dépassées, même pendant de courtes périodes, de l'ordre de 10 à 15 minutes. Cela signifie que,

pour une substance donnée, les fluctuations de concentration, qui se produisent inévitablement dans le temps et dans l'espace au cours des expositions professionnelles, doivent obligatoirement se situer en dessous de la valeur adoptée pour la limite tolérable. Cette dernière représente donc ce que les hygiénistes industriels gouvernementaux des Etats-Unis appellent une *«ceiling value»* (valeur plafond). La substance est alors affectée de la lettre C dans les listes de limites tolérables, recommandées ou proposées, publiées annuellement par ces experts (threshold limit values ou TLV).

Peuvent être classés dans cette première catégorie:

a) les agents *d'irritation* des muqueuses de l'arbre respiratoire, tels que le chlore, le phosgène, l'acide chlorhydrique, le chlorure de vinyle, le trifluorure de bore, l'hydrogène sulfuré, l'anhydride sulfureux, l'ammoniac, le peroxyde d'azote et le formaldéhyde.

b) les composés à propriétés narcotiques, tels que les solvants halogénés.

c) les agents générateurs *d'asphyxie,* par des mécanismes pouvant d'ailleurs être différents, tels que l'oxyde de carbone et l'acide cyanhydrique.

d) les substances provoquant des phénomènes de sensibilisation, tels que le diisocyano-2,4-toluène (toluylène-2,4-isocyanate) et même des produits, tels que le benzène, susceptibles d'être stockés rapidement dans l'organisme ou d'exercer, à la suite de l'absorption d'une dose même unique, des effets graves à plus ou moins long terme. Cette dernière éventualité se présente pour divers toxiques à effets retardés, tels que, en dehors de cancérogènes comme les produits de la classe des nitrosamines, l'herbicide Paraquat qui est un chlorure de bipyridinium à tropisme pulmonaire, les organophosphorés à action neurotoxique et les dérivés du méthyl, et de l'éthyl-mercure à action également neurotoxique. Ce sont, comme les a si bien appelés Barnes, des poisons qui frappent et s'en vont (Poisons which hit and run).

Il convient de remarquer ici que, jusqu'à présent, toutes les valeurs adoptées par les experts soviétiques sont, ainsi que je l'ai déjà indiqué, des valeurs plafonds. Il en est de même des MAC recommandées par le Comité Z 37 de «l'American Standards Association», devenues des valeurs plafonds à la suite de la nouvelle définition des MAC donnée par ce Comité en 1957 [26], la conséquence étant un abaissement de la concentration moyenne tolérable, en raison des inévitables fluctuations que nous avons mentionnés. A mon avis, une telle attitude n'est pas logique, car elle ne tient pas compte de la nature des potentialités toxiques.

2. Substances dont les effets principaux sont des *effets cumulatifs* provoqués par des expositions répétées à des concentrations de l'ordre de celles pouvant être rencontrées dans la pratique (travail de 6 à 8 heures par jour, pendant 5 à 6 jours par semaine).

Pour les substances de ce groupe, les limites de concentration tolérables doivent être considérées comme des valeurs moyennes intégrées par rapport au temps (*«time weighted averages»* des experts anglo-saxons), ce qui cadre avec la définition donnée dans les résolutions du Symposium de Prague et correspond, en gros, à celle des *«threshold limit values»* des hygiénistes industriels gouvernementaux des Etats-Unis.

A titre d'exemples d'éléments ou de composés se rattachant à cette catégorie, on peut citer:

a) l'arsenic et les métaux lourds (plomb, mercure, cadmium ...),

b) les fluorures,

c) les insecticides organo-halogénés (DDT, lindane, aldrine, dieldrine, heptachlore, chlordane ...),

d) le dinitro-o-crésol,

e) certains solvants, comme le sulfure de carbone ou l'alcool méthylique.

Dans leur cas, les concentrations nécessaires pour provoquer, après une courte durée d'exposition, des symptômes toxiques, soit dans l'immédiat, soit après une phase de latence, peuvent être considérablement supérieures aux valeurs moyennes adoptées comme limites tolérables. Ces valeurs peuvent donc supporter une marge relativement ample de fluctuations dans le temps et dans l'espace au cours de la journée de travail, à condition que ces fluctuations se compensent. La marge admissible de fluctuations dépend de toute une série de facteurs, parmi lesquels figurent, en dehors de la nature du toxique considéré, les taux des concentrations nécessaires à la production d'effets de toxicité aiguë, la fréquence avec laquelle se produisent des dépassements de la valeur moyenne tolérable, la durée de ces dépassements, le caractère plus ou moins cumulatif de la substance. Les décisions à cet égard relèvent du jugement des experts. Si l'on considère, par exemple, le cas de l'exposition au plomb, il n'a pas de risque à ce que la concentration de cet élément soit, dans l'air des locaux de travail pendant une heure, 8 fois supérieure à la valeur moyenne, si, pendant les 7 autres heures de la journée, cette concentration est nulle.

En revanche, dans le cas des solvants volatils liposolubles à propriétés narcotiques, comme le trichloréthylène ou le sulfure de carbone, la marge admissible de dépassement de la valeur moyenne est, de toute évidence, beaucoup plus réduite.

Tableau 2   Quelques valeurs de limites tolérables proposées pour des éléments, des composés, ou des produits, susceptibles de manifester une *activité cancérogène* (Hygiénistes industriels gouvernementaux des Etats-Unis, 1969)

	ppm	*mg/m³*
Béryllium		0,002
Acide chromique et chromates (en CrO_3)		0,1
Nickel-carbonyle	0,001	0,007
Dérivés du sélénium (en sélénium)		0,2
Dérivés de l'uranium naturel (en uranium)		0,2
Microbrouillards d'huiles minérales		5

3. Substances dont l'effet principal est une action *cancérogène*.

Pour les substances de ce groupe, pour les raisons que nous avons mentionnées dans les généralités sur la nature des risques, il n'est pas possible, dans l'état actuel de nos connaissances, de fixer, en toute sécurité, des limites tolérables. Cela signifie que tous les efforts doivent tendre à les exclure des ambiances de travail. Cette exclusion avait été recommandée, en ce qui concerne la diméthylnitrosamine et ses homologues, la benzidine, la β-naphtylamine et la β-propriolactone, par le Symposium de Paris en avril 1963. Une telle recommandation ne saurait, à mon avis, avoir son plein effet que si les composés en question ne sont plus employés et sont remplacés par d'autres substances fonctionnellement équivalentes (acide amino-2-naphtolsulfonique-1, dit de Tobias, à la place de la β-naphtylamine par exemple). Sinon, il me paraît indispensable, en raison de l'inapplicabilité, à la fois sur le plan scientifique et sur le plan pratique, du concept de la tolérance 0, de fixer des limites pratiques de concentration tolérable; aussi basses que possible bien sûr, en tenant compte de la limite de sensibilité des méthodes de contrôle analytique. Cet important problème a été étudié activement sur le plan international par un Comité de l'Union Internationale Contre le Cancer, chargé de l'étude des aspects quantitatifs de la cancérogénèse et travaillant en étroite liaison avec le Centre International de recherches sur le Cancer, rattaché à l'Organisation Mondiale de la Santé, ainsi qu'avec la Section de Toxicologie et d'Hygiène industrielle de la Division de Chimie appliquée de l'Union Internationale de Chimie Pure et Appliquée [47].

Déjà, une telle attitude a été adoptée dans le passé pour un certain nombre d'éléments ou de composés cancérogènes ou potentiellement cancérogènes, comme le montre le *Tableau 2* ci-contre donnant les valeurs retenues par les hygiénistes industriels gouvernementaux des Etats-Unis (valeurs relevées dans la liste publiée à la suite de la réunion de Mai 1969 à Cincinnati — Tableau 2).

Comme toute classification, celle adoptée par le Symposium de Paris présente un caractère quelque peu arbitraire; beaucoup des composés à évaluer présentent des potentialités toxiques, dont certaines les rattachent à la catégorie 1 et d'autres à la catégorie 2.

Le Symposium de Paris avait, en outre, estimé possible, sur ma proposition [34], d'un groupe soient classés dans celui pour lequel la limite de concentration tolérable assure le maximum de sécurité. Après réflexion, il me paraît bien préférable, pour assurer ce *maximum de sécurité*, d'essayer d'établir, pour de tels toxiques, 2 catégories de limites tolérables: a) des *valeurs moyennes*, et b) des *valeurs plafonds*.

Des exemples spectaculaires de l'intérêt d'une telle approche sont ceux des solvants chlorés notamment du trichloréthylène, du perchloréthylène et du trichloro-1,1,1-éthane, de l'ozone, des vapeurs nitreuses et de l'oxyde de carbone.

Le Symposium de Paris avait, en autre, estimé possible, sur ma proposition (34), d'adopter, à titre provisoire, une liste de limites tolérables pour un petit nombre de toxiques constituant un premier «nucléus d'entente» à l'échelle internationale, en tenant compte, entre autres, des points d'accord pratique entre les valeurs adoptées par les organismes de différents pays, notamment Etats-Unis et U.R.S.S. (Tableau 3).

Mais certains participants avaient estimé prématuré l'établissement, même provisoire, d'une telle liste de valeurs, en considérant qu'il convenait de procéder, au préalable, à une évaluation critique rigoureuse des informations d'ordre expérimental ou épidémiologique.

Pour cette raison, le Symposium avait recommandé la constitution de petits groupes d'experts dont la tâche, s'effectuant sous l'égide de la Sous-Commission d'études des limites tolérables de la Commission Internationale permanente de Médecine du Travail, serait de rassembler et de procéder à l'évaluation critique des informations toxicologiques concernant certaines classes bien définies de toxiques industriels ou agricoles.

Subséquemment, és qualité de Président, j'ai proposé à la Sous-Commission, qui a approuvé (Vienne, Septembre 1966), de constituer dix groupes, auxquels j'ai ajouté, plus récemment, un 11ème sur la méthodologie d'évaluation toxicologique. Certains de ces groupes ont déjà abouti à des conclusions qui ont été discutées et approuvées lors des réunions de la Sous-Commission à l'occasion du Congrès International de Médecine du Travail qui s'est tenu à Tokyo en Septembre 1969.

C'est ainsi que le groupe sur les dérivés minéraux et organiques du mercure, animé par Friberg (Suède), a organisé, à l'Institut Karolinska de Stockholm en Novembre 1968, un Symposium international, dont les comptes-rendus ont été publiés dans *Archives of Environmental Health*[6]. La discussion de la masse

d'informations rassemblées a conduit aux recommandations suivantes quant aux limites admissibles:

a) pour les dérivés du *méthyl-* et de *l'éthyl-mercure*, aucune valeur dans l'air n'a été proposée, mais il a été admis que le taux de mercure dans

Tableau 3 *Limites admissibles* recommandées, à la majorité et a titre provisoire, par le 2ème Symposium International sur les concentrations maximales tolérables pour les substances toxiques dans l'industrie (Paris, Avril 1963). Elles ne s'appliquent pas à la surveillance la pollution de l'air en général, mais seulement aux *ambiances professionnelles*

A-Gaz et vapeurs	ml/m^3	mg/m^3
Chlore	1	3
HCl	5	7
Ozone	0,05	0,1
SO$_2$	4	10
H$_2$S	10	15
NH$_3$	50	35
AsH$_3$	0,05	0,2
Alcool n butylique	100	300
Butylamine	5	15
Diisocyano-2,4-toluène	0,02	0,14

B-Poussières, fumées, brouillards	
Acide sulfurique	1
Acide chromique et chromates (en CrO$_3$)	0,1
Oxyde de Zn (fumées)	5
Vanadium (en V$_2$O$_5$) poussières	0,5
fumées	0,1

Groupe II	
Poussières, fumées et brouillards	
Fluorures	2,5
Beryllium (oxydes et sels exprimés en Be)	0,002
Oxydes de Cd (fumées)	0,1
Chlorodiphényle (42% Cl$_2$)	1
Naphtalènes chlorés (plus de 5 atomes par mol)	0,5
Métadinitrobenzène	1
Dinitrotoluène	1
Trinitrotoluène	1
Parathion	0,1

Groupe III

Diméthylnitrosamine et homologues

Benzidine 〕 à exclure des ambiances
β-naphtylamine 〕 de travail
β-propiolactone

le sang total ne devait pas dépasser 10 µg/100 ml, cette valeur devant être considérée comme une valeur plafond qui ne devrait pas être dépassée pour une exposition continue de 8 heures par jour à une concentration de 0, 01 mg/m³ de méthyl ou d'éthylmercure dans l'air.

b) pour la vapeur de mercure dans l'air: 0,05 mg/m³.

c) pour les sels minéraux du mercure et les sels de phénylmercure et de méthoxyéthylmercure dans l'air: 0,10 mg /m³ (en mercure).

C'est ainsi également que le groupe sur le *plomb* minéral animé par R. Zielhuis, a organisé à Amsterdam, en Novembre 1968, un Symposium international, dont les comptes-rendus sont en cours de publication dans *les Annals of Occupational Hygiene* et seront peut-être également publiés dans les *Archives of environment health*. La discussion des nombreuses informations présentées a conduit a la conclusion que le taux sanguin du plomb était en corrélation avec le taux absorbé et qu'il convenait d'adopter un taux maximal admissible de 70 µg/ 100 ml de sang. De cette valeur, les limites maximales admissibles suivantes ont été déduites:

Plomb dans l'air (valeur moyennes intégrée par rapport au temps pour une exposition de 40 heures par semaine): 150 µg/m³.

Plomb dans l'urine: 130 µg/litre.

Acide delta aminolévulinique (ALA) dans l'urine: 10 mg/litre.

Coproporphyrine dans l'urine: 300 µg/ litre.

Il a été fortement recommandé de poursuivre des *études épidémiologiques* sur des groupes de sujets exposés au plomb, en mettant en œuvre, en association avec les examens cliniques, dont l'importance a été reconnue primordiale, des méthodes

adéquates de détermination du plomb dans l'urine et surtout dans le sang, de l'acide amino delta amino lévulinique et des coproporphyrines dans l'urine et de la protoporphyrine dans les érythrocytes. C'est ainsi, enfin, que le groupe sur *l'oxyde de carbone*, animé par A. Grut (Danemark), après avoir travaillé par correspondance, s'est réuni Londres en Octobre 1968. L'examen critique des informations rassemblées a révélé que l'exposition répétée à de faibles concentrations d'oxyde de carbone provoquait une augmentation de la perméabilité vasculaire et une réduction du volume plasmatique, ainsi que des perturbations dans le métabolisme du cholestérol conduisant à une augmentation marquée des dépôts de ce composé au niveau de l'intima des vaisseaux. La limite maximale admissible, recommandée dans le cas sujets sains effectuant un travail normal, est celle adoptée par l'Association des Hygiénistes Industriels gouvernementaux des Etats-Unis, soit 50 ppm, ce qui correspond à 8 à 10% de l'hémoglobine transformée en carboxyhémoglobine, COHb. La limite doit être abaissée dans les cas des sujets effectuant un dur travail.

Les seuils de danger suivants, à considérer dans le cas de situations accidentelles exceptionnelles, ont été adoptés:
   400 ppm pour une heure d'exposition
1000 ppm pour 20 minutes d'exposition.

L'interêt de la constitution de ces groupes et du travail qu'ils ont déjà effectué a été fortement souligné par une réunion mixte d'experts de l'OMS et de l'Organisation Internationale du Travail (OIT) sur les limites tolérables qui s'est tenue à Genève en Juin 1968 (46), et a recommandé qu'ils restent en étroite liaison avec les deux organisations.

Lors de cette même réunion, a été réexaminé le premier nucleus d'entente internationale et une liste de zones de *concentration non dangereuses* pour 24 produits chimiques a été recommandée pour adoption sur le plan international (Tableau 4).

Je tiens à souligner, une fois encore, que les valeurs adoptées représentent seulement des ordres de grandeurs destinés à servir de guides pour la prévention. C'est pourquoi, d'ailleurs, il est question de «zones de concentrations».

C'est également comme valeurs guides que doivent être considérées les *seuils de danger* («Emergency exposure limits» ou Eels) proposés, pour un certain nombre de toxiques industriels: peroxyde d'azote, diméthyl-1,1,-hydrazine, trichloro-1,1,1-éthane et, plus récemment, Pentaborane 9, conjointement par le Comité de Toxicologie de l'«US National Research Council» et l'Association américaine d'Hygiène Industrielle.

L'établissement de ces valeurs qui sont des valeurs plafonds définies en relation avec la durée d'exposition, a pour objectif de fournir des bases pour l'appréciation des dangers de brèves *expositions accidentelles*.

Elles représentent des concentrations pouvant être tolérées, sans risque d'effets nocifs majeurs pour la santé, pendant un temps déterminé, sans que, pour autant, soit exclue l'éventualité d'effets nocifs mineurs, tels qu'irritation des muqueuses, atténuation du confort ou même effets toxiques bénins à caractère réversible (diminution momentanée de la vision, de la capacité de jugement et de coordination, par exemple).

Ces limites diffèrent fondamentalement de celles précédemment envisagées.

Ce sont des concentrations maximales pouvant être tolérées exceptionnellement pour une exposition brève et ne se produisant qu'une fois ou très rarement dans la vie d'un ouvrier. Elles concernent donc des expositions accidentelles qui ne doivent pas être répétées, sauf si le retour

Tableau 4   Zones de *Concentrations non Dangereuses* Recommandees pour adoption sur le Plan International (Comité mixte OIT/OMS de la Médecine du travail, Genève, Juin 1968)

Substances	*Concentrations maximales acceptables commes non dangereuses* (mg/m³)*
Acide Chlorhydrique	5–7
Phosgène	0,4–0,5
Hydrogène sulfuré	10–15
Gaz sulfureux	10–13
Acide sulfurique et anhydride sulfrique	1
Ozone	0,1–0,2
Ammoniac	20–35
Arsine-hydrogène arsénié	0,2–0,3
Ethanol	1000–2000
Acrylate de méthyle	20–35
Nitrobenzène	3–5
Dinitrobenzène	1
Dinitrotoluène	1–1,5
Trinitrotoluène	1–1,5
Parathion	0,05, 0,1
Iode	1
Beryllium et ses composés (en B)	0,001–0,002
Molyndène, composés solubles, poussières (en Mo)	4–5
Vanadium (en V₂Oᵦ)	
poussières	0,5
vapeurs	0,1
Ferro-vanadium	1
Oxydes de zinc (vapeurs)	5
Zirconium et ses composés (en Zr)	5
Dérivés chlorés du diphényle	1
Dérivés chlorés de l'oxyde de diphényle	0,5

* Les chiffres cités sont considérés dans certains pays comme des valeurs maximales pour une exposition de courte durée (valeurs plafonds) et, dans d'autres, comme des valeurs moyennes intégrées par rapport au temps.

du sujet exposé à un état absolument normal peut être démontré par des méthodes physiologiques et des examens médicaux adéquats, en un mot si la preuve peut-être apportée de la complète réversibilité des effets.

L'application de ces limites impose donc une surveillance médicale particulièrement rigoureuse des sujets exposés.

Tableau 5   Seuils de danger pour le peroxyde d'azote, NO_2

Durée de l'exposition	ml/m³ (ppm)	mg/m³
5 minutes	35	66
15 minutes	25	47
30 minutes	20	38
60 minutes	10	19

Je crois utile d'indiquer, à titre d'exemples, les seuils de danger proposés pour le peroxyde d'azote NO_2 (*Tableau 5*), en rappelant que la limite tolérable généralement admise est de 5 ppm, soit, dans les conditions normales de température et de pression, sensiblement 9 mg/m³, cette valeur devant être considérée comme une valeur plafond.

J'ai tenu à mentionner ce concept des *seuils de danger,* car il s'insère dans une approche encore plus générale adoptée en 1963 par un Comité d'experts de l'OMS sur les polluants atmosphériques [41, 11], en ce qui concerne les critères et indices de la pureté de l'air respiré par la population générale. Cette approche a consisté à définir les *indices de pureté de l'air* par 4 niveaux de concentrations, durée d'exposition et effets correspondants:

«*Niveau I.* La concentration et la durée d'exposition sont égales ou inférieures aux valeurs pour lesquelles, dans l'état actuel de nos connaissances, aucun effet direct ou indirect (y compris une modification des réflexes ou des réactions d'adaptation ou de protection) ne peut être observé.

Niveau II. Les concentrations et les durées d'exposition sont égales ou supérieures aux valeurs pour lesquelles on observera probablement une irritation des organes des sens, des effets nocifs sur la végétation, une réduction de la visibilité ou d'autres effets défavorables sur le milieu.

Niveau III. Les concentrations et les durées d'exposition sont égales ou supérieu-

res aux valeurs pour lesquelles il y aura probablement soit une atteinte des fonctions physiologiques vitales, soit des altérations risquant d'entrainer des maladies chroniques ou une mort prématurée.

Niveau IV. Les concentrations et les durées d'exposition sont égales ou supérieures aux valeurs pour lesquelles il y aura probablement maladie aiguë ou mort prématurée dans les groupes culnérables de la population.»

Il est bien évident que la fixation de tels niveaux, équivalant, dans une certaine mesure, à l'établissement d'une courbe *dose-action,* est susceptible d'apporter des bases d'un grand intérêt pour l'appréciation des risques d'exposition des ouvriers. Une *classification des effets biologiques* en 4 niveaux du même type a été élaborée par le Comité mixte OIT/OMS de la Médecine du Travail, réuni, à Genève en Juin 1968, pour étudier les niveaux admissibles d'exposition professionnelle aux substances toxiques véhiculées par l'air [46]. Mais cette classification ne recouvre pas exactement celle précédemment mentionnée, car, outre que l'exposition professionnelle aux substances toxiques diffère considérablement de l'exposition à la pollution atmosphérique, il n'y a pas à prendre en considération, lorsqu'on définit des niveaux d'exposition professionnelle, des facteurs tels que les effets nocifs de la pollution sur les plantes. En conséquence, dans une classification des *niveaux d'exposition professionnelle,* les zones de concentration sont fatalement différentes de celles qu'on utilise pour classer les *niveaux de pollution atmosphérique générale.*

Finalement, la classification suivante a reçu l'assentiment de nombreux membres du Comité.

Catégorie A (expositions non dangereuses)

Expositions qui, dans l'état actuel de nos connaissances, ne provoquent aucune modification décelable de l'état de santé ou de l'aptitude physique des personnes exposées, à un moment quelconque de leur existence.

Catégorie B

Expositions qui peuvent provoquer des effets rapidement réversibles sur la santé ou l'aptitude physique, sans entrainer d'état morbide précis.

Catégorie C

Expositions qui peuvent entraîner une maladie réversible.

Catégorie D

Expositions qui peuvent provoquer une maladie irréversible ou la mort.

Ces catégories correspondent à une exposition de huit heures par jour, pendant cinq jours par semaine, répétée, en principe, pendant de longues périodes. Toutefois, dans le cas de certaines substances, il suffit d'expositions brèves à des concentrations élevées pour provoquer une irritation intense, l'intoxication ou la mort et il convient d'en tenir compte lors de l'établissement des limites admissibles. En outre, dans diverses circonstances-périodes de travail plus longues ou plus courtes que celles indiquées ci-dessus, charge thermique anormalement élevée, malnutrition parmi les ouvriers, etc., il importe de recueillir toutes les données voulues avant de décider si les quatre catégories précitées restent valables.

Le groupe de travail sur les *gaz irritants* (NO_2, SO_2, O_3) de la Sous-Commission sur les limites admissibles pour les substances toxiques dans les ambiances professionnelles, de la Commission internationale permanente pour la Médecine du Travail, animé par Stokinger (USA), a appliqué ce concept des niveaux d'action biologique. Quatre niveaux ont été en conséquence proposés pour chacun des trois gaz étudiés. Les valeurs de zones de concentration qu'il est préférable de re-

commander sont évidemment celles dépourvues d'effets significatifs directs ou indirects. Les valeurs retenues par le groupe précité sont:

0 à 1 ppm pour SO_2
0 à 0,5 ppm pour NO_2
0 à 0,1 ppm pour O_3

Cette recommandation s'est accompagnée de celle de ne jamais dépasser les valeurs des zones de concentrations pouvant éventuellement provoquer des effets discrets et réversibles. Les valeurs proposées par le groupe à cet égard sont:

> 1—5 ppm pour SO_2
> 0,5 à 5 ppm pour NO_2
> 0,1 à 1 ppm pour O_3

Possibilités actuelles d'Application du Concept de Limite tolérable aux Agents chimiques de Pollution de l'air des Villes et des Environnements Industriels

J'aborde là le point crucial de ce rapport. Il faut bien souligner, tout d'abord, que les *limites* ou niveaux *admissibles* d'exposition *professionnelle* aux substances toxiques véhiculées par l'air ne sauraient s'appliquer aux polluants pouvant se rencontrer dans l'air respiré par la *population en général*.

Il y a à celà 2 raisons principales:

1) alors que les travailleurs sont exposés de façon discontinue (6 à 8 heures par jour, pendant 5 à 6 jours par semaine) avec possibilité de récupération, la population générale est, le plus souvent, exposée de façon relativement continue.

2) alors que les ouvriers sont presque toujours des adultes soumis à des visites médicales d'embauche et à un contrôle médical périodique, permettant d'exclure les sujets en mauvaise condition physique, la population générale comporte non seulement des adultes bien portants, mais encore des enfants, des vieillards, des femmes enceintes, des malades et des su-

jets porteurs de tares, de déficiences ou d'affections non apparentes, qui peuvent être beaucoup plus sensibles.

Il suffit, pour comprendre l'importance de cette remarque, de rappeler que, lors des épisodes aiguu de pollution par brouillards toxiques à Londres, en Novembre-Décembre 1952 et en Décembre 1962, les morts se sont produites chez des sujets âgés ou porteurs d'affections pulmonaires ou cardiovasculaires.

On peut logiquement déduire de ces considérations que les valeurs à adopter en ce qui concerne les limites admissibles pour les polluants pouvant se rencontrer dans l'air des cités et des environnements industriels doivent être, surtout les valeurs moyennes intégrées par rapport au temps, considérablement plus basses que dans le cas de l'air des lieux de travail.

C'est en général sur la base de ce principe que des chiffres ont été proposés dans divers pays pour certains des principaux polluants.

A titre d'exemples, aux Etats-Unis, les essais d'évaluation préliminaire ont conduit les experts à proposer:

a) dans le cas du *plomb* minéral, un chiffre de 10 µg/m³ comme concentration moyenne intégrée dans une période d'un mois [44], soit une valeur 20 fois plus basse que celle, un peu trop élevée à mon avis, adoptée par les Hygiénistes industriels gouvernementaux comme limite admissible dans l'air des locaux de travail.

b) dans le cas du *béryllium* (sels et oxydes dans l'environnement des rampes de lancement de fusées), un chiffre de 0,01 µg/m³, soit 200 fois plus faible que celui adopté par les hygiénistes industriels gouvernementaux comme limite admissible dans l'air des locaux de travail (0,002 mg/m³).

c) dans le cas des *aldéhydes,* des chiffres situés en dessous des valeurs suivantes:

— 0,1 ppm pour le formaldéhyde

— 0,01 ppm pour l'acroléine

— 0,2 ppm pour les aldéhydes totaux exprimés en formaldéhyde [42].

En Juillet 1964, la revue américaine «Chemical Engineering News» a publié le compte-rendu d'une discussion entre les représentants du «Public Health Service» et ceux de l'industrie pétrolière, dans lequel il est fait état, dans le cas de *l'anhydride sulfureux* (SO_2), de limites de 0,1 ppm pour l'Etat de Floride et de 0,3 ppm pour celui de Californie, mais de tels chiffres, qui constituent des valeurs moyennes, n'ont de signification que si l'ampleur des fluctuations de concentration dans le temps et dans l'espace est connue et des valeurs plafonds fixées.

En Union Soviétique, d'après Riazanov [17], des valeurs ont été adoptées pour 40 polluants. Elles sont rassemblées dans le Tableau 6.

Je dois avouer que, dans le cas de certains polluants, *l'oxyde de carbone,* par exemple, le respect des valeurs indiquées me parait devoir comporter, en raison des émissions inévitables, notamment par les gaz d'échappement des véhicules automobiles même en limitant au minimum la concentration d'oxyde de carbone présente, une sorte de conditionnement de l'air des cités.

Les problèmes de fixation des *normes de pureté de l'air* des cités sont en fait très délicats. Cette fixation comporte en effet obligatoirement un compromis entre les souhaits des hygiénistes et des toxicologues qui ont à établir les *seuils de nocivité* et les *impératifs sociaux et économiques* limitant les possibilités de réduction des taux des polluants à l'émission; l'exemple des hydrocarbures aromatiques polycycliques à potentialité cancérogène est, à cet égard, spectaculaire. L'idéal serait de ne pas en avoir dans l'environnement; malheureusement, ils se forment, notamment par des réactions de pyrolyse, dans de multiples circonstances et on ne peut éviter complètement leur présence.

En outre, l'intervention, à un degré variable au cours du temps, de toute une série de facteurs, notamment ceux de nature météorologique ou climatique, rend très difficile l'établissement de corrélation précise entre les taux des polluants à l'émission et l'évolution de leurs concentrations dans l'air. En conséquence, il faudrait fixer, pour chaque polluant, des *seuils de nocivité* en fonction de la *durée d'exposition*. C'est la tendance que traduit un volumineux document consacré à l'étude des critères de qualité de l'air pour les dérivés oxygénés du soufre, publié en Janvier 1969, par la «National Air Pollution Control Administration» du «Public Health Service» des Etats-Unis. D'autre part, sont à prendre en considération non seulement les *effets sur* la santé et le bien-être de *l'homme,* mais encore ceux sur le milieu et notamment sur la *végétation.* Ceci s'est traduit, dans le cas de *l'éthylène* par exemple, par la proposition de limites admissibles différentes pour les zones rurales (0,25 ppm pour une période d'une heure, 0,05 ppm pour une période de 8 heures) et les zones industrielles (1 ppm pour une heure, 0,10 ppm pour une période de 8 heures), les chiffres adoptés pour les zones rurales comportant l'absence d'effets nocifs sur les plantes les plus sensibles, ce qui n'est pas le cas pour ceux retenus pour les zones industrielles [43].

Dans le même contexte, la très haute sensibilité de certains végétaux aux effets nocifs des *dérivés minéraux de fluor* a conduit à proposer, dans le cas des zones agricoles, des limites admissibles, exprimées en acide fluorhydrique, aussi basses que

1 ppb, 2 ppb, 3,5 ppb, 4,5 ppb
(parties par billion vol/vol)

Tableau 6   *Limites admissibles pour les polluants atmosphériques* en Union Soviétique (en mg/m³)

Polluant	Valeur-Plafond	Valeur moyenne
Acétone	0,35	0,35
Acétophénone	0,003	0,003
Acroléine	0,3	0,1
Acétate d'annyle	0,1	0,1
Aniline	0,05	0,03
Arsenic et dérivés minéraux (en arsenic)		0,003
Benzène	2,4	0,8
Essence de pétrole	5	1,5
Acétate de butyle	0,1	0,1
Sulfure de carbone	0,03	0,01
Oxyde de carbone	6	1,0
Chlore	0,1	0,03
Chloroprène	0,25	0,08
Chromates (en CrO₃)	0,0015	
Diméthylformamide	0,03	0,03
Dichloréthane	3	1
Dinyl(diphényle + oxyde de diphényle)	0,01	0,01
Poussières non toxiques	0,5	0,15
Acétate d'éthyle	0,1	0,1
Fluor et ses composés (en fluor)	0,03	0,01
Formaldéhyde	0,05	0,05
Acide chlorhydrique	0,05	0,015
Hydrogène sulfuré	0,008	0,008
Plomb minéral (en Pb)		0,0007
Sulfure de plomb (en Pb)		0,0017
Manganèse et ses dérivés (en Mn)	0,03	0,01
Mercure		0,0003
Méthanol	1,5	0,5
Acétate de méthyle	0,07	0,07
Méthacrylate de méthyle	0,1	0,1
Oxydes de l'azote	0,3	0,1
Phénol	0,01	0,01
Anhydride phosphorique	0,15	0,05
Suies	0,25	0,05
Styrène	0,003	0,003
Acide sulfurique	0,3	0,1
Anhydride sulfureux	0,5	0,15
Toluène diisocyanate	0,05	0,02
Acétate de vinyle	0,2	0,2

comme concentrations moyennes pour des périodes respectives de 1 mois, 1 semaine, 24 heures et 12 heures [45].

L'organisation Mondiale de la Santé s'est activement occupée, surtout depuis une dizaine d'années, des problèmes de santé publique résultant de la pollution atmosphérique et notamment de ceux concernant les *critères et les indices de pureté de l'air*. J'ai rappelé, dans la partie de cet article consacrée à l'étude du problème des limites admissibles dans les ambiances professionnelles, que l'approche adoptée par un Symposium interrégional réuni à Genève en Août 1963, et subséquemment, par un Comité d'experts OMS, en Octobre de la même année, avait consisté à définir, de façon très générale, les indices de la pureté de l'air par 4 niveaux de concentrations et de durées d'exposition en relation avec les effets produits, non seulement sur l'homme, mais encore sur les animaux, la végétation et le milieu en général.

Mais ce sont là des principes généraux qu'il s'agit de mettre effectivement en oeuvre, dans le but d'établir, sur la base d'informations toxicologiques adéquates, notamment sur les effets insidieux à long terme, des seuils de toxicité pour les polluants chimiques de l'air.

Cela signifie qu'il y a une masse énorme de recherches toxicologiques à effectuer, car, à mon avis, la fixation, même provisoire, de *normes de pureté de l'air* ne saurait, malheureusement, dans l'état actuel de nos connaissances, intervenir sur la base du critère idéal souhaité par les toxicologues, correspondant au niveau I des experts de l'OMS, c'est à dire l'absence démontrée de nocivité.

Nos informations purement toxicologiques sont, en effet, encore beaucoup trop incomplètes. Si l'on veut fixer de telles normes, à titre provisoire, il faut tenir compte des données déjà existantes et procéder à des jugements de valeur, avec acceptation de risques calculés, en considérant les éventuelles conséquences sociales et économiques, néfastes pour le bien-être des populations, pouvant résulter d'attitudes trop rigoureuses.

Cette attitude de compromis sur le plan toxicologique ne saurait dispenser les gouvernements, les autorités sanitaires et les responsables des pollutions de mettre tout en oeuvre pour réduire, autant que faire se peut, les émissions de polluants chimiques potentiellement toxiques.

Pour l'avenir, à mon avis, c'est un impératif de faire tous les efforts nécessaires pour combler les lacunes de nos connaissances toxicologiques. Etant donnée la multiplicité des problèmes à étudier, il s'impose de coordonner les recherches à l'échelle internationale, en fixant des ordres de priorité s'appliquant, par exemple, aux polluants soufrés, aux oxydes de l'azote, à l'oxyde de carbone et aux hydrocarbures aromatiques polycycliques.

L'Organisation Mondiale de la Santé qui a déjà tant fait pour le rassemblement des informations, la diffusion de recommandations générales et l'étude des méthodes de contrôle analytique de la pollution de l'air, se doit d'agir dans ce sens, comme elle l'a d'ailleurs déjà fait dans d'autres domaines.

Conclusion

Ma conclusion sera brève. Nos *connaissances* sur les seuils de nocivité des agents de pollution de l'air des villes et des environnements industriels sont, pour la plupart d'entre eux, encore très *incomplètes*. Nous sommes en particulier dans l'incertitude quant aux effets possibles à long terme pouvant insidieusement résulter, pour la santé des populations, de l'exposition à cet air pollué. Cette incertitude est d'autant plus inquiétante pour les toxicologues et les hygiénistes que certains polluants se sont révélés indubitablement nocifs, même à concentrations très minimes, ce qui a conduit, au moins dans certains cas, les autorités législatives à prescrire des mesures tendant à *réduire l'émission* de tels polluants. Il s'impose donc, à mon avis, de façon impérative, dans ce domaine de la toxicologie comme dans beaucoup d'autres, de stimuler et d'aider les recherches destinées à *fixer les seuils de toxicité* et les niveaux inoffensifs, même pour des *expositions très prolongées*.

C'est la condition primordiale pour l'établissement de mesures vraiment efficaces de protection de la santé des populations, qui n'ont, malheureusement, aucun moyen de choisir l'air qu'elles respirent.

Bibliographie

[1] Amdur, M. O., L. Silverman, P. Drinker: Archives Industrial Health 6: 305–313, 1952

[2] Amdur, M. O.: International Journal Air Pollution 1: 170–182, 1959

[3] Ball, W. L.: American Industrial Hygiene Association Journal 20: 257–363, 1959

[4] Bauer, K. H.: „Das Krebsproblem." Ed. Springer, Berlin–Göttingen, Heidelberg, 1905

[5] Fairchild, E. J., S. D. Murphy, H. E. Stokinger: Sciences 130: 861, 1959

[6] Friberg, L.: Archives of Environmental Health 19: 891–906, 1969

[7] Kotin, P., H. L. Falk, C. J. McCammon: Cancer 11: 476–481, 1958, cf. égt. Kotin, P., H. L. Falk, M. Thomas: Cancer 9: 905–909, 1956, cf. égt. Kotin, P., H. L. Falk: Cancer 9: 910–917, 1956

[8] La Belle, C. W., J. E. Long, E. E. Christofan: Archives Industrial Health 1: 297–304, 1955

[9] Lawther, P. J., R. E. Waller: Trans. Assoc. Ind. Med. Off. 9: No 1, 1958

[10] Lawther, P. J.: Proceedings Royal Society of Medicine 51: 262, 1958, cf. égt. Ref. 11, cf. égt. Lawther, P. J.: J. Just. Fuel 36: 341, 1963

[11] Lawther, P. J., A. E. Martin, E. T. Wilkins: L'épidémiologie de la pollution de l'air; Rapport sur un symposium, Genève, OMS. (Cahiers de Santé Publique No 15), 1963

[12] Matzen, R. N.: American Journal Physiology 190: 84, 1957

[13] Nakahara, W., F. Fukuoka: Naturwissenschaft 47: 44, 1960

[14] von Oettingen, W. F.: Public Health Bulletin No 290, 1944

[15] Pemberton, J., C. Goldberg: British Medical Journal 11: 567–570, 1954

[16] Reid, R. D.: Proceedings Royal Society of Medicine 49: 767–771, 1956

[17] Riazanov, V. A.: Bulletin Organisation Mondiale de la Santé 32: 389–398, 1965

[18] Roe, F. J. C., M. C. Lancaster: British Medical Bulletin 20: 127–133, 1964

[19] Schell, L. D., O. J. Dobrogorski, J. T. Mountain, J. L. Svirbely, H. E. Stokinger: Journal Applied Physiology 15: 67, 1958

[20] Scott, W. E., E. R. Stephens, P. L. Hants, R. C. Dofer: Pr. Am. Petr. Ind. 37: 171, 1957

[21] Stephens, R. E.: Journal Franklin Institute 263: 349, 1957

[22] Stokinger, H. E., W. D. Wagner, O. J. Dobrogorski: Archives Industrial Health 14: 158–162, 1956

[23] Stokinger, H. E., W. D. Wagner, O. J. Dobrogorski: Archives Industrial Health 16: 514–522, 1957

[24] Stokinger, H. E.: International Journal Air Pollution 2: 313–326, 1960, cf. égt. (5)

[25] Stokinger, H. E.: American Industrial Hygiène Association Journal 13: 8–19, 1962

[26] Stokinger, H. E.: Archives of Environmental Health 4: 115, 1962

[27] Svirbely, J. L., O. J. Dobrogorski, H. E. Stokinger: American Industrial Hygiène Association Journal 22: 21–26, 1961

[28] Teisinger, J.: Pure and Applied Chemistry 3: 253–267, 1961

[29] Truhaut, R.: "Les substances chimiques, agents de cancers professionnels", Archives de Maladies Professionnelles 15: 431-467, 1954

[30] Truhaut, R.: "Sur l'évaluation des risques de cancérisation pouvant résulter de l'incorporation volontaire ou fortuite d'agents chimiques aux aliments", Revue d'Hygiène et de Medicine Sociale 9: 8, 667–685, 1961

[31] Truhaut, R.: "Sur les risques pouvant résulter de la pollution de l'air des villes et sur les moyens de lutte à mettre en oeuvre", Revue Pollution Atmosphérique 4: 3–19 et 148–186, 1962

[32] Truhaut, R.: "Additifs aux Aliments: Les risques de nocivité pouvant résulter de leur emploi inconsidéré. Les méthodes de prévention", Bulletin de la Société Scientifique d'Hygiène Alimentaire 50: 4, 5, 6, 77–185, 1962

[33] Truhaut, R.: "Evaluation of carcinogenic activity of food additives and contaminants", Archives of Environmental Health 7: 351–358, 1963

[34] Truhaut, R.: (Paris, 1er–6 Avril) "Les résultats obtenus dans l'étude des principaux toxiques industriels-Divergences et points d'accord à l'échelle internationale", Comptes-rendus du IIIè Symposium Internationnal sur les limites tolérables pour les substances toxiques dans l'industrie, Ed. Institut National de Sécurité, 9 avenue Montaigne Paris 103–117, 1963

[35] Truhaut, R.: "Toxicité à long terme et pouvoir cancérogène", Actualités Pharmacologiques, Ed. Masson Paris, 15ème Série 257–306, 1963

36 Truhaut, R.: "Le problème des limites tolérables pour les substances toxiques dans l'industrie-Divergences et points d'accord à l'échelle internationale", Comptes-Rendus du 14è Congrès International de Médecine du Travail de Madrid 16–21 Septembre 1963, Excerpta Medica II: 146–158, 1964

37 Truhaut, R.: (Strasbourg, 24 Juin–1er Juillet), "Seuils de nocivité", Comptes-Rendus Conférence Européénne sur la pollution de l'air du Conseil de l'Europe 78–118, 1964

38 Truhaut, R.: (Buenos Aires, 14–21 Novembre), "Aperçus sur les seuils de nocivité et les limites tolérables pour les substances potentiellement toxiques dans les atmosphères de travail et dans l'air des cités", Comptes-Rendus 1er Congrès Mondial sur la Contamination de l'air 1–50, 1965

39 Comptes-Rendus du Symposium International sur les limites tolérables pour les substances Toxiques dans l'industrie, Prague, Avril 1959, 1961, Pure and Applied Chemistry 3: 1, 2, 8–373

40 Comptes-Rendus du IIè Symposium International sur les limites tolérables pour les substances toxiques dans l'industrie, 1963 (Paris, 1er–6 Avril), Ed. Institut National de Sécurité, 9 Avenue Montaigne Paris

41 Les Polluants Atmosphériques: Série des Rapports techniques de l'OMS; No 271, Genève, 1964

42 Community Air Quality Guides: American Industrial Hygiene Association Journal 29: 505–512, 1968

43 Community Air Quality Guides: American Industrial Hygiene Association Journal 29: 627–631, 1968

44 Community Air Quality Guides: American Industrial Hygiene Association Journal 30: 95–97, 1969

45 Community Air Quality Guides: American Industrial Hygiene Association Journal 30: 98–101, 1969

46 Niveaux Admissibles d'exposition professionelle aux substances toxiques véhiculées par l'air: Sixième rapport du Comité mixte OIT/OMS de la Médecine du Travail, Série des rapports techniques No 415, 1969

47 The quantification of Environmental Carcinogens: Ed. Shubik, P., D. B. Clayson, B. Terracini. U.I.C.C. Technical Report; Séries 4: 1970

Factors to be Considered in the Evaluation of the Toxicity of Pesticides to Birds in Their Environment

E. E. Kenaga

Agricultural Department, The Dow Chemical Company, Midland, Michigan 48640, USA

Summary. The most useful toxicological unit for evaluating pesticide effects on birds is the *milligrams of pesticide* intake per *kilogram of body weight per day*. Since data used for the calculation of one or two terms of this unit are often unknown, the principles necessary for estimating the amount of daily food consumption by birds, the body weight of representative species or ages of birds and the level of pesticide residues on wildlife food are summarized.

Representative data related to food consumption and body weights of different species and ages of birds are tabulated. Based on the wide range of body weights and species feeding habits, the smallest birds may consume at least ten-fold more food in terms of milligrams per kilogram of body weight per day than the largest birds. Considering also the variation of pesticide residues on different types of plants as representatives of bird food particles, the daily dietary intake of pesticide by birds (expressed as mg/kg/day) may vary at least 100-fold in environments treated with the same dosage of pesticide.

Evaluation of the toxicity of pesticides in the food of birds is most realistic if the results of subacute or chronic dietary feeding studies are available. The above information and the physical, chemical and biological properties of the pesticide are important factors which are related to the quantity and type of residues in the environment and to the type of toxicological tests needed to properly simulate or interpret environmental exposure of birds to pesticides under use conditions.

Zusammenfassung. Dieser Beitrag behandelt bedeutende ökologische Themen hinsichtlich der Bewertung der Auswirkungen von Pestiziden auf Vögel. Die Definition der täglichen Aufnahme einer Verbindung mit der Nahrung, die ohne Gegenwirkung toleriert werden kann, entspricht klassischen toxikologischen Untersuchungen an Tieren. Auf Vögel übertragen beziehen sich die Einheiten notwendiger Informationen auf mg/kg/Tag, d. h., die Menge eines Pestizides in Milligramm, die ein Vogel entsprechend seinem Körpergewicht pro Tag vertragen kann. Der Vergleich zwischen dem Pestizidanteil, den ein Vogel in Laboruntersuchungen aufnimmt, und demjenigen, den er in der freien Natur verträgt, liefert die Grundlage, um die Wahrscheinlichkeit abschätzen zu können, ob ein Pestizid Vögeln schaden könnte oder nicht.

Introduction

This paper deals with some of the more important ecological segments which relate to evaluation of the effect of pesticides on birds. Definition of daily dietary intake of a compound which can be tolerated without adverse effect represents the classical approach to animal toxicological studies. Translated to birds, the key units of needed information relate to the *milligrams of pesticide consumed per kilogram of body weight of the bird per day*. This term is commonly expressed as mg/kg/day. Comparison of the level of pesticides tolerated by birds in laboratory studies with the level expected in field situations provides a basis for estimating the probability of whether or not a pesticide might be harmful to birds.

An examination of the pesticide literature regarding bird toxicity shows that the test method criteria are often incomplete or unreported. In tests run in the laboratory, a typical report of results of a dietary feeding study is often given in terms of the ppm of pesticide necessary to obtain an LC_{50} (a concentration in the diet resulting in 50% mortality of the organism) without giving the daily ingestion rates or weights of the bird. A typical field test may show that X number of pounds of pesticide were applied per acre with a given observed bird mortality rate. In the above laboratory example, two of the three required criteria stated above for determining mg/kg/day are missing, and in the field example all of them are missing!

The scope of this paper is therefore concerned with the more important particular ecological factors which relate to defining the mg/kg/day consumed by terrestrial birds in field and forest environments. The purposes of this paper are:

1) To provide information from which the body weight of particular birds and the related daily dietary intake of food and pesticide residues can be estimated.

2) To define with a degree of quantitativeness the level of pesticide residues on various categories of plants. With this quantitative base, the level of pesticide residues for specific plants or insects can be estimated with an acceptable degree of accuracy.

3) To define the types of food representative of various sizes and shapes of those consumed by birds in a way that reasonable estimates of the resultant dietary ingestion of pesticide residues can be made.

4) To emphasize the importance of conducting toxicological studies with birds which simulate the levels and changes in levels of pesticides which occur in the environment.

5) To illustrate the principles in evaluating the probability of pesticide residues being a hazard to birds.

Relationship of Food Consumption to Body Weight of Birds

Body Weights of Various Ages and Species of Birds

The size and weights of the more than 600 species of birds in North America as well as those around the world are extremely variable. The smallest species of adult hummingbird weighs around 3 grams while small song birds such as warblers, vireos, chickadees, finches, and small sparrows weigh 10—25 grams. Intermediate sized species of adult song birds such as some fringillids, thrushes, and blackbirds weigh 25—75 grams. Larger birds such as the jays, doves, robins, starlings, and grackles weigh 75—150 grams. Small predatory birds, many aquatic birds and small upland fowl such as small hawks and owls, teal, some quail, and some shorebirds weigh 150—400 grams. Certain ducks, some grouse, large hawks and owls weigh 400—1000 grams. The large species of birds such as some herons, ducks, grouse, geese, eagles, and turkeys weigh over 1000 grams. Thus, adult birds have about a 1000-fold variation in body weight. Examples of variations in body weight of species of adult birds are shown in Tables 1 and 2 (Nice 1939, Kendeigh 1969a, b).

Table 1   Relation of food consumption to body weight (after Nice 1939)

Bird Species	Adult Weight (Grams)	Food Eaten Per day (Grams)	Per cent of Body Weight Eaten Per day
Blue Tit (Parus caeruleus)	11	3.3	30.0[1]
Great Tit (Parus major)	18	4.4	26.0[1]
European Goldfinch (Carduelis carduelis)	13	2.3	17.5[2]
Robin (European) (Erithacus rubecula)	16	2.35	14.7
Chaffinch (Fringella coelebs)	22	2.9	13.2[2]
Mourning Dove (Zenaida macroura)	100	11.2	11.2[2]
Song Thrush (Turdus philomelos)	89	8.8	9.8[1]
Dunlin (Erolia alpina)	114	8.9	8.5[1]
Bobwhite (Colinus v. virginianus)	170	15.2	8.8[2]
Lapwing (Vanellus vanellus)	195	15.3	7.8[1]
Kestrel (Falco tinnunculus)	200	15.4	7.7
Blackbird (European) (Turdus merula)	118	8.4	7.3
Little Owl (Athene noctua)	164–172	9.1– 9.4	5.5
Tawny Owl (Strix aluco)	442–475	22.1–23.7	5.0
Common Buzzard (Buteo buteo)	855–900	38.5–50.0	4.5
Domestic Fowl (Gallus domesticus)	1800	61	3.4[2]

[1] Dry weight (estimated as 40 per cent of the weight of the mealworms used as food).
[2] Seeds and grains.

There are also variations in the *adult* weights of individuals of the same species depending upon sex, breeding status, climate, length of daylight, availability of food and other factors. However, the greatest difference in weights within a species is the result of growth.

The *young* of large precocial species of birds such as the bobwhite *(Colinus virginianus)* are as light in weight as *adults* of many small species of altricial birds. Altman and Dittmer (1964) show the hatching weights of bobwhite, Pekin duck *(Anas platyrhynchos domesticus)*, grayleg goose *(Anser anser)*, domestic chickens *(Gallus domesticus)*, and domestic turkeys *(Meleagris gallapavo)* to be 4, 59, 77, 32—37, and 45—50 grams, respectively. Many other examples are available. Weights of some precocious species such as game and domestic fowls may increase 40—275 times over their natal weights (calculated from Altman and Dittmer 1964).

These differences in weights are obvious when brought to one's attention, however, in much of the literature concerning the toxicity of chemicals to birds, body weights are not given. Body weights are needed in order to determine standard toxicological measurement units such as LD_{50} data in terms of mg/kg and the acceptable daily intake (ADI) of pesticides or other dietary additives in terms of mg/kg/day. If weights of birds are not given it is usually possible to estimate their weights quite closely if their age and species are known, on the basis of the published literature.

Quantity of Food Consumed by Birds of Various Ages and Species

The food consumption rates of birds in the laboratory are rather easily measured when they are eating standardized mash food or grain. However, many species of

Table 2 Relationship of food consumption to body weight (below about 35 °C) (after Kendeigh, 1969a, b)

Bird Species	Weight (Grams)	Grams Food Eaten per Day* (Hrs Photo-period)	Per cent of Body Weight Eaten per Day	Per cent Increase in Existence Metabolism at 0 °C Over 30 °C
Yellow-bellied Seedeater	9.3	5.58 (15)	60.0	162
(Sporophila nigricollis)		4.41 (12)	47.5	
Blue-black Grassquit	9.4	4.69 (15)	49.9	137
(Volatinia jacarina)		3.98 (12)	42.4	
Variable Seedeater	10.7	5.30 (15)	49.5	122
(Sporophila aurita)		4.71 (12)	44.0	
Zebra Finch	12.1	5.44 (12)	45.0	172
(Taencopygia castanotis)				
Field Sparrow	13.2 ♀	4.84 (15)	36.7	
(Spizella pusilla)		4.38 (10)	33.2	
Field Sparrow	13.9 ♂	4.97 (15)	35.8	116
(Spizella pusilla)		4.65 (10)	33.5	
Common Redpoll	14.0	4.98 (7)	35.6	71
(Acanthis flammea)				
Hoary Redpoll	15.0	4.99 (7)	33.2	96
(Acanthis hornemanni)				
Tree Sparrow	19.0	7.11 (15)	37.4	81
(Spizella arborea)		5.95 (10)	31.4	
House Sparrow	25.2	7.39 (15)	29.3	54
(Passer domesticus)		7.09 (10)	28.2	
White-throated Sparrow	27.4	7.91 (10)	28.9	98
(Zonotrichia albicollis)				
Dickcissel (Spiza americana)	29.6 ♀	9.45 (15)	31.9	143
		7.89 (10)	26.7	
Dickcissel (Spiza americana)	31.6 ♂	9.24 (15)	29.2	
		8.39 (10)	26.5	
Green-backed Sparrow	37.0	10.14 (15)	27.4	125
(Arremonops conirostris)		9.79 (12)	26.5	
Blue-winged Teal	309 ♀	32.33 (12)	10.46	121
(Anas discors)	363 ♂	35.33 (12)	9.73	132

* Calculated from information furnished by Dr. S. C. Kendeigh and based on kcal energy used per bird day multiplied by a metabolic conversion factor of 67–80%, depending on the species, then divided by 4.35 kcal/g food value to get gross weight of food eaten per day of chicken mash.

birds have never been reared in the laboratory. Wild birds eat a variety of species of plant and animal foods in their daily diet, and the proportion of these foods vary with season and location of the bird. The total weight of food eaten by wild birds is difficult to determine even though the variety of species of organisms eaten is sometimes known. The lack of such data makes it necessary to have a method of estimating these intake rates as accurately as possible, for calculating mg/kg/day data to be used with various species of birds.

Small birds eat less than large birds, however, in general the smaller the bird, the greater the amount of food it eats (by dry weight) on the basis of percent of its body weight. This is in keeping with the increased energy output related to heat loss, necessary because of the increased surface area to body weight ratio of the smaller bird. Small birds also often eat smaller food particles than large birds and so must find and eat more of them.

Examples of the amounts of food eaten by birds of different species and their food habits, based on weight of the bird, have been studied by various people. These data include song birds, game birds, passerine and non-passerine, insect eaters, seed eaters, carrion eaters and predators. Nice's (1939) data (Table 1), when plotted on log-log graph paper using body weight versus percent of body weight eaten per day, form the basis for a reasonably close fitting straight line (Fig. 1). The amount of food, on a dry weight basis, consumed by birds weighing 20, 100, or 1,000 grams is 18, 9.2 or 3.6 percent of body weight, respectively, based upon this line.

Kendeigh (1969a) assembled data from the literature concerning the food energy requirements of a number of species of birds, comparing the effect of temperature changes. Kendeigh (1969b) also furnished existence metabolism data on birds feeding during different lengths of photo periods as well as a formula for converting these data to grams food eaten per day. From this food consumption infor-

mation, Table 2 has been constructed which gives the percent of body weight eaten per day for various species of caged birds. These data when plotted on log-log graph paper (Fig. 1) result in a line parallel to that of Nice's, however, the values are about two times greater. For example, using data derived from Kendeigh's converted caloric intake information, a bird weighing 20, 100, or 1,000 grams wold eat 33, 17, and 6.7 percent of its body weight per day. The length of daylight (photo period) decreases or increases these values slightly, depending on whether the photo period exposure time is shorter or longer. A fairly significant increase in existence metabolism at 0° C compared to 30° C is noted in various species of birds, independent of size. The greatest increase occurs with tropical birds. Data with northern fringillid species such as redpolls show that the increase in energy necessary to maintain birds in winter versus summer is nearly two-fold. The food habits of different species of birds vary considerably. Larger species of birds which have an abundance of food, may be able to engorge themselves on occasions and have a more variable daily intake than smaller species whose metabolism require them to search more constantly for food. Small passerine birds that feed on small insects and seeds must eat fairly regularly througout the day because they often pass food through their digestive system in less than one hour and thus cannot fast for long after their alimentary canal is empty.

It is recognized that a two- or three-fold difference in the daily rate of food consumption could occur in the diet of a bird, depending on the moisture content and the caloric and nutritive values of its food. Most of the food consumption values reported in the literature have been for dry food (not over 10—15 percent

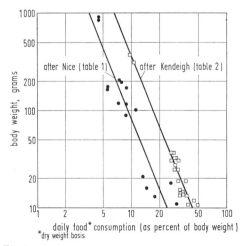

Fig. 1.

moisture) or have been corrected for moisture or caloric content. There are few if any data available on the weight of the total diet of wild insectivorous species of birds. Birds eating a high percentage of food with a high moisture content may consume greater quantities of pesticide than birds with diets containing a low moisture content because of the increased bulk needed for equivalent nutritive value. Individual variation in the daily rate of food consumption in the field may be due to weather conditions or lack of food. Even under laboratory conditions, Stickel et al. (1965) found that woodcocks *(Philohela minor)* ate a mean value of 68 to 77 percent of their body weight per day in earthworms (wet weight). Individual woodcocks in the group varied from 11 to 143 percent. Such large differences might occur for a few days, but would probably average close to the mean over a period of a week. Since earthworms may contain 83 percent water by weight (French et al. 1957), the variation in percent of food eaten in dry weight per day would be about 1.9 to 24.3 percent, or a mean of 12 to 13 percent over several months.

Caution must be used in calculating the amount of food eaten by pen reared birds such as ducks and pheasants since they will waste a considerable amount. The food wastage must be either measured or prevented to get accurate consumption data.

Birds of each species change weight rapidly as they grow from nestlings to adults. A high energy conversion bird such as the domestic turkey decreases its consumption rate from 13.6 to 4.0 percent of body weight per day between hatching and 28 weeks of age (Table 3). Similar data on the bobwhite are also given in Table 3. Many passerine species complete their entire weight growth in two weeks. In all cases, the growing bird decreases

Table 3   Relationship of food consumption to body weight and age in the Bobwhite and the Broad-Breasted Bronze Turkey

Age (Weeks)	Average Weight (Grams)	Per cent of Body Weight Eaten per Day[3]
Bobwhite[1]		
0– 1	11.2	26.3
1– 2	18.6	23.4
2– 2.5	31.7	15.7
Domestic Turkey[2]		
0– 1	88	13.6
1– 2	184	13.1
2– 3	338	11.3
3– 4	533	9.1
4– 8	1,315	7.8
12–16	4,944	5.6
16–28	7,507	4.0

[1] From a report of The Dow Chemical Company.
[2] Modified from the Salisbury Poultry Product Use Guide, Salisbury Literature. Charles City, Iowa.
[3] Turkey mash not corrected for dry weight.

its food intake when measured as percent of its body weight. Data on various species and sizes of birds illustrate the principle that there is a general linear relationship within and between species of birds based on the weight of the bird and dry weight of the food consumed. The dry weight of the food in itself is not important except as an indicator of the caloric and nutritive value of the food which all birds need for growth, reproduction and existence. While actual food consumption rates are desirable for calculation of pesticide intake, it is helpful to have a method for making an approximation for the many species of birds for which the consumption rates are unknown. The data in Tables 1 and 2 furnish a working basis, where the species or weight of the bird is known, for estimating the percent of their body weight eaten in food per day. In general, on a dry-weight basis, the daily food eaten by large species is less than 10 percent of their weight, compared to small species or small immature individuals of a large species, which eat up to 40 percent. By use of the data of Nice and Kendeigh these food-body weight relationships can be estimated within approximately a two-fold factor (Fig. 1).

Size, Shape, and Variety of Foods Eaten by Birds

Birds show varying degrees of preference to specific foods in their diet. A species which is so adapted to its diet that it is restricted to a very few food items, or to one, is said to be stenophagous (Welty, 1963). If a species eats a wide variety of foods it is called euryphagous or omnivorous. These terms are relative. A bird may be stenophagous at a certain stage in its life history, or at a certain time of year, and euryphagous at other times. These changes may be related to such things as food availability, migration, weather, age, health, and sex. Birds may be arbitrarily divided into phyllogenetic groups on the basis of their principle food source. Examples are as follows:

Feeding Habits	Birds (examples)
Stenophagous Predaceous on vertebrate animals	Certain hawks, owls, kingfisher,
Flower nectar	Hummingbirds
Specific kinds of snails	Everglade Kite (Rostrhamus sociabilis)
Euryphagous Invertebrates, such as a large variety of insects	Certain warblers, vireos, flycatchers, swallows, swifts, nighthawks, shorebirds
Seeds of many types	Many sparrows, doves
Invertebrates and seeds	Certain blackbirds, thrushes, mimics, grouses, fringillids, waxwings, woodpeckers, upland game birds
Invertebrates, vertebrates, and plants	Certain ducks, herons, gulls, crows

The necessary protein in the diet of many species comes mostly from various insects and other invertebrates; from vertebrates such as fish, frogs, and mammals; and from plant seeds. Few birds eat the foliage of plants as a large portion of their diets .

In general, the size of seeds, insects, and other foods eaten by the first three groups of euryphagous bird species, are fairly small, averaging perhaps the size of a grain of wheat or smaller, or the size of a small to medium sized insect. These groups of birds are most numerous from the standpoint of species and individuals among the terrestrial birds.

Various plant foods such as seeds and fruits eaten by birds, are often covered by remnants of the flower or are enclosed in pods, shells, etc., which are not themselves eaten by birds, but protect the bird food from part or all of the pesticide residue originally applied. Many seeds and fruits of plants are dispersed away from the original source by various natural forces such as wind, popping open, sticking onto animals, or being transported by them, and other ways which tend to decrease the availability of the original pesticide residue applied to the plant. The feeding habits of the specific birds helps determine the exposure to the pesticide. If for example, the bird shucks off the seed cover containing the pesticide residue and eats only the seed which has no pesticide residue, then obviously the bird is not ingesting much of the residue occurring on the whole plant. In addition to these factors which reduce residue intake, birds have a tendency to randomize their selection of food by visiting different areas and eating various sizes and shapes of food in their diet.

Estimation of Residues of Pesticides in or on Plants and Insects

The main purpose for determining the amount and kind of *residues of pesticides on bird food* is to assess their acute and chronic toxicological effects on birds. The principle acute effects of pesticides on terrestrial wildlife food are those occurring immediately after application of the pesticide to the environment and for most pesticides they are the highest residue levels encountered by birds. The chronic effects are those related to long-term residues. Since pesticide residues on terrestrial plants decline with time, the principle chronic toxicological problem is with the level of residue remaining after several weeks.

The amount of residues of various pesticides on most of the specific food items in the diet of birds has not been determined by analytical methods. In order to develop some principles which would apply as guidelines to the highest residue levels expected on various plants, immediately after application and after 5—6 weeks, Hoerger and Kenaga (1971) reviewed literature from a number of pesticide journals. These references included ppm residues of 21 representative pesticides in or on 36 different crops from 22 literature sources. These residue values were obtained as the result of analyses from many different application dosages in terms of pounds per acre. For the purposes of comparison, all of these residues were unitized on the basis of ppm residue per one pound of pesticide applied per acre. The residues reported here usually represent the pesticide applied and not the degradates or metabolites. While this is not usually important immediately after application, obviously degradates or metabolites do occur with time, and need to be taken into account, at least from the toxicological standpoint.

The pesticides reported in these pesticide journals are herbicides such as 2,4-D, dalapon, dicamba, picloram; insecticides such as the more persistent of the chlorinated hydrocarbons, that is, DDT, endosulfan, dieldrin, endrin, aldrin, chlordane, toxaphene; phosphate insecticides such as parathion, malathion, diazinon, demeton, phosdrin, carbophenothion, dimethoate, EPN, dioxathion; and other pesticides. Thus, a wide variety of chemical structures are represented.

For use in translation to situations where no residue data is available, it is important to estimate the "worst" or most rigorous situation that might occur, in order to establish an upper limit of residue potentiality. For this purpose the *highest residues* of all pesticides were selected by Hoerger and Kenaga (1971), from each of the references. These highest (or upper limit) residue levels ranged from as high as 240 ppm at one pound per acre immediately after application down to as low as undetectable amounts after several weeks, as shown in Table 4. The variations in upper limit residues on various crops with various pesticides occur principally because of weathering, metabolism, chemical breakdown, and vegetative factors. The typical residues are an average figure for the *highest* residues for all pesticides studied on the various crop groups.

Three important vegetative factors significantly affect the level of pesticide residues during application of a pesticide to plants. These factors are the weight of the vegetation per unit area; the degree of penetration of the pesticide through various layers of plants vegetation; and the surface area/mass ratio.

The upper limit residues on various crops vary greatly. Crops having similar levels of residues were classified together into one vegetative group. By this means several vegetative groups were formed as shown in Table 4.

The plant categories shown in Table 4 represent in part, a decreasing ratio of surface area to mass ratio and illustrates that such ratio changes may result in nearly a thirty-five fold difference in ppm of pesticide residue, on various plant species or plant parts, immediately after application of a given pesticide dosage. Grass and other foliage presents a large surface dimension to a thin volume and low weight (large surface area to mass ratio), as do small food particles.

On the basis of information in Table 4, it is obviously important to know what sizes and shapes of foods are present in the environment which provide the diets of the species of birds being evaluated for toxicity due to pesticide residues, since this helps determine the upper limit of residues to be expected immediately after application and is very useful for acute toxicological tests. It must be remembered that most residue values will fall far below the upper limit values.

Very little information concerning *pesticide residues on insects,* resulting from the use of a given pesticide control program in the field, is available in the literature. Target and nontarget insects are often the major sources of food for some species and some life stage of birds. Ini-

tial residues on insects are probably in the same order as those on plants of similar surface area to mass ratios.

El Sayed et al. (1967) analyzed 15 species of insects for insecticide residues, samples of which were collected from July to November from cotton fields heavily sprayed with DDT and other insecticides. The various representative insects analyzed for pesticide residues included species of the orders Lepidoptera, Odonata, Coleoptera, Homoptera, Ephemeroptera, and Orthoptera. No detectable pesticide residues occurred in five species. The most omnipresent and the highest concentrations of insecticide residues found were *p,p' DDT* and its metabolite *p,p' DDE* which ranged in total from undetectable amounts up to a maximum of 13 ppm.

El Sayed et al. (1967) used topical applications of DDT to larvae of the bollworm, *Heliothus* sp., and by residue analyses showed only DDE to be present after 5.5 days. (If the *Heliothus* larvae weighed 0.1 grams, the DDT dosage cited [40 micrograms] would be equivalent to 400 mg/kg [400 ppm]. Thus, in 5.5 days, no detectable DDT was present and only 5 ppm of DDE, or less than 2% of the calculated original dosage applied was found as DDE.) Such an example points out the great potentiality for *biodegradation of insecticide residues in insects.*

Table 4 *Upper limits* and *typical* limits of residues of pesticides on differing categories of plants.

	ppm Residue on the Basis of a Pesticide Dosage of 1 lb Per Acre			
	Immediately After Application		6 Weeks After Application	
	Upper Limit	Typical Limit	Upper Limit	Typical Limit
Range Grass (short)	240	125	30	5
Grass (long)	110	92	20	1–5
Leaves and Leafy Crops (vegetables and fruit)	125	35	20	<1
Forage Crops (alfalfa, clover)	58	33	1.0	<1
Pods Containing Seeds (legumes)	12	3	1.5	<0.1
Fruit (cherries, peaches, grapes, citrus)	7	1.5	1.5	<0.2

This biodegradation by insects surviving the pesticide application should result in a corresponding decrease in pesticide residue consumption by insect eating birds, at various times after application of the pesticide.

Most of the factors which affect the decline of residues on plant surfaces are also operative for insect surfaces and so inert residues may be estimated on the basis of insect species having a surface to mass ratio similar to those of the equivalent plant type listed in Table 4 (for example, in size a beetle might be equivalent to a grain of wheat).

The Interchangeable Use of mg/kg/day and ppm Residue Data for Toxicity Studies with Pesticides, Based on the Body Weight and Food Intake of Birds

Mg/kg/day toxicity information, and the species and weight of the bird, indicating the percent by weight of *daily dietary intake,* can be used to help predict the safe level of ppm residue of pesticide in the diet of a particular species or size of bird. This ppm data can be matched with the estimated or actual pesticide residues from the proposed pesticide dosage to see if a safety margin exists. An example of the variation in mg/kg/day intake from 50 ppm pesticide in diets of birds of various weights or species, or the variation in ppm in the diet to obtain an intake of one mg/kg/day is shown in Table 5.

An example of the variation in the mg/kg/day pesticide intake of various sizes of birds based on the upper limit residue values in Table 4 resulting from a 1 pound per acre pesticide treatment, are shown in Table 6. In most cases the ppm and mg/kg/day figures for toxicants in the environment will be considerably below those high residues shown in applications of pesticides on the first day, as

Table 5 Mg/kg/Day and ppm Conversion units used as equivalents in pesticide dietary intake calculations for birds of varying weights

	Body Weight of Bird (Grams)		
	20	100	1,000
Food Ingested Per Day			
Grams[1]	3.6	9.2	36.0
Per cent of Body Weight[1]	18.0	9.2	3.6
Diet Containing 50 ppm of Toxicant			
Mg Toxicant Ingested per Day per Bird	0.18	0.46	1.8
Mg/kg/Day of Toxicant Ingested	9.0	4.6	1.8
1 Mg/kg/Day Toxicant Ingested			
Mg Toxicant Ingested per Day per Bird	0.02	0.10	1.0
Ppm in Diet to Obtain 1 Mg/kg/Day	5.6	10.9	27.8

[1] Adapted from data presented by Nice (1939). See Fig. 1 and Table 1.

Table 6 Estimation of the mg of toxicant/kg of body weight/day intake by birds of varying sizes resulting from eating different foods from an area treated uniformly with an application of 1 pound of toxicant per acre.

Illustrative Examples of Ppm[1] In or On Different Types of Food Eaten by Birds	Mg/kg/Day Ingested by Different Sized Birds		
	20 gm 18⁰/₀[2]	100 gm 9.2⁰/₀	1,000 gm 3.6⁰/₀
240 (sparse foliage)	43	22	9
58 (dense foliage, insects)	10	5.3	2.1
10 (seeds, fruit, large insects)	1.8	0.9	0.4

[1] Ppm pesticide residue immediately after application based on maximum values cited.
[2] Per cent of body weight ingested in dry food per day.

shown in Table 6. On subsequent days or weeks after application where biological magnification in chain or life food is not involved, these mg/kg/day values may decrease rapidly depending on the stability and residual nature of the compound, its application, and various environmental factors.

Interpretations and Uses of Toxicological Results Obtained by Various Means of Administering the Pesticide to the Bird

Many interpretations of the *toxicological effects of pesticides* on organisms in the environment are made by use of laboratory test methods. Test methods which measure toxicity by injection into various parts of the bird anatomy and/or their eggs do not represent the normal pesticide intake of the bird as ingested in its diet.

Following the lead of mammalian toxicologists who study the acute hazards of pesticides to man, the *acute oral LD$_{50}$* (dosage causing 50% mortality) *toxicity test method* is often used in birds as a basis for comparison between various pesticides. Results from the use of such a test method may often be the only standardized toxicological basis of comparison available between different pesticide compounds and is often used to categorize compounds as "highly toxic", "moderately toxic", "of low toxicity", and "essentially non-hazardous". The categories require definitive numbers which are usually given in terms of mg/kg. These numbers are sometimes used as the basis for state or national regulation or laws for restricting the use of pesticides. However, the acute oral LD$_{50}$ or EMLD (estimated minimum dosage causing mortality) values frequently do not provide a realistic assessment of the true toxicity of compounds to wildlife under the conditions of the proposed pesticide application.

The administration of the acute oral test is artificial and is not representative of the normal method of consumption of food by birds because most of their diets normally contain small amounts of pesticide residues in terms of ppm and *the intake occurs at various intervals during the day*, in constrast to the acute oral test in which intake occurs once a day and in

which the pesticide is undiluted or present in a concentrated formulation at the time of ingestion. The effect of the acute oral route of administering pesticides to the bird is an additional unknown factor because of the handling and excitement engendered by such treatments. Data from more meaningful test methods are needed in place of the acute oral test. Although literature data comparing pesticides on other standard test methods is limited, a comparison of the toxicity of several pesticides by use of three test methods on two species of birds can be used to illustrate this point in this paper. These methods include the toxicity results of single acute oral (AO) dosage, thirty repeated acute oral (AO[30]) dosages, and *dietary feeding (DF) studies;* using the adult Mallard duck *(Anas platyrhynchos),* and the adult Coturnix quail *(Coturnix coturnix japonica),* and insecticides of completely different chemical and physical properties, such as DDT, dieldrin, dimethoate, DOWCO 179 (0,0-diethyl 0-[3,5,6-trichloro-2-pyridyl] phosphorothioate [the active ingredient of DURSBAN Insecticide]), and DOWCO 139 (4-dimethylamino-3,5-xylyl-methylcarbamate [the active ingredient of ZECTRAN Insecticide]). The single acute oral dosage and the thirty daily acute oral dosages were administered to the birds in gelatin capsules through glass tubing. The dietary feeding studies, while not completely analogous to each other in the number of days fed and the dietary concentration of the pesticide, are similar enough for general comparative purposes. These data are transformed from ppm to mg/kg/day mostly from data of Tucker and Crabtree (1970) and the USDI (1964). The insecticides are compared individually on the basis of the ration of mg/kg/day for each test method.

The comparison of the three test methods as indicators of toxicity to birds in the environment is emphasized in the column "Index of Chronicity" in Table 7. This index is a ratio derived from the following equations:

$$\frac{LD_{50} \text{ mg/kg/day (AO)}}{LD_{50} \text{ mg/kg/day (DF)}}$$

or

$$\frac{LD_{50} \text{ mg/kg/day (AO)}}{(EMLD) \text{ mg/kg/day} - (AO\ [30])}$$

The single acute oral LD_{50} is used here as the unit of comparison and is assigned an index number of one. *Index numbers greater than one* indicate the degree of *chronic toxicity* as judged by the test method used for comparison. The larger the index number, especially those above one, the more likely the insecticide is to cause chronic toxicity. Index numbers less than one tend to indicate the relative lack of chronic toxicity. The index of chronicity for dietary feeding tests ranges as high as several hundred with *DDT* and as low as 0.01 with *DOWCO 139*. The decreasing order of chronic toxicity of the compounds based on data in Table 7 is DDT > *dieldrin* > *dimethoate* > *DOWCO 179* > DOWCO 139. The decreasing order of *acute oral toxicity* of the same compounds is DOWCO 139 > dimethoate > DOWCO 179 > dieldrin > DDT, which is almost the exact opposite in the order of compounds listed. By this illustration it is seen to be very important that the proper test method is used as the principle criteria for evaluating safety.

The acute oral LD_{50} of DDT is quite high and when compared with other insecticides makes DDT appear to be quite safe. (It must be remembered, however, the DDT is not very soluble in the various aqueous media of the gut, being low in water solubility and reasonably fat soluble, and in addition, is not very easily metabolized.) Thus, in the short passage of time going through the gut of the bird, in order to produce toxic symptoms, DDT must be absorbed by the body rapidly. Stickel et al. (1965) treated woodcocks with massive single oral doses of DDT and *heptachlor*. Residue data from birds treated with DDT suggested that woodcocks absorbed relatively little of the DDT that was administered. Quantities of butter oil or corn oil, used to formulate heptachlor, when fed to woodcocks were passed in about three hours, thus very likely allowing very little absorption of heptachlor. Thus, the acute oral LD_{50} data for Mallard ducks shown in Table 7, may be used to considerably underestimate the chronic toxicity of DDT and other compounds of similar physical and chemical properties, due to poor adsorption of the pesticide into the body during the normal passage time allowed for digestion and voidance by the duck.

Tucker and Crabtree simulated chronic dosage effects by use of thirty successive single daily dosages, applied in the same way as the single oral dosage. Results from this method tends to indicate the chronic nature of DDT toxicity as shown by the decrease in dosage from > 2240 down to 50 mg/kg/day. While the LD_{50} and EMLD are used for comparison, the relative toxicity of one compound to the other is still valid. With use of the thirty day repeated acute oral test as with the single acute oral dosage, the passage of the partially unchanged, and unabsorbed DDT through the gut may give a reduced toxic response which results in underestimation of the toxicity of DDT to birds.

The physical and chemical properties of DOWCO 139, which account for a low index of chronicity, are entirely different than those of DDT. Williams et al. (1964) studied the metabolic products of DOWCO 139 in dogs. Krishna and Casida (1966), and Oonithan and Casida (1968) studied metabolic products in rats. In both animals DOWCO 139 was found to be rapidly metabolized. Such meta-

Table 7   The relative toxicity of some insecticides to birds correlated with acute and chronic toxicity test methods used for safety evaluation

Pesticide	Test Method	LD_{50} Mg/kg/Day	LC_{50} ppm in Diet	No. of Days Treated	Index of Chronicity
		Mallard Duck			
DDT	AO[1]	>2240	–	1	1
	AO (30)[2]	50[2]	–	30	> 45[2]
	DF[3]	4–6	200	6–8	>280
	DF	4[4]	100	365	>560
Dieldrin	AO	381	–	1	1
	AO (30)	5[2]	–	30	76[2]
	DF	19–20	1000	28–30	19–20
		10	500	24	38
Dimethoate	AO	41.7	–	1	1
	DF	16.0	5000	21	2.6
DOWCO 139 (ZECTRAN)	AO	3	–	1	1
	AO (30)	1.25	–	30	2.4[2]
	DF	13.0	5000	16	0.23
		9.0	2500	25	0.33
		6.5	1000	27	0.46
DOWCO 179 (DURSBAN)	AO	70–80	–	1	1
	AO (30)	>2.5[2]	–	30	> 30[2]
	DF	36[4]	361	5	≈2
		Coturnix quail			
DDT	AO	841	–	1	1
	DF	76–175	500	7–48	4.8–11.1
		28–83	250	49–57	10.1–30
Dieldrin	AO	69.7	–	1	1
	DF	17	100	8	4.1
		14	50	66	5.0
		4	20	67	17.4
DOWCO 139	AO	3.2	–	1	1
	DF	114[5]	1000	3	0.03
		25–250[5]	500	3–5	0.01–0.13
DOWCO 179	AO	15.9	–	1	1
	DF	50[4]	500	28	0.32

[1] Single acute oral application by stomach tube. LD_{50} (AO).
[2] Based on 30 daily acute oral applications-estimated minimum lethal dosage (AO 30).
[3] Birds fed ad lib on dietary treatments. LC_{50} (DF).
[4] Estimated on the basis of bird weight.
[5] Young birds instead of adults.

bolic degradation may account for the ability of the Mallard duck to tolerate a dosage equivalent to a large proportion of the single acute oral LD_{50} daily for thirty days. However, birds normally eat pesticides as residues in part or all of their diet, usually at rates below 100 ppm so that the most likely laboratory method of simulation of field residue consumption is via dietary feeding. In the dietary feeding studies with DOWCO 139, the birds eat an amount equivalent to two to four-fold the amount of the single acute oral LD_{50} every day for 16—27 days for the same resultant mortality. The dietary feeding test results shown in Table 7 emphasize the ability of the Mallard duck and Coturnix quail to detoxify DOWCO 139 even when fed at higher levels in the diet than with DDT.

From data in Table 7 it is seen that the acute oral LD_{50} test, when used as a relative indicator of toxicity to birds, results in gross *underestimation* of the long-term effect of DDT, which has a high index of chronicity and results in gross *overestimation* of the effect of DOWCO 139, which has a low index of chronicity. When the acute oral LD_{50} is used in legislation as the basic toxicological unit for measurement and restriction of pesticides which may affect wildlife, it is obvious that DOWCO 139 could be maligned and DDT could be exemplified. Because of the potentially misleading results from acute oral tests, and because birds obtain most of their pesticide residue intake from their natural diet, it is recommended that the *dietary feeding test is the best method of measuring the pesticide residue consumption of birds* concerning effects of pesticide residues in food in the environment and that other current laboratory test methods are not satisfactory substitutes for it. The duration of such a feeding study will depend upon the toxicity and magnitude of pesticide residues in the environment as influenced by the physical, chemical and biological properties of the pesticide.

Suggestions for Conducting Laboratory Dietary Feeding Toxicological Tests with Birds

Declining Residues in the Environmental Food

Laboratory tests with caged birds are used to help to predict field pesticidal effects. In published laboratory studies the concentration of the pesticides in the bird diet is kept at a constant level; such tests are used as part of the evaluation

throughout the exposure period. When such tests are used as part of the evaluation of the *long-term effects of pesticides on birds* in the field it should be taken into consideration that for pesticides where biological magnification in food chains is not involved, that the *pesticide residues on plant foods often decline to less than 1 ppm* in less than a week with the use of many pesticides and to less than 10 ppm in a few weeks even with many uses of the so-called persistent *chlorinated hydrocarbon insecticides* as shown in Table 4. Such pesticide residues on plants often decline rapidly, even with frequently repeated applications, and are not found at the same unchanging residue concentrations day after day, as are used in most laboratory toxicology studies. The use of such laboratory toxicological tests in field toxicity predictions should be considered in light of this added safety factor, or the pesticide concentration in the laboratory diet should be reduced to match the disappearance of the pesticide residue in natural bird food in the environment where the pesticide is applied.

Metabolites in Environmental Food

The ratio of the residues of a pesticide to its metabolites or degradates in bird food varies with time and depending upon the environmental stress conditions to which the food is subjected. In laboratory tests the pesticide diet is usually prepared by using dry food, thus, the rate of change into metabolites or degradates may be considerably less rapid or different than in the birds natural foods. Some analytical knowledge of the rate of these changes in wild bird food is helpful in simulating such dietary tests in the laboratory.

Food acceptance

The quantity of pesticide residues consumed in the diet of birds is a function of ppm of pesticide in the diet and the amount of food eaten. When the rate of daily dietary consumption is reduced by birds eating treated food compared to untreated food it is important to determine whether such reduction is due to toxicity or repellancy. *Measurements of weights of birds and of food consumed* by untreated control birds are absolutely necessary for comparison and determination of these factors. Other more subtle tests may be necessary to determine sublethal effects.

Size of Bird

Because of the wide difference in mg/kg/day intake of pesticides between small and large birds, it would seem desirable to *use small birds* such as the young of large species or small species of adult birds for measuring the pesticide residue that birds will tolerate in their diet.

Conclusions

1) Body weights of birds can be estimated reasonably to within ± 20 % of the actual weight when not known exactly, by use of available literature values.

2) The quantity of food eaten per day by birds can be reasonably estimated when not known exactly. In general, species weighing over 100 grams eat less than 20 % of their body weight in food per day, and those under 10 grams may eat as much as 60 % in dry weight food per day. Based on a linear *relationship of body weight to food intake*, percent of body weight eaten per day can be estimated within a 2-fold factor, as shown in Fig. 1.

3) The *upper limits of* the amount of *residues of a pesticide* on various *species, shapes, and sizes of plant food* and insects immediately after application can be reasonably estimated. In general,

the highest residues may be expected on treated food particles which have the highest surface area to weight-volume ratios and greatest exposure to the pesticide application. The upper limits of pesticide residues on plants and their *decline with time* on various types of crops, as correlated from data on many pesticides, is given in Table 4.

4) Based upon the information developed in this report on a unit basis, a 1 pound per acre dosage of a non-cumulative pesticide could result in a 35-fold difference (7—240 ppm) between *maximum pesticide residues* in different sized natural bird food particles. Also a 20-gram bird of nearly any species will likely eat 5 times as much food in mg per kg of body weight per day as a 1,000-gram bird of nearly any species. Thus, it is possible that a 20-gram bird could eat 100 times or more pesticide than a 1,000-gram bird, in terms of mg/kg/day. Variations in concentrations and dosages of pesticides and variations in food consumption rates of different weights of birds can be estimated as shown in Table 6 to obtain specified mg/kg/day data. As has been shown, the maximum residues estimated here rarely occur in nature and therefore mitigating circumstances of stability of the residues, food habits of the birds, etc. would result in corresponding reduction of this estimate.

5) The estimated or calculated mg/kg/day of intake of pesticides from residues on bird food should be matched with toxicological responses from similar dosages in laboratory or field test results with birds, preferably *by use of ad lib dietary feeding tests,* rather than *acute oral,* injection or other less correlatable *tests,* not as suitable for environmental interpretation. Subacute and *chronic* laboratory *toxicity tests* are

often conducted with diets containing a constant concentration rate of pesticides over a period of days, weeks, or months. Interpretation of such laboratory tests should take into consideration the frequently large and quick *rate of decline* in pesticide residues which occur on natural bird food, or the dietary intake of birds in such tests should be adjusted accordingly. In addition, the test methods should provide for the study of metabolites of the pesticide or other derived molecules if any occur. *Food acceptance* of the pesticide in the diet of birds and the relative toxicity of the pesticide to representative species of birds must also be taken into account.

References

Altman, P. L., D. S. Dittmer: Biology Data Book. Biological Handbooks. Fed. Amer. Soc. Exper. Biol. Washington, D.C., p. 101, 1964

El Sayed, E. I., J. B. Graves, F. L. Bonner: Chlorinated Hydrocarbon Insecticide Residues in Selected Insects and Birds Found in Association with Cotton Fields. J. Agr. Food Chem. 15 (6): 1014–1017, 1967

Hoerger, F. D., E. E. Kenaga: Pesticide Residues on Plants. Correlation of Representative Data as a Basis for Estimation of their Magnitude in the Environment. Environmental Quality. Academic Press, New York 1: 9–28, 1972

Kenaga, E. E., W. K. Whitney, J. L. Hardy, A. E. Doty: Laboratory Tests with *Dursban®* Insecticide. J. Econ. Ent. 58 (6): 1043–1050, 1965

Kendeigh, S. C.: Personal Communication. 21 March 1969 a

Kendeigh, S. C.: Tolerance of Cold and Bergman's Rule. The Auk 86: 13–29, 1969 b

Krishna, J. G., J. E. Casida: Fate in Rats of the Radiocarbon from Ten Variously Labelled Methyl and Dimethylcarbamate C^{14} Insecticide Chemicals and Their Hydrolysis Products. J. Agr. Food Chem. 14 (2): 98–105, 1966

Nice, M. M.: The Biological Significance of Bird Weights. Bird-Banding 9 (1): 1–11, 1938

Oonnithan, E. S., J. E. Casida: Oxidation of Methyl and Dimethylcarbamate Insecticide Chemicals by Microsomal Enzymes and Anticholinesterase Activity of the Metabolites. J. Agr. Food Chem. 16 (1): 28–44, 1968

Salisbury Laboratories: Salisbury Poultry Product Use Guide. Charles City, Iowa 1968

Stickel, W. H., D. W. Hayne, Lucille F. Stickel: Effects of Heptachlor-contaminated Earthworms on Woodcocks. J. Wildlife Management. 29 (1): 132–146, 1965

Tucker, R. K., D. G. Crabtree: Handbook of Toxicity of Pesticides to Wildlife. Fish and Wildlife Service, USDI, Resource Publication No. 84. Denver Wildlife Research Center, Denver, Colorado 1970

United States Department of the Interior: Pesticide-Wildlife Studies. A Review of Fish and Wildlife Service Investigations During the Calendar Year. Circular 199. pp. 98–106. Fish and Wildlife Service, USDI. Washington, D.C. 1964

Welty, J. C.: The Life of Birds. Knopf, New York 1963

Williams, E., R. W. Meikle, C. T. Redemann: Identification of Metabolites of Zectran Insecticide in Dog Urine. J. Agr. Food Chem. 12 (5): 457–461, 1964

The Effect of Air-Polluting Gases on Plant Metabolism

I. Ziegler

Botanisches Institut der Technischen Universität, Forschungsgruppe für Biochemie der Gesellschaft für Strahlen- und Umweltforschung mbH, 8 München 2, Arcisstraße 21, West Germany

Summary. Among the air-polluting gases, SO_2, ozone, peroxyacetylnitrate (PAN) and fluorine are those whose action is studied most. This review tries to show the connection between the well-known macroscopic symptoms, on the one hand, and the primary points of attack at the enzymatic level, the changes in the plant's metabolism, and the microscopic and electronmicroscopic results, on the other. PAN and ozone, which originate through the action of sunlight on auto-exhausts, cause the strong oxidizing character of this type of smog. Their primary point of attack seems to be their oxidizing effect on protein SH-groups. PAN in special oxidizes the SH-groups of a photoreducible disulfide containing chloroplast protein, thus blocking photosynthesis. SO_2, which originates from combustion of coal and petroleum as well as from roasting of sulfur-containing ores, causes the reductive character of this type of smog. SO_2 has a special position among the air-polluting gases because it can be incorporated without damaging effect into the normal sulfur metabolism up to a certain level. After exceeding this limit, it causes a rapid depression of photosynthesis. Its primary point of attack thereby, however, is scarcely known. F^- is bound as a salt in the cell wall or in the cell vacuole and is thereby prevented from its damaging effect on metabolic processes up to a certain level. Upon exceeding this, it acts mainly on the enzymes of carbohydrate metabolism. The action of air-polluting gases on the permeability of cell membranes will deserve special interest in the future. In a few examples it is shown in which way the collapse of cell compartmentation causes the loss of regulatory mechanisms of the cell. The influence of internal (genetic conditions, physiological age etc.) and external (light, temperature, humidity etc.) factors on the general metabolism, and, in this way, on the sensitivity of the plant to air-polluting gases, is shown.

Zusammenfassung. Unter den Schadgasen sind bislang SO_2, Ozon, Peroxiacetylnitrat (PAN) und Fluor in ihren Wirkungen auf Pflanzen relativ am eingehendsten untersucht. Es wird in der vorliegenden Übersicht versucht, zwischen den primären Angriffspunkten im enzymatischen Bereich, den Veränderungen im Gesamtstoffwechsel sowie den elektronenmikroskopischen und mikroskopischen Befunden einerseits und den eingehend studierten makroskopischen Schadsymptomen andererseits eine Verknüpfung zu finden. PAN und Ozon, die durch Einwirkung des Sonnenlichts auf Autoabgase

entstehen, sind die Hauptbestandteile eines Smogtyps von stark oxidieren-
dem Charakter. Ihr primärer Angriffspunkt dürfte in der Oxidation von
SH-Gruppen liegen, wobei PAN speziell die SH-Gruppen eines photoredu-
zierbaren Disulfid-Proteins der Chloroplasten oxydiert und dadurch die
Photosynthese hemmt. SO_2 verleiht dem aus Kohle- und Erdölverbrennung
sowie aus der Erzröstung stammenden Smog einen reduzierenden Charakter.
Es nimmt eine Sonderstellung ein, da es bis zu einem gewissen Grad in den
normalen Schwefelstoffwechsel eingeschleust werden kann und erst bei Über-
schreitung einer Toleranzgrenze vor allem eine Assimilations-Depression
hervorruft. Sein Angriffspunkt hierbei ist jedoch noch weitgehend unbe-
kannt. Auch F⁻ bleibt in niedrigen Konzentrationen als Salz in der Zellwand
oder der Vakuole gebunden und damit dem schädlichen Einfluß auf den
Stoffwechsel entzogen. Nach Überschreiten der Toleranzgrenze hemmt es
vorwiegend die Enzyme des Kohlenhydrat-Stoffwechsels. Eine stärkere Be-
achtung wird in Zukunft die Wirkung der Schadgase auf die Permeabilität
der Zellmembranen verdienen. An einigen Beispielen wird gezeigt, wie
der Zusammenbruch der Zellkompartimentierung den Verlust von Regu-
lationsmechanismen der Zelle verursacht. Einflüsse innerer (Erbanlagen,
physiologisches Alter u. a.) und äußerer (Licht, Temperatur, Feuchtigkeit
u. a.) Faktoren auf den Stoffwechselzustand und damit auf die Empfindlich-
keit der Pflanzen gegenüber Schadgasen werden besprochen.

The Composition of Air

The Composition of Uncontaminated Air

The medium, in which the leaves of higher plants perform their fundamental processes of assimilation and respiration, consists mainly of nitrogen (78.09% or 780 900 ppm[*]), oxygen (20.94% or 209 400 ppm), argon (0.93% or 9 300 ppm), and carbon dioxide (0.03% or 315 ppm). The remaining 0.01% is composed of neon (18 ppm), helium (5.2 ppm), methane (1—1.2 ppm), as well as krypton, nitrogen oxide, hydrogen, xenon, nitrogen dioxide, and ozone (each ≤ 1 ppm)[1]. In addition, there are varying amounts of water vapor. This "natural atmosphere" is steadily changing under the influence of civilization. These changes deeply affect the plant kingdom, since in

[*] The information about the gaseous components is always valid for ppm per volume.

plants not only their dissimilatory processes but especially their assimilation are dependent on the environmental air. Moreover, the individual plant cannot escape from an unfavorable environment as animals can.

The Air-Polluting Compounds

The type and concentration of air-contaminating compounds have already been documented in numerous publications and review articles. In West Germany, a network of stations controlling air-polluting material is run by a Commission of the Deutsche Forschungsgemeinschaft (DFG). This consists of 5 monitoring units in "normal air" and 2 units in industrial areas with polluted air (Gelsenkirchen and Mannheim). These register SO_2, CO_2, fluorine compounds and the total dust content of the air; in addition, the radioactivity and the pollen and spore content of the air is recorded. The results of these measurements are documented by communications from the Secretary of the DFG.

For a detailed review, see articles by Georgii[1] and Tebbens[2]. The following information summarizes very briefly the compounds and the concentrations with which we have to deal.

In respect to the degree of dissemination and the surface area covered, two groups can be destinguished. First, the almost ubiquitous pollution caused by both types of "smog", the "SO_2-smog" (also called "London smog") and the "ozone smog" (also called "Los Angeles smog"). Next are the emissions, mostly from industrial plants, that intensively affect the flora through their high concentration and specificity but which are usually locally restricted.

The Ubiquitous Pollution from Smog

Both "smog-types" are fundamentally different in composition, origin, and effect (Table 1). As a result of its origin from auto exhaust, the "Los Angeles smog" is originally composed of CO, NO and NO_2, as well as a large number of unsaturated and aromatic hydrocarbons. The CO-level of a large city like Los Angeles averages about 10—12 ppm; on "smog-days", up to 60 and can even rise locally and transiently to 120 ppm[2,4]. The rush hour traffic in the large cities

of Western Europe, as in Bochum, also causes maxima up to 88 ppm[3]. The concentration of NO usually amounts to somewhat less than 1 ppm, but can reach 2 ppm on smog-days[2,4], and 3.93 ppm for a maximal concentration[5]. The concentration of the hydrocarbons, which usually averages 2 ppm in Los Angeles, can rise for some hours to 4—9, and for several minutes even to 20—40 ppm. By thin layer chromatography, gas chromatography, and UV- and fluorescence spectroscopy[6], numerous substances could be identified and their list already includes more than 80 compounds[2]. About 60% of the total hydrocarbons is composed of the C_1—C_7 members of the aliphatic series, including single and double unsaturated compounds.

Among these are propylene with 14.29% and ethylene with 9.26%. Among the aromatic hydrocarbons, benzene (6.11%), toluene (8.2%), and p-xylene (6.25%) are predominant[7]. Other hydrocarbons, such as cresol, phenol, phenanthrene, perylene (from diesel engine), or benzanthracene, as well as chlorinated hydrocarbons and lead tetraethylacetate are found only in traces, but because of their partially strong poisoning effect they are nevertheless important contaminating factors.

Some of them do not belong to the gaseous air pollutants in the strictest sense, but rather to the aerosols and dusts (see below). A list of such components has been compiled by Corn[8] and Korte et al.[9].

Table 1   Characteristics of "SO_2-smog" ("London smog") and "ozone-smog" ("Los Angeles smog")[1]

Characteristics	Los Angeles smog	London smog
Air temperature	25 to 35 °C	−3 to +5 °C
Relative humidity	< 70%	> 80%
Type of inversion	down draft inversion	radiation inversion
Wind velocity	< 2 m/sec	< 2 m/sec
Most common occurrence	July–October	November–January
Reaction character	oxidation	reduction

Sunlight (especially the UV-part) affects the primary components of smog. As a result of *photochemical reactions,* ozone and a series of peroxyacylnitrates (RCC (:O)OONO$_2$), of which *peroxyacetylnitrate* (PAN) is the most important compound, originate from the mixture of olefines and nitrogen oxides. Besides these, aldehydes, especially formaldehyde, are formed. The reaction sequence, which leads to these secondary products, is dependent on many locally different factors, like the composition and concentration of the primary exhaust gases or the energy of the irradiated light. These reactions have been thoroughly studied in the last few years[10]. While these secondary reactions overwhelmingly lead to ozone and PAN, the "Los Angeles smog" takes on a strong oxidizing character. The total amount of oxidants on smog-days reaches 0.75 ppm[2]. The total concentration of aldehydes in cities fluctuates between 0.04 and 0.18 ppm, but can rise to 0.27 ppm[1].

The *"London smog"*, on the other hand, has a reducing nature because of its *SO$_2$-content*. It arises from the combustion of coal and petroleum.

Therefore, besides domestic uses, its sources are all industrial complexes, which utilize coal or petroleum for energy production. An extensive amount of SO$_2$ is produced by a variety of industrial processes, such as the roasting of sulfur-containing ores. The SO$_2$-concentrations for Frankfurt on smog-days average 0.2—0.3 ppm and sometimes reach such values as 0.55 ppm over a period of hours. An average of 0.6 ppm was established for Bochum during the winter of 1962[3]. During the same time in December of 1962, the SO$_2$ level reached 1.4 ppm in Gelsenkirchen[1]. Especially with this smog type, the gaseous components are accompanied by the particles of fog and soot (see below).

The vast expanded air pollution of cities and industrial areas is composed of both smog-types in varying extents. Through the use of sulfur-free coal and petroleum, the air contamination from SO$_2$ has diminished in the last few years especially in the industrial Ruhr district[11]. But at the same time, the smog-type originating from auto exhausts has increased.

Locally Restricted Air-Polluting Compounds

In contrast to the two ubiquitous smog-types, there are locally limited air pollutions of which the *fluorine compounds,* especially hydrogen fluoride, play the most important role. They originate from aluminum and superphosphate factories, as well as industrial plants that process stone and earth. The fluorine content of the air in this environment reaches 0.018, and in some industrial districts, 0.08 ppm[2].

The pollution of the air by HCl from the alkali industry used to be serious, but since HCl can easily be recovered, it contributes only to a slight extent to the air contamination today.

In place of this, the *emission of HCl-vapor* from the burning of rubbish has increased, since the proportion of chlorine-containing synthetics, especially PVC, has grown there. The average load of a large city amounts to 0.016—0.095 ppm, based on Cl^{-4}.

The air pollution from industrial plants which release NH$_3$ or H$_2$S is negligible. Nitrogen oxides are liberated to some extent in the chemical industry by nitric and sulfuric acid production, as well as by the nitration of organic compounds. Their contribution, however, is spatially limited and is not so significant in comparison with that originating from the auto exhausts (see above). The same is true for the release of ethylene from the

coke industry, as well as for phenol, pyridine, and its derivatives, and many
other heterocycles in tar vapor, which originate from the roofing industry. A survey of the 3,4-benzpyrene content of the
air in Düsseldorf has been furnished by
Schlipköter[3]. It reaches a maximum value
in winter of 838 µg/1000 m³.

Aerosols and Particulates

A wide and general use of the pesticides
(fungicides, insecticides, herbicides) in
the last few years has led to an enrichment in the soil, water, and air. Insecticides are found to be 40 pp 10^{12} in the
London air, as well as 231—402 pp10^{12}
in rain water[9]. Together with the hydrocarbons they are not considered gaseous
air pollutants, but rather components of
the aerosol. Increasing particle size finally results in the various dusts and dirts.
Their physical characteristics are reported in detail by Katz[4] and Corn[8]. While
in open country the dust sediment yielded
a maximum of ≤ 0.65 gm/m²/day in 1967/
68, it increased, for example, for the area
between Duisburg and Dortmund to
0.66—1.3 gm/m²/day and reached a value
of 1.3 gm/m²/day in certain centers of
this area[11].
In these dusts are found such metals as
lead, zinc, iron, cadmium, etc., according
to the type of industry[8]. An important
source of the lead content of the air is
lead tetraethyl; in this form about 225 x
10^6 kg of lead are released into the air
by auto exhausts in the USA. About 50%
falls within 30 m of the street on the
bordering land; the rest is distributed
over a larger area[12].
Table 2 may illustrate better than the
concentrations mentioned above to what
extent the total atmosphere is charged
with exhaust.
The following survey will deal only with
the effect of the air pollutants which are

Table 2   Pollutants emitted in USA in 1968 (in
millions of metric tons)[157]

Parti-culates	S-oxides	N-oxides	CO	Hydro-carbons
8.1	22.1	9.1	1.7	0.6

assimilated by the plants from the air
with the help of the stomata or by cuti-
cular absorption, but not with those taken
up from the soil by root absorption. It is
largely limited, therefore, to the gaseous
compounds and to the components of
aerosols.

The Mechanism of Action of Air-Polluting Compounds on Plants

To trace the mode of action between the
primary points of attack and the complex
symptoms in the whole plant, the components of air pollution will be grouped
here not according to their common occurrence, but rather acording to their
chemical character.

Ozone and Peroxyacetylnitrate (PAN)

Effect on Enzymes and Coenzymes
Ozone and PAN distinguish themselves
through their strong oxidizing effect. The
redox-potential of ozone is $+ 2.07$ V; that
of PAN exceeds $+ 0.4$ V[13]. PAN oxidizes
the cystein moiety of glutathione; the
oxidation of GSSG then partially proceeds to cysteic acid[13, 14, 15]. The active
form of ribonuclease is not inactivated by
PAN, but its presence hinders the ability
of the enzyme to recover its activity when
it is fumigated in the reduced (SH-) form.
The susceptibility of many proteins (e. g.
hemoglobin or papain) depends largely
on the number of SH-groups which are
available to the access of PAN. This in
turn, changes with the pH-dependent
configuration[16]. Catalase, peroxidase,
and urease are substantially less sensitive

to ozone than papain; nevertheless, an average of 0.3 mM is enough for 50% inactivation. Ozone is therefore about 30 times more effective than H_2O_2[17].

The *inactivating effect* of PAN (and the somewhat weaker effect of ozone) on phosphoglucomutase and on the UDPG-utilizing, polysaccharide synthesizing system of the cell wall is probably due to its oxidative effect on SH-groups[18, 19]. Also the inactivation of NADP-dependent isocitrate-dehydrogenase, glucose-6-phosphate-dehydrogenase, and of the malate dehydrogenase seems to be due to this oxidative effect[14]. The reduced pyridine nucleotides and of course all reduced members of photosystem II are also oxidized by both compounds[20, 21]. Mudd[20] points out that the oxidation products formed by the action of ozone on NADAPH and NADH, can, in addition, inhibit the enzyme reaction for which the reduced coenzymes would be needed. PAN also oxidizes indole-3-acetic acid. One oxidation product was identified as 3-hydroxymethyloxindole[22].

Effect on Complex Metabolic Processes

The *inhibition* of phosphoglucomutase and of the polysaccharide synthesizing system by gassing with PAN in vitro corresponds to the drastically reduced incorporation of $^{14}CO_2$ into cellulose and glucan, when sections of Avena coleoptiles are incubated in the presence of 2.4 ppm PAN[23,24]. If the sections are treated with PAN in vivo, the extracted enzymes prove to be almost as strongly inhibited as by in vitro incubation with the toxic material. As a consequence, the cell elongation is completely stopped.

Besides, this growth of oat coleoptiles is blocked by oxidation products of indole-3-acetic acid which resulted from the action of PAN (see above)[22].

In vivo incubation with ozone results in a much weaker inhibition of the polysac-charide synthesis and the cell elongation. The phosphoglucomutase subsequently extracted from the coleoptiles treated in vivo is not blocked[19]; the polysaccharide synthesizing enzyme system is only slightly affected.

Nevertheless, week-long fumigation of citrus fruits with 0.25 ppm ozone shows a remarkable decrease in starch content, while the amount of reducing sugar is increased[25].

If a suspension of tobacco mitochondria is gassed with ozone (2.4 µmol/sec), after 15 seconds the oxidative phosphorylation is completely inhibited and the O_2 uptake is blocked to 70%. Also mitochondria, which are isolated from in vivo fumigated plants (1 ppm/1—5 hours), show a 75% inhibition of the oxidative phosphorylation and a 30% inhibition of the O_2 uptake[26]. Mitochondria from fumigated tobacco leaves (0.7 ppm/1 hour) exhibit an even greater inhibition of the O_2 uptake with succinate, than as with NADH as a substrate for the oxidation. This may be further evidence that the respiration inhibition is a consequence of the oxidative effect on the SH-groups of the enzymes involved[27]. If glutathione or ascorbic acid are added to isolated cabbage or spinach mitochondria before the ozone treatment, they can to a high degree prevent this oxidative effect and thereby the respiration inhibition (with citrate as substrate). They can only partially cancel the inhibition of respiration when their addition follows the ozone treatment[28].

The final increase in respiration when the leaf tissue already shows visible damage, seems to be much more complex (cf. p. 198). Furthermore, the oxidation of NADPH by PAN or ozone has the consequence, that the incorporation of ^{14}C-acetate into the lipids of the chloroplasts is reduced by half, while the NADPH-independent incorporation into the citric acid remains unaffected[29].

The *effect* of both oxidants *on photosynthesis* is especially interesting. If isolated chloroplasts of Phaseolus are gassed, it makes no difference for ozone or PAN whether this pretreatment takes place in the dark or in the light; the Hill-reaction and the photophosphorylation are extensively inhibited [30, 31]. This damage can not be reversed by following with a N_2 gassing, washing of the chloroplasts, or by the addition of glutathione. This indicates that the inhibition by gassing of isolated chloroplasts is directly dependent on an irreversible oxidation of functional SH-groups in the enzymes, beyond the disulfide stage, or on an oxidation of components concerned with the photosynthetic electron transport system [31].

The results are different when intact leaves are fumigated instead of isolated chloroplasts. *Ozone inhibits* the *photosynthesis* (O_2-development, NADP-reduction) of the chloroplasts isolated after treatment not only when they are exposed to light during fumigation, but also (and even to a greater extent) when the leaves were in the dark 48—72 hours prior to the fumigation. The *inhibition of photosynthesis by PAN* (particularly the photophosphorylation) is, on the contrary, absolutely dependent on the presence of light during and after the fumigation of the intact leaves [32, 33]. There is good evidence (cf. p. 200), that this light-dependent damage of the photosynthesis by PAN in vivo, is primarily caused by oxidizing the SH-groups of a disulfide-containing photoreducible chloroplast protein. If predarkened Phaseolus leaves are exposed to light, then in the course of 2 hours the amount of SH-groups (determined by titration increases in the same proportion (to 30%) as the sensitivity to PAN increases; for its part, fumigation with PAN reduces their number [21].

This photoreduction of a disulfide-protein is closely connected with the process of photophosphorylation [34]. Asahi and Masaki [35] partially purified such a protein with disulfide-bonds, which are reduced by exposure to light.

The action spectrum of the damage caused by the presence of PAN, however, does not completely correspond to the absorption spectrum of the chlorophyll. Maximum morphological damage of Phaseolus vulgaris in the presence of PAN is obtained at wavelengths of 370, 420, 480, and (with a small peak) at 640 nm [33]. It is possible that auxiliary pigments are involved or that either photosystem II or photosystem I + II are operating, depending on the wavelength. Thus, besides its oxidative effect on photoreduced SH-groups, PAN may find in vivo quite a number of active sites for an oxidative effect in the course of photosynthetic electron transport. These points of attack may vary with wave length [21].

Fumigation of Phaseolus vulgaris or of spinach chloroplasts with PAN and subsequent measurements of the $^{14}CO_2$ fixation even drastically reduces the rate of fixation, but without an alteration of the fixation pattern [36]. After 15 minutes of ozone fumigation (1 ppm), $^{14}CO_2$-fixation has been reduced to 34%. As a consequence, pulse labeling results in an impaired incorporation of ^{14}C into the free amino acids and the amino acids of the protein hydrolysates [37].

However, further metabolism not only seems to be altered quantitatively but also qualitatively before visible damage appears. By the action of natural smog or ozone, formic acid, acetic acid, and lactic acid accumulate in spinach [38] and, leucine, valine, phenylalanine, and other amino acids in Phaseolus [37]. The accumulation of γ-aminobutyric acid in Phaseolus shows the *complex effect* of fumigation on *further metabolism*. This γ-aminobutyric acid originates from glutamic acid (which correspondingly decreases), as a result of the enhanced *membrane permeability*. This allows the free glutamic acid in the mitochondria and the chloroplasts to diffuse into the cytoplasm, where glutamic acid decarboxylase is found [37]. There is no detailed knowledge of the correlation in the processes of these long range changes (fumigating 1 hour or more). It is possible, as in the case of fluoride treatment (see below), that an activation of the phosphoenol-pyruvate-carboxylase and an increase of the reductive amination occurs.

Leaves impaired by ozone or auto exhaust finally show a red-brown discoloration, which, for example, in Rumex crispus is accompanied by the appearance of anthocyanin (cyanidin glucoside [39]), or in Phaseolus by the accumulation of caffeic acid [40]. These visible phenomena (cf. p. 195), which are also caused by plant pathogens or other "stress factors" are not yet understood.

Effect of the Structure of the Cells

A central point of attack for the metabolic changes as well as for the following *morphologic changes* may be the oxidizing effect of both compounds on the $C=C$-double bonds of the unsaturated fatty acids in the cell membrane [41]; in addition, the SH-groups, which are necessary for the biosynthesis of the long chain fatty acids in chloroplasts, are destroyed. The incorporation of ^{14}C-acetate into the lipids is blocked [29] (see above).

Felmeister et al.[42] demonstrated the detrimental effect of a mixture of NO_2, trans-2-butene, and air on artificial membranes made from lecithin and unsaturated phospholipids.

The first visible result of the effect on membrane permeability is the osmotically irreversible swelling of the mitochondria which occurs after 50—60 seconds gassing of a suspension of tobacco mitochondria (1.7—3.4 mol/sec ozone); at the same time, protein and nucleotides are released [43]. Thus, this change in permeability immediately follows the primary effect, which is the inhibition of the oxidative phosphorylation (see above). Further effects of the change in permeability [44] may include the increased uptake of ^{14}C-glucose by sections of fumigated citrus leaves (0.15—0.25 ppm ozone), their higher Q_{O_2}, and their increased $^{14}CO_2$ development [45] or the appearance of γ-aminobutyric-acid (see above) as a result of the collapse of the cell compartmentation. These important regulation mechanisms of the cell metabolism and their interruption by air polluting gases will certainly deserve more attention (see p. 198).

The first damage after ozone or PAN treatment observed by the electron microscope, is a granulation of the chloroplast stroma. There the granula gradually arrange themselves in continuous rows. In the second phase there is a general disintegration of the thylakoid system and the cell membrane [46, 47]. As yet the results do not allow a correlation between the metabolic ability, the remaining compartmentation of metabolism, and the electron microscopic picture during the course of fumigation. The final aggregation of the cell contents in the center of the cell [47], the decomposition of the chlorophylls[48,49], and the accumulation of phenols (see above), finally result in the morphologically visible damage pattern. Its appearance and its variation in dependence on the length and intensity of the fumigation, the age, and past life of the plant have been reported in many publications and reviews [50, 51].

It is remarkable that ozone and PAN, which prove to be very similar as oxidants in their primary effects, lead to such different damage patterns in the plant. PAN especially harms the cells of the spongy parenchyma and causes, thereby, a glossy, bronze discoloration of the lower surface of the leaf; while ozone leads to a spotting and decoloration of the upper surface of the leaf through a major destruction of the palisade parenchyma.

Hydrogen Fluoride

The Fate of Fluorine in the Plant

The hydrogen fluoride of the air penetrates mainly through the stomates, but can also enter through the bark (lenticells?) and be stored there[7]. As experiments with ^{18}F have shown, it is absorbed onto the leaf surface, from where it can again be washed off[52]. Especially at night time when the stomates are closed, the damaging gas is absorbed to a relatively large extent onto the surface of the plant[53]. A spontaneous reliberation can result from evaporation[54]. The fluorine ion collects finally in the cell vacuole as Ca- or Al-salt or in the cell wall[55]. The content of fluorine in the plant averages 7—16 ppm (based on dry weight). However, there are plants such as Camellia which normally contain 57—355 ppm fluorine, and even as much as 1370 ppm. The fluorine can be present in such a form as fluoracetate or fluoroleic acid[52].

The *sensitivity* of a given species *to HF* also varies greatly; i.e. the extent to which fluorine can be deposited in the cell without damaging it (see p. 199).

It is not known in which compounds the F^- supplied by fumigation is incorporated. In dicots these compounds are transported to the margin of the leaves; and in monocots, to their tips.

Thus, the damaging concentrations accumulate there, while the remaining parts of the leaves remain relatively unaffected[50]. From 20 days of constant fumigation of the bush bean, the normal F-content of 1.0—6.9 ppm (based on dry wt.) can rise to 107.3 ppm[56]. The distribution

of the accumulated fluorine between the inner tissues of the leaf and the surface seems to vary in different plants. In the relatively resistent cotton, there is evidently a continual exchange between both of these areas[57]. A further transport of the fluorine in the conducting tissue, as in the case of sulfur (see below) does not seem to take place[53, 58].

Effect on Enzymes

The best known effect of the F^- *ion* may be the *inhibition of enolase*[59]. There are, nevertheless, a series of enzymes which are inhibited by F^- as a result of the binding of Mg^{++} or a divalent heavy metal[52]. Ordin and Skoe[60] found a strong inhibition of the phosphoglucomutase if Avena coleoptiles were incubated with NaF. UDPG-pyrophosphorylase remains largely unaffected[61].

An *activation of glucose-6-phosphate dehydrogenase* was established by Arrigoni and Marré[62], when extracts from the tips of germinating Pisum sativum plants were incubated with NaF. Addition of a low concentration of NaF to extracts of Glycine max activates the phosphenolpyruvate dehydrogenase[63]. This agrees with the fact that dark fixation of $^{14}CO_2$ in HF-fumigated plants is enhanced[63].

Since fluoride is separated from the plasmalemma by its accumulation in the vacuole and in the cell wall (see above), it has little influence on the metabolism up to a certain level. However, homogenization of the tissue destroys the compartmentation of the cells and brings the fluoride in contact with the enzymes. This has to be taken into account and the enzyme inhibition found in vitro may not always be concluded for the in vivo conditions.

Effect on Complex Metabolic Processes

The inhibitory effect of F^- on the phosphoglucomutase may be the reason that after fluoride fumigation of Glycine max there is less saccharose but more reducing sugar[64]. And it may explain why the IAA-stimulated Avena coleoptiles incorporate only a little ^{14}C-glucose into the cellulose and into the components of the water and acid soluble cell wall fraction (see above). This results in the strongly inhibited growth by the presence of 0.01 NaF[18, 60, 65]. The synthesis of starch in pine needles also decreases from HF-fumigation[66].

Besides this, fluoride causes *growth retardation* by slowing cellular multiplication and expansion in IAA-stimulated corn seedling roots. Thereby total RNA content is related to the growth rate[67]. Further analysis has shown that the amount of both free and bound ribosomes is reduced. Especially in bound ribosomes the RNA : protein ratio is modified[68]. In contrast to earlier results[69] the base composition, however, does not seem to be changed[68].

The strong inhibition of the enolase by fluoride has the consequence that the decomposition of fructose diphosphate is fluoride sensitive in seeds, where exclusively the glycolytic pathway operates[70]. In contrast the decomposition of fructosediphosphate is little affected in the leaves, where it proceeds largely to glucose-6-phosphate, phosphogluconate and then by the pentosephosphate cycle[71]. Fluoride fumigation of Chenopodium murale or Polygonum orientale (5–6 days with 6 ppm or treatment of the leaves with KF [2 days, 5×10^{-3} M]) leads to substantially more ^{14}C being released from the C_1 of glucose than from the C_6; i.e. the pentosephosphate pathway is almost exclusively used[72].

In all these experiments the *respiration was stimulated*. This originates from the fact that the permeability for glucose is enhanced in the leaf disk incubated with fluoride. In some cases, e.g. beans, the O_2 uptake can be greatly increased already before a visible damage occurs[56].

McNulty and Lords[73] trace back the increased respiration from fluoride, however, to an interference with the regulation by means of the PO_4-ADP-ATP system.

Nevertheless, the respiration is generally not seen to increase before necrosis starts in the adjacent tissue[55]. This may depend on an unspecific reaction due to the tissue lesion (see p. 198). HF-fumigation results in a large accumulation of malonic acid, malic acid, and citric acid at the time of the beginning necrosis. It is not known whether this is due only to an increased activity of the phosphoenolpyruvate-carboxylase (see above), or whether there is also more phosphoenolpyruvate produced as a result of increased diffusion or respiration or from inhibition of a phosphatase.

The acids are finally converted into amino acids, especially into aspartic acid (and aspargine), glutamic acid (and glutamine), and serine[64].

The *photosynthesis* does *not* seem to be directly *affected* by a low HF concentration (1—12 pptm). Rather, in most cases, the loss of the photosynthetic O_2 development parallels the appearance of visible necrosis. The bordering green tissue remains unaffected[50]. However, higher concentrations (500—1000 pptm) reduce the photosynthesis about 40% before visible damage appears[50].

Finally, the *synthesis of chlorophyll a* and *b* and even that of the protochlorophyll is *inhibited* by fluoride fumigation. As soon as Mg is incorporated into the ring structure, there is no more inhibition; i.e. the light dependent transformation of protochlorophyll to chlorophyll proceeds normally. However, since the carotene formation also ceases, McNulty and Newman[74] assume that the decomposition of the chloroplast membrane gives rise to the inhibition of the pigment synthesis (see below).

The Effect on the Structure of the Cell

The simultaneous inhibition of carotene and chlorophyll synthesis by F^- causes McNulty and Newman[74] to suggest that the *chloroplast structure* is decomposed. As of yet there are no detailed electron microscopic studies of the changes in the chloroplasts or of the structural changes accompanying the plasmolysis of the mesophyll[7]. The numerous macroscopic damage characteristics are described by Thomas[58] and Brandt and Heck[51].

Sulfur Dioxide

The Fate of SO_2 in the Plant

Expecially in the case of SO_2, fumigation of isolated enzymes and cofactors says little about its action on the metabolism in vivo. The reason for this may be that sulfur is itself a component of the plant and that it is much more involved in normal metabolism than is fluoride.

There is no doubt that the SO_2, penetrating chiefly through the stomates is first dissolved in the tissue during formation of H_2SO_3. The dissolution proceeds gradually. Therefore, SO_2 can be recovered by distillation of freshly fumigated leaves or it may even be freed spontaneously[75, 76]. Thomas et al.[77], as well as Weigl and Ziegler[78] have shown that the acidic effect of the H_2SO_3 caused by low fumigation intensity (e.g. sugarbeet leaves 60 days/5.2 h daily/0.186 ppm[77]), can be buffered, whereby the buffer capacity of the leaves gradually decreases. In tissue cultures of animal cells, Thompson and Pace[79] showed that the buffer capacity increased with the serum content of the medium. It may be assumed, that also in the leaves the proteins are responsible for the buffering.

The sulfite ion is, at least partially, oxidized to sulfate. This possibly occurs with the help of the sulfite oxidase, which was found by Arrigoni[80] in pea mitochondria; it seems to be stimulated by ADP, and cytochrome c is reduced thereby. It is unknown how far this oxidation is done enzymatically and to what extent SO_3 is directly oxidized according to the redox condition prevailing in the cell. Using $^{35}SO_2$ it was found that after 2 hours of fumigation of spinach, about 18% and after 7 hours, 43% of the incorporated total activity was present as SO_4^{-}[78]. Besides in SO_4^{--}, the sulfur isotope is incorporated into organic compounds. These are cystein, glutathione and especially a compound unidentified until now. After a 2 hour fumigation this compound constitutes 10%, and after 7 hours, 15% of the total incorporation. It also appears when spinach leaves are provided with $^{35}SO_4$ through the leaf stalks or if Chlorella suspensions have assimilated in a solution containing $^{35}SO_4^{--}$[78]. Infiltration of Mung bean leaves with ^{35}S-labelled sulfate yields labelled cystine, methionine, and cysteine-sulfinic acid, as well as other unidentified compounds[81]. Thomas et al.[77] and Steward et al.[82] obtained cysteine and methionine in the protein hydrolysate of alfalfa by fumigating with $^{35}SO_2$ as well as by the application of $Na_2^{35}SO_4$.

The sulfate ion, from which the normal-S-metabolism starts, is known to be activated in a two step reaction[83]:

1) $$ATP + SO_4 \xrightarrow{\text{sulfurylase}} \text{adenosine-5-phosphosulfate (APS)} + PP_i$$

2) $$ATP + APS \xrightarrow{\text{APS kinase}} \text{3-phospho-adenosine 5-phosphosulfate (PAPS)} + ADP$$

PAPS is then reduced with the help of the SH-form of a disulfide protein ("fraction c [SH₂]"). In yeast, Neurospora, and others, the SH-form originates from the disulfide by the action of "Enzyme A" and NADPH[83, 84].

In irradiated chloroplasts the reduction is bound to the photosynthetic electron transport. Thereby sulfite, sulfide, cysteine and other S-containing compounds are formed. A sulfate reduction of 3 μmol/h/mg chlorophyll is obtained[85].

The reduction of sulfate to sulfite in the light either needs intact chloroplasts or a reconstituted system, which contains broken chloroplasts, the chloroplast extract and a phosphorylating system ($ATP + P_i$ + ferredoxin + NADP + glutathione)[85]. This indicates that ATP is required to activate sulfate by sulfurylase and APS kinase, yielding PAPS (see above). Both enzymes are present in the chloroplasts extract[85, 86]. The reducing agent either is NADPH or glutathione (or another sulfhydryl compound) and ferredoxin is seemingly required for the photosynthetic reduction of NADP[85].

For sulfate reduction in the dark, besides an ATP-regenerating system, reduced glutathione (or another sulfhydryl-compound) is required. NADPH cannot replace glutathione which itself seems to be the reducing agent in the dark. A reduction rate of 30 μmol/mg chlorophyll/h is obtained with this system[85]. From these results one may conclude that the sulfate reduction of green plants, taking place in the dark[58] is not achieved by a sulfhydryl-containing enzyme, as it is in yeast, Neurospora and others.

Thomas[58] and Asahi[86, 87] had already shown that the reduction to the sulfide step can also proceed if sulfite is the starting material instead of sulfate. According to Asada et al.[88], this sulfite reductase is a cytochrome-containing protein. The photoreduction of sulfite is also coup-

led to the photosynthetic electron transport by the need for reduced ferredoxin. It proceeds not only in intact chloroplasts but also in broken ones without further supplement needed for sulfate reduction[85]. In contrast to Asada et al.[88] in the system of Schmidt and Trebst[85] methylviologen cannot replace ferredoxin.

The results suggest that in chloroplasts, like in some microorganisms, the PAPS is first reduced to sulfite by a sulfhydryl compound like glutathione and then the sulfite is reduced to sulfite by a ferredoxin dependent sulfite reductase[85]. A sulfhydryl containing enzyme as in yeast or Neurospora (see above) does not seem to participate.

Thereby the reduction of PAPS needs 2 molecules of a sulfhydryl compound (HS-X), e.g. glutathione. In the first step a thiosulfonate arises, which in turn is reduced by a second molecule of glutathione[85].

$$1) \quad PAPS + HS\text{-}X \xrightarrow{\text{sulfotransferase}} \text{-}S\text{-}X\text{-}S\text{-}SO_3 + PAP$$

$$2) \quad X\text{-}S\text{-}SO_3H + HS\text{-}X \xrightarrow{\text{thiosulfone reductase}} HSO_3H + X\text{-}SS\text{-}X$$

There may be no doubt that SO₂ is incorporated in this way into the normal sulfur metabolism. ³⁵S is thus concentrated first in the leaf. Microautoradiographic investigations[78] show a distinct accumulation especially in the guard cells of the stomates.

Much less is found in the epidermis and even less in the mesophyll. It is established that sulfur supplied by the leaf[89], is distributed probably as sulfate through the whole plant by the conducting tissue[77,90, 91]. Microautoradiography also shows in addition to the phloem a marked accumulation of the activity around the xylem, probably in the parenchyma sheaths[78]. Finally, the sulfur isotope is found wherever proteins are accumulated, such as in cambium, in fruit, and in seeds[58]. At length it is also incorporated into the secondary plant metabolism and, for instance, it was found in the isothiocyanate moiety of the thioglycoside glucobrassicine of fumigated Brassica[92].

These results may explain the observation that weak fumigation with SO_2 over a longer period of time can cause an increase in yield[93]. The upper limit of the positive effect depends on many factors, like plant species, age of plant, nitrogen supply, etc.[51], and may lie around 0.2—0.3 ppm. The activation of photosynthesis in alfalfa and other plants[94], may also be connected with the intake of sulfur into the metabolism.

The Damaging Effect of SO_2 on Metabolism

The Effect of SO_2 on Enzymes and Coenzymes

Disulfide proteins are converted by H_2SO_3 to thiosulfonic acids and thiols. Thus disulfide proteins are decomposed by cleavage of the S-S bonds in the polypeptide chains[95]. Diastase[96], peroxidase[97], and catalase[98] are supposed to be damaged immediately from the action of SO_2. Also the cleavage of glycosidic bonds is inhibited by SO_2[94].

Bisulfite compounds like α-hydroxypyridinemethanesulfonate in concentrations of 10 mH inhibit phosphoendpyruvate carboxylase as well as NADH dependent malate dehydrogenase[99]. The action of sulfite on the β-carboxylation pathway of photosynthesis therefore deserves special interest.

As a nucleophilic agent even in very low concentrations (2.5 μg/ml) sulfite combines with NAD to give a hydro-pyridine-4-sulfonic acid. This in turn combines specifically with lactate- or malate-dehydrogenase in a stoichiometric amount to form an almost undissociated complex. Thereby, the hydrogen transfer is inhibited[100, 101].

Thiamine is cleaved by sulfite into the pyrimidine and thiazole moieties[102]. It is not known whether there is an influence of SO_2 fumigation on cocarboxylase in vivo or if an inhibition of decarboxylation of α-keto acids (pyruvate decarboxylase!) or of the cleavage of α-hydroxy- and α-keto acids (transketolase!) occurs. In vitro experiments have shown that sulfite treatment splits "active acetaldehyde" formed by the binding of acetaldehyde to thiamine pyrophosphate, into the pyrophosphate ester of the thiazole and the pyrimidine moiety. The acetaldehyde group seems to be associated with the thiazole part[103].

The Effect on the Metabolism and the Composition of the Plant

When the buffer capacity of the organism is exhausted, as well as the possibility for incorporation of the sulfur into organic compounds, the assimilation, respiration, and transpiration are affected before visible damage can be observed. The threshold value for this is determined by the product of fumigation time and intensity. Many attempts were made to calculate this product[51].

It is easy to see from the fate of SO_2 in the plant (p. 192), that the capacity of the plant to incorporate SO_2 into the metabolism in vivo depends on many partial processes, which are to some extent still unexplained; and moreover, it depends on so many factors that can influence these steps, that such a mathematical treatment can only furnish approximate values. Short, sporadic fumigation with high SO_2 concentrations especially makes the meaning of the product of time and intensity relative, because of the momentary overload of the system.

As the first consequence of the SO_2 damage, there is a depression of the assimilation. Most of the measurements were made with the infrared recorder. In numerous cases it was shown that the degree of the assimilation inhibition as well as the ability to regain a normal assimilation level after fumigation has stopped, are dependent upon the concentration of SO_2[94, 104, 105].

The assimilation of lichens is especially sensitive to SO_2, which caused their extinction in the cities[106]. Increasing length of SO_2 fumigation has the consequence that increasing light intensities must be employed to reach the compensation point. After 3 days of fumigation, Parmelia sulcata showed no more net assimilation, even after irradiation with 9000 lux[106].

The mechanism of inhibition of photosynthesis is still unknown. It is not yet unequivocally decided whether the effect of SO_2 (or that of H_2SO_3) depends completely or only partly on an acidification of the tissue respective of a specific cell compartment[78, 50]. It is also unknown to what extent an altered redox-state may contribute to the inhibition.

Former hypotheses concerning the effect of SO_2 on the assimilation were based on the fact that the "chloroplast iron" (Chloroplasteneisen") combines irreversibly with SO_2, and thereby "loses its catalytic effect"[107, 108, 109]. The appearance of water soluble iron[108] proportional to the duration of the SO_2 action is also thus explained. Whether a reaction of the "chloroplast iron" (which, as we know today, is found in ferredoxin and in the cytochromes) with SO_2 leads to an inhibition of the assimilation, has not been investigated since that time.

It is also uncertain whether SO_2 acts directly on the photosynthetic reaction sequence or whether the membranes are attacked first. This could secondarily result in a damage of the assimilation processes[110].

The decomposition of chlorophyll to phaeophytin[111, 112, 113] may proceed too slowly to explain the sudden decline of the photosynthesis. In contrast to the NO_2 effect (see below) carotene is conserved[112]. The increase of respiration intensity, which is observed in many cases along with progressive damage[105], may be nonspecific and a general damage symptom (see p. 198).

The analysis of the fumigated leaves shows an increase in sulfur content even before metabolic damage appears. In particular, the amount of sulfate increases. Numerous investigations tried to find a correlation between the appearance of the first visible damage and the sulfur content, which can rise 2—5 fold[50, 58].

The amount of sulfur tolerated by the plant varies from species to species. Vogl[114] emphasized that the S-content of Pinus is not proportional to the degree of damage. But it is unknown whether the sulfur content and the onset of impairment of the metabolism are actually correlated with each other.

As a further result of the damaging effect, there is reported, for example, a rise in the Ca and Mg content, and occasionally in the amount of K or Si in spruce[115], or a reduction of the terpene production[116]. The increased cloudiness of the hot water extract from spruce needles after SO_2-fumigation, is supposed to be caused by an increase in the amount of leaf waxes[117] (not by easier extractibility?).

Effect on the Structure

The damage is microscopically expressed, especially in plasmolysis symptoms of chlorophyll containing palisade cells and spongy parenchyma. Thereby, the chloro-plasts swell, are deformed, and finally the mesophyll collapses[51]. There are no pub-lications yet on the relationship between the physiological effect of SO_2 and the microscopic or submicroscopic structure. The various damage characteristics are described by Thomas[58] and Brandt and Heck[51].

Effect on the Transpiration

A very complex subsequent reaction that is important for the plant metabolism, is the *effect of SO₂* fumigation *on transpira-tion*. While damage from ozone results in only a slight decline in the transpira-tion[118], it increases by SO_2 gassing during the first hour. Then it declines rapidly (in spinach about 50%[78, 105, 119]). The lethal increase in evaporation leads to withe-ring after which the destruction of the osmotic potential is followed by advanced desiccation. The transpiration damage is caused by a paralysis of the stomates. Re-markably, during the preceding phase of physiological S-incorporation, ^{35}S is al-ready accumulating in the guard cells[78] (see above). It is still unknown, why the guard cells accumulate ^{35}S, in which com-pound they incorporate it and how these events result in their paralysis.

Hydrochloric Acid, Chlorine, Nitrogen Oxides, and Ammonia

Plants may tolerate a concentration of *HCl* up to 10 ppm for several hours, whereas the upper limit for *Cl₂* is <0.45 ppm. The symptoms are similar, especial-ly with gas, to those caused by SO_2[58]. Thus, the chlorine content of the plant can increase to more than double its ori-ginal amount[7]. The reactions which cause the depression of assimilation are not yet known.

Incubation of Chlorella with *nitrous acid* ($2.5 \times 10^{-4}M$) in the light reduces the pho-tosynthetic CO_2-incorporation. The elec-tron transport is, however, not affected; rather, one of the enzymes of the CO_2-reduction cycle is inhibited. Hiller and Bassham[120] assume that this inhibition occurs between the diphosphates of fruc-tose and sedoheptulose and the corres-ponding monophosphates. From these ex-periments with Chlorella very limited conclusions can be drawn as to the action of gaseous NO_2 on plants. For these con-ditions only a few, rather unspecific symptoms have been reported to date. While 5 ppm *nitrogen oxides* are already toxic for bean plants, many other species are able to tolerate 20 ppm. After expo-sure to minor concentrations, the plants show reduced growth. Reduction of the leaf size and weight results in an increase of the chlorophyll content, based on fresh weight[5]. Highly toxic concentrations cause a decomposition of chlorophylls to pheophytins. In contrast to the SO_2 effect, the carotene is also decomposed[112]. It is not known if plants can use NO_2, within the toxic limit, as a source of nitrogen[58]. Plants are able to tolerate a concentration of < 8.3 ppm *NH₃* for several hours with-out visible damage. Higher concentra-tions inhibit respiration; only the oxi-dative decomposition of succinate seems to be resistant. The oxidation of NADH is especially decreased, while the NAD-reduction appears to be slightly increa-sed[121]. Besides this, the pH change of the cell sap may be a central point of attack. This causes precipitation of pro-tein and tannins; the latter results in a black brown coloration of the epidermis and mesophyll[7].

Carbon Monoxide

Light-dependent *carbon monoxide* fixa-tion in barley leaves and cucumber was found to a small extent. Labelled carbon of ^{14}CO appears mainly in serine and glycine. Addition of dinitrophenol causes a fixation pattern which more closely resembles the fixation pattern of typical Calvin cycle[122, 123].

In general, CO seems to be much more harmful for animals and man than to plants because of its binding on the hemoglobin[124]. The cytochromoxidase is probably the only CO-binding pigment of the plant mitochondria[125]. In young leaves it inhibits respiration, in older ones it can even have a stimulating effect. Ducet and Rosenberg[126] explain this to the effect that the cytochrome oxidase of older leaves is present in excess, and by partial binding of it, the oxidation-reduction state of the respiratory chain is altered. This results in an inhibition of phosphorylation. As with dinitrophenol, the uncoupling leads in turn to an increase in respiration. With respect to the damaging effect of air-polluting CO, these experimental results are probably without much meaning.

In Chlorella, CO inhibits the extra oxygen evolution which is associated with nitrate reduction. Possibly it is the nitrite which is the photosynthetic inhibitor when nitrate reduction is blocked by CO. At low light intensity, CO prevents the stimulation of photosynthesis caused by nitrate or ammonia in nitrogen starved cells. These phenomena probably are due to the inhibitory effect of CO on the cytochrome oxidase of the Krebs cycle which is necessary for N-utilization[127 a].

Epinasty, chlorosis, and anesthetic effects on Mimosa pudica are attributed to a high concentration (500—1000 ppm) of CO[58]; the way in which the CO works is, however, still unexplained.

In contrast, the N_2-fixation is highly sensitive to CO and is inhibited by traces of it. This is the same for the N_2-fixing system of the non-symbiontic bacteria, as well as for the symbiontic root nodule bacteria[127 b], although only the latter contain hemoglobin. A possible diminution of the nitrogen supply of the plants caused by CO as air pollutant, has not yet been studied.

Hydrocarbons

Espinasty, inhibition of growth, as well as acceleration of fruit ripening and leaf fall are known to be caused by air-polluting components. Olefins, especially ethylene, were ascertained to be the causative agents. Orchids are especially sensitive to ethylene and react with wilting and loss of the perigon[128, 50]. Ethylene holds a special position among the damaging gases, since it is a hormone and is also produced by the plant itself. Its half maximal effect normally ranges between 0.1—0.2 ppm; its mechanism of action is described in detail by Pratt and Groeschl[129].

Aldehydes, whose concentration exceed 0.2 ppm, injure petunias, for example[130]. The mechanism of action is still unknown. Aromatic as well as heterocyclic hydrocarbons, for instance benz(a)pyrene are very important as carcinogenic compounds or as allergenes for man, but they seem to have little or no effect on plant metabolism[131, 132]. There is also little known about the effect of insecticides and fungicides on the higher plants. Aldrin, dieldrin, and endrin are metabolized to hydrophilic compounds, which still have not been identified[9].

Heavy Metals and Particulates

On heavily travelled highways, growing grass can contain 3000 ppm lead; vegetables from gardens situated near the street, 10—700 ppm[12].

It is not known to what degree the lead is assimilated by the roots and what portion in absorbed by the leaves. Heavy metals can be absorbed directly through the cuticle[133]. Lead strongly inhibits the oxidation of succinate in isolated corn mitochondria. This takes place especially in phosphate deficient plants, since lead is precipitated in the presence of higher levels of phosphate and is thereby removed from the metabolism[12]. The assimilation of pine needles also decreases with a high lead content[7]. Garber[7] also gives a detailed description of the damaging effect of metal containing particulates. However, there has been no physiological and metabolic analysis to date. Cement dust results in a strong reduction of photosynthesis[134]. This may be partially due to the reduced access of light to the dust covered leaves, but mainly it could be a result of the strong acid affect and a general destruction of the leaf tissue[135].

The Metabolism of Visibly Damaged Plant Parts

The principal damage symptoms are the *chloroses* (loss or decrease of the chlorophyll) and the *necroses* (collapse of the tissue)[51]. They are the result of metabolic changes as well as of accumulation or decomposition of plant constituents. Chloroses or necroses are also caused by other types of damage to the plant; e.g. they may be caused by virus attack, mechanical injury or by the effect of heat[55]. The known metabolic damage produced by gases, which result in the above-mentioned leaf injuries, were summarized on pp. 183—194. In no case, however, do we know of a definite metabolic agent which causes necrosis. Take pipicolinic acid as an example. It appears in necrotic (not in chlorotic) tissue both by fluoride damage and after virus attack. In healthy tissues it is not present. Feeding healthy plants with this compound has no effect,

however. Pipecolinic acid is therefore only an accompanying effect, not the causative agent of necrosis[64].

A general characteristic of visible damaged tissue seems to be the *inhibition of assimilation*, which is accompanied by an increase of respiration. But even attack by parasites, e.g. Erisyphe graminis, causes the same symptoms[136]. As after fluoride damage (see above), the affected tissue shows a reduction in the use of the glycolytic pathway in favor of that of the pentosephosphate pathway.

A *breakdown of the permeability barriers* may be the reason for most metabolic changes in damaged tissue. For instance, the total amount of NADH, NADP, and NADPH in the leaf is not changed after attack by Erisyphe graminis. Separation of the cell components shows, however, that the NADP, previously limited to the chloroplasts[134], is found to 60% in the non-plastid parts of the cells of damaged tissue. There it is needed as cofactor for the pentose-phosphate dehydrogenase which starts the pentosephosphate cycle (with erythrocytes[138], with chloroplasts[139]). "False" regulations of enzyme activity certainly are an important factor in this respect, but have been scarcely considered to date.

Modification of the Damaging Effect by External and Internal Factors

The effects of air-polluting compounds investigated under defined culture conditions in the laboratory are, in practice, frequently modified by additional factors. Some essential points shall be briefly summarized in the following section.

The Synergistic Effect of Air-Polluting Gases

As shown on pp. 183ff., there are usually several toxic gases combined with one another in the polluted air. In Germany, the most important fact for the action of air pollution might be the synergism of SO_2-smog from domestic and industrial uses and of the oxidizing smog from auto-exhaust. For example, tobacco tolerates ozone to 0.05 ppm; simultaneous presence of SO_2, even 2 pphm ozone can cause visible damage[140]. Although a reducing and an oxidizing agent are present here, they do not neutralize each other's action, but rather enhance it. The SH-compounds like glutathione are possibly the common point of attack (see p. 187 and p. 192). They are oxidized by ozone and, because they are lacking, the plant is not able to incorporate low concentrations of SO_2 into its normal metabolism. Thus SO_2 immediately has a deleterious effect.

The Varying Sensitivity of the Plants to Air-Polluting Gases

Species Differences

There is a wide scale of sensitivity to each gas, which ranges from extremely sensitive to relatively resistent species.

Cotton, for example, tolerates HF fumigation averaging 0.74 ppm for 20 hours without decrease in assimilation and without visible damage. On the contrary, in some fruit trees the assimilation is reduced to about one third after 4 hours fumigation with an average of 0.052 ppm and a large part of the leaf surface is damaged[50].

Using mosses and lichens as most sensitive plants, a scale of annual average levels of SO_2 has been estimated[141].

It is easy to see that evergreen plants, such as conifers, are more sensitive to SO_2 damage than the deciduous species, since they are exposed to damaging material throughout the year. Besides this, an important factor is certainly the gas absorption capacity of the leaves[50]. An extensive causal analysis has, however, not yet been made.

Race Differences

Even within one species there are large differences in the *sensitivity*. As an example there is an ozone-sensitive tobacco strain Bel-W 3, which is damaged by 0.10 ppm, whereas the insensitive strain Bel-B, tolerates a concentration up to 0.22 ppm[142]. Irradiation preceding fumigation reduces the damaging effect of ozone due to the increase of ascorbic acid. However, this is not the reason for the difference in sensitivity of the two strains since the ascorbic acid level is the same for both[143]. Only in one case was the cause for the resistence of a strain followed one step further. In the onion, a dominant gene (or a gene complex) causes the closing of the stomata as a result of the ozone effect. The stomata of the ozone-sensitive strain, on the other hand, remain open even under the influence of ozone[51]. Since the guard cells are controlled by the osmotic pressure and this in turn depends on the concentration of the soluble carbohydrates, the question of which point in the cell is attacked by ozone might be interesting. The influence of genetic factors on the resistance is discussed in[51]. Breeding of relatively resistant spruce and pine races was attempted by grafting individuals that were proven phenotypically as SO_2-resistent onto SO_2-sensitive plants[144].

The Influence of Age

The degree of *sensitivity* is less dependent on the age of the whole plant than on that of the *particular leaves*. Normally, the sensitivity increases until complete unfolding of the leaves and then falls again[51].

By the damage from PAN, a relation between sensitivity and age of the leaves is found to exist. By titration of the PAN-sensitive, photoreducible SH-groups in the chloroplasts (see p. 185), it could be established that these increase in young leaves parallel to the PAN sensitivity up to the ninth day, and then in older leaves both decrease again[21]. Due to the various modifications of development by external factors, especially by light intensity, the physiological age is very often more important than the actual age in days.

The Influence of External Factors

External factors, such as *photoperiod, light intensity, temperature,* or supply with *mineral substances* may predispose the plant's sensitivity in a very complex way before the onset of fumigation or they may influence the toxic effect within a short time during the fumigation itself. In some cases there is a known connection between the preceding life history and the sensitivity to the damaging gas. Thus, for example, an ample supply with mineral nitrogen can increase the resistance of Medicago sativa to SO_2[145]. With respect to the fate of SO_2 in the plant (p. 189), one might assume that an increased supply of nitrogen enables the plant to increase incorporation of sulfur into the metabolism of the plant. In many cases, however, the amount of nitrogen in plants is positively correlated with increase in sensitivity[146]. In Nicotiana tabacum the most intensive damage by ozone was found to occur in plants with optimum nitrogen content (2.5—5% nitrogen content in the leaves[147]).

While tobacco leaves with a medium sugar content (3.4-4.5 mg/gm fresh weight) react the most sensitively to ozone[148], the sensitivity in beans is reduced by elevated sugar levels. This may explain the reduced sensitivity of many plants to ozone damage which is caused by previous irradiation[13].

Changes in sugar content, however, only result in altered sensitivity if they are above or below a certain concentration span[148]. Its margin in turn is not only dependent on the irradiation, but also on the further metabolism of the assimilates as well as their transport. Thus external factors may influence the sensitivity in a very complex way. This variety explains the many different and contradictory results obtained in respect to the influence of external factors on the plant's sensitivity[13].

The increase in sensitivity to PAN parallels the content of photoreducible SH-groups in the plant (see p. 187). Therefore, it increases with the duration and intensity of the preceding irradiation to a maximum[21]. In this case the influence of the preceding life history as well as that of the physiological age (see p. 199) are relatively well explained. Leaves of Nicotina glutinosa which had just completed differentiation and were therefore characterized by high metabolic activity, proved to be the most sensitive to a "smog atmosphere"[149].

Also the leaf content of vitamins and growth factors is regulated by the cooperation of the developmental stages and external factors. Irradiation of 22—23 hours causes 2—3 fold increase of the ascorbic acid level; thereby the ozone sensitivity of the plant is reduced[143]. In this case, the reducing action of the vitamin and its protecting effect may be closely related. It is unknown in which way the concentration of growth factors may influence the action of ozone[150].

External factors which change the plant's sensitivity to air polluting gases, especially to SO_2 within a very short time, act primarily on the stomates. Water supply, light intensity, and temperature are the most important modifiers of stomatal opening. In this way a relative humidity of 29% causes an extensive resistance of Medicago sativa to SO_2[145]. Vogl[114], however, emphasized that SO_2 is still absorbed in the dark where stomates are known to be closed.

The complex *relationship between external factors,* stomates opening, and the penetration of air polluting gases, which could only be briefly summarized here, were investigated by Heck et al.[151] and they are described in detail by Dugger and Ting[13], Thomas[58], and others.

Measures against the Damaging Effects of Air-Polluting Gases

As in other fields, the possibilty of therapy implies an accurate knowledge of the damaging effect. The scant knowledge of the mechanism of action of air-polluting gases corresponds to the still very limited curative possibilities. They apply chiefly to two points. First, it is attempted, through application of *antioxidants,* to lower the damage of the oxidizing auto-smogs[152, 153, 154]; to decompose ozone with the help of charcoal, diatomaceous earth, or similar compounds before its penetration into the plant[155], or to keep F^- ions on the leaf surface by spraying with *fluoride binding chemicals*[156]. The breeding of resistant strains, from crossing, results in an accumulation of gene-dependent physiological characteristics and thereby causes an accumulation of resistance increasing factors. This may be the method of choice. In this case it is advantageous that the "pest" is not of a biogenic nature; i. e. it itself does not produce mutants, which then adapt to the new resistent forms. The use of positive metabolic characteristics with respect to resistance to air pollution, however, needs more detailed knowledge of the action of air polluting gases on the plant metabolism.

Addendum

Interest in environmental research increased rapidly during the past year. Of the numerous publications dealing with the action of air polluting gases on plants, only a few will be referred to here. They will be confined to those concerned with the basic action of such gases on the plant's biochemical or physiological processes.

1. Ozone and PAN

The fact that these compounds oxidize protein sulfhydryl groups[13, 14] could be directly demonstrated by increased formation of disulfides by action of ozone on *Phaseolus*[158]. This effect was found to be responsible for the detrimental effect on growth and ribosomal RNA. Fumigation of *Phaseolus* with 0.35 ppm ozone specifically decreases the population of chloroplastic but not of cytoplasmic ribosomes[159]. Polysomes and monosomes of isolated chloroplast ribosomes are dissociated by ozone in the same manner as by other sulfhydryl-reagents like p-chloromercuribenzoate[160, 161]. The different susceptibility of chloroplastic and cytoplasmic rRNA to the oxidizing effect of ozone probably is due to the fact that the tertiary structure of chloroplastic ribosomal protein makes more SH groups accessible for the oxidizing action[160].

The inhibition of glycolipid biosynthesis in spinach chloroplast preparations[162] also seems to be due to the oxidizing effect of ozone. As do other sulfhydryl reagents, 1—2 µmoles ozone/min block the formation of di- and trigalactosyldiglyceride from uridine 5'-pyrophosphate-galactose. Ozone may oxidize the enzyme sulfhydryl groups themselves, or the inhibition may be a secondary effect caused by oxidation of unsaturated fatty acids; the peroxide produced reacts further with the sulfhydryl compounds[162].

The inhibitory action of PAN on the polysaccharide synthesizing system of the cell wall in *Avena* coleoptiles[18, 19] can be partially overcome by auxin[163]. Probably the formation (or activation) of glucan synthetase is thereby stimulated and the PAN-inactivated system replaced[163]. Not only increased starch degradation[25] but also impairment of polysaccharide synthetase seems to cause the increase in total soluble carbohydrates and reducing sugars found in *Pinus* seedlings exposed to 5 pphm ozone for 5—22 hours[164]. Much better knowledge of normal variation in ascorbic acid level however is needed for the evaluation of seemingly elevated amounts of this compound in fumigated plants[164].

Earlier findings[165] that ozone induces higher levels of peroxidase activity were extended to include its action on the pattern of peroxidase isozymes[166]. After fumigation of *Phaseolus vulgaris* with 20—25 pphm ozone for 90—180 min, almost all bands obtained by polyacrylamide gel electrophoresis show an increase in intensity. Even apparently unaffected plants show this increase in activity which is most expressed in leaves with symptoms of injury. Investigations concerning the effects of activation of secondary plant metabolism[39, 40] have been carried out on *Pinus ponderosa*[167]. Accumulation of methyl-chavicol indicates that the oxidizing air polluting gases disturb the carbohydrate-shikimic acid metabolism.

At the cellular level, accumulation of chloroplasts and carbohydrates in the peripheral portions of the mesophyll cells and disruption of the homogenous distribution of protein and nucleic acids precede visible damage in ponderosa pine[168]. As is the case with fluoride and SO_2 there is evidence that also ozone penetrates mainly through the stomata. Low sensitivity of *Pinus* species is correlated with a low number of stomata per unit mesophyll and ozone injury is associated with substomatal cavity cells[169].

2. Hydrogen Fluoride

In single cell cultures of *Glycine max* it was shown that the physiological incorporation of fluoride ion (10^{-3}M) results in the formation of fluoroacetate and fluorocitrate[170]. In young leaves of *Dichapetalum toxicarum* conversion of F^- to fluoroacetate is probably intimately related to photosynthesis; later the acetate is converted to long chain fluoro fatty acids[171]. In fluor-sensitive species such as *Funaria hygrometrica* and *Hypogymnia physodes*, exposure to HF levels as low as 65 ppb for 12 hours causes disintegration of chloroplasts and plasmolysis[172].

3. Sulfur Dioxide

According to the pK_s value of $HSO_3^- \rightleftharpoons H^+ + SO_3^{--} = 7.0$, the dissolution of SO_2 in the plasma yields an equilibrium of HSO_3^-/SO_3^{--} of about 1:9—1:9,8[173]. If present in low doses the incorporation of the sulfate ion into the normal metabolism[58, 81] and its stimulatory effect on the yield of plants[93] not only includes reduction to S^{--}[83, 85], but also oxidation to SO_4^{--}[78]. This oxidation takes place in irradiated, isolated chloroplasts, delivering electrons to photosystem II. SO_3^{--} at concentrations from 0.25—5 mM stimulates the photoreduction of both $K_3(Fe(CN)_6)$ and NADP[174]. Consequently, $^{14}CO_2$ incorporation (up to 1 mM SO_3^{--}; see below) is enhanced[174]. At these stimulatory SO_3^{--} concentrations, the changes in fixation pattern towards an increase of sugar-monophosphates at the expense of sugar-diphosphates and phosphoglyceric acid may be due to the activation of fructosediphosphatase by reduced ferredoxine[175], the level of which is probably increased by the stimulated electron flow[174].

Whereas the stimulatory effect of SO_3^{--} on the Hill reaction increases with its concentration (reaching a plateau at 3 mM

SO_3^{--}). $^{14}CO_2$ incorporation decreases at concentrations > 1 mM[174]. This level corresponds to the range at which the inhibitory action on ribulosediphosphate carboxylase starts[173].

Kinetic studies of SO_3^{--} action on ribulosediphosphate carboxylase (EC 4.1.1.39; Rudph-carboxylase) and phophoenolpyruvate carboxylase (EC 4.1.1.31; PEP-carboylase) revealed some general features of SO_3^{--}/HCO_3^- interaction[173, 176]. With respect to Mg^{++} both enzymes are non-competitively inhibited by SO_3^{--}. The K_i-values (84,5 mM SO_3^{--} for PEP-carboxylase, 9,5 mM SO_3^{--} for Rudph-carboxylase) demonstrate the much higher sensitivity of the latter enzyme. This relative sensitivity also holds with regard to their substrates PEP and Rudph respectively. PEP-carboxylase even at SO_3^{--} concentrations as high as 10 mM shows no inhibition with respect to PEP (and thus the K_i-value practically would be ∞) whereas Rudph-carboxylase is inhibited non-competitively with a K_i-value of 14 mM SO_3^{--}. With respect to HCO_3^- as the common substrate, both enzymes are inhibited competitively, indicating that replacement of HCO_3^- by SO_3^- is a more general feature of SO_3^{--} action. The K_i-values (27 mM SO_3^{--} for PEP-carboxylase, 3 mM SO_3^{--} for Rudph-carboxylase) reveal that this competitive inhibition is much more sensitive than the non-competitive one. Moreover, comparison of K_m-values of HCO_3^- and K_i-values of SO_3^{--} indicate that the replacement of HCO_3^- by SO_3^{--}, common to both enzymes, is facilitated by decreasing affinity of Rudph-carboxylase to its substrate HCO_3^{-}[176].

From these results it may be concluded that the competitive inhibition of Rudph-carboxylase by SO_3^{--} will predominate at low SO_3^{--} and low internal CO_2 concentrations.

Although SO_3^{--} is a very potent inhibitor of malate dehydrogenase[100, 101], malate formation in enzyme preparation of *Zea mays* is not at all or only slightly inhibited[176]. This is due to the fact that the NADH-dependent activity is abundant in *Zea mays* (a NADPH-dependent malate dehydrogenase is also present) and PEP-carboxylation is by far the rate limiting step. Consequently, as far as is indicated by enzymatic studies, CO_2 fixation by action of PEP-carboxylase + malate dehydrogenase in plants with the dicarboxylic acid pathway of photosynthesis[177], seems to be an evolutionary advantage towards SO_2 resistance, as CO_2 refixation by Rudph-carboxylase in the bundle sheath chloroplasts takes place under elevated CO_2 tension.

The increased sensitivity of lichens towards sulfite[106] is reflected by the complete inhibition of $^{14}CO_2$ incorporation at SO_3^{--} concentrations as low as 0.4 to 0.85 mM[178]. The differing sensitivities within the lichens studied were correlated with their known sensitivities to atmospheric pollution.

The inhibitory action of bisulfite compounds like α-hydroxypyridine methane sulfonate (HPMS) on PEP-carboxylase and NADH-dependent malate dehydrogenase[99] were examined with respect to their action on photosynthetic $^{14}CO_2$ fixation in isolated chloroplasts[179] and leaf strips[180]. HPMS concentrations of 0.7 and 0.5 mM applied to chloroplasts and leaf strips respectively reduce CO_2 fixation to about 50%. The ATP level at 5 mM HPMS is reduced in the light and in the dark[180]. However, Rudph-carboxylase, ribosephosphate isomerase and carbonic anhydrase are not affected by 1 mM HPMS[179]. Furthermore, the marked differences in inhibition patterns of HPMS[99] and SO_3^{--}[176] on PEP-carboxylase strongly indicate that HPMS and SO_3^{--} act differently on the enzymes of photosynthetic CO_2 fixation. They emphasize that HPMS action cannot serve as a model for SO_3^{--} action [180], even though the over-all reaction of CO_2 fixation is blocked by both to about the same extent.

Long-term fumigation of pea-seedlings affects the amino acid metabolism. First the concentrations of glutamate, then that of glutamin and γ-aminobutyric acid are increased in shoots as well as in roots [181]. These phenomena are connected with an increase in the activity of glutamate dehydrogenase in the direction of reductive amination of α-ketoglutarate [182]. Glutamate dehydrogenase in pea-seedlings exists in multiple molecular forms which have identical molecular weights and seem to be conformers [182]. SO_2 fumigation changes the pattern of the multiple molecular forms separated by polyacrylamide gel electrophoresis, but renaturation after urea-denaturation results in the formation of one identical activity band in control and fumigated plants [182]. Thus the action of ozone on peroxidase multiple forms [166] is different from that of SO_3^{--} on glutamate dehydrogenase. The importance of pH-dependent enzyme conformation for SO_3^{--} action also seems to be indicated by the fact that horse-radish peroxidase catalyzed oxidation of indole-3-acetaldehyde is inhibited by 10^{-4} M SO_3^{--} at acidic pH, whereas it is stimulated at pH 7.8 [183]. The enhancement at pH 6 is based on the tendency of sulfite ions to react with free radicals, induced by a peroxidizable substrate [184].

The tendency of SO_3^{--} toward free radical reactions at concentrations of 10^{-3} — 10^{-2} M leads moreover to the cleavage of phosphodiester bonds of DNA, thereby causing inactivation of transforming activity of B.subtilis DNA [185]. At higher concentrations of bisulfite (1 M), modified polymers of polyuridylic acid are generated in which up to 95% of the uracil

residues have been converted to 5.6-dihydrouracil-6-sulfonate residues. However, even 2.6% of bisulfite modification abolishes 46% of phenylalanine coding ability in an E.coli cell free protein synthesis system. Thus biological damage by SO_3^{--} may partially be caused by inactivation of messenger RNA [186].

4. Carbon Monoxide

Recent data suggest that biological production of CO on a global level may be larger than that by combustion engines [187, 188] even though the latter may cause exaggerated local CO concentrations. One of its main sources seems to be swamps; the methane produced by them is permanently converted to CO in the atmosphere [188].

Recently its uptake and metabolism in bean leaves, at concentrations of 200 to 360 ppm has been investigated for the first time in detail and clarified on its fundamentals [189]. ^{14}CO uptake occurs in light and in darkness. In darkness it is converted to $^{14}CO_2$ and released. In light ^{14}CO is incorporated into serine which is then converted to sucrose. The rate of CO-fixation is roughly proportional to its concentration but is not related to rates of photosynthesis, although photosynthetic reducing power is also necessary for CO fixation.

This CO fixation by vegetation is in the range of 12—120 kg/km²/day and thus in the same magnitude as that by soil [189].

5. Heavy Metals

Most investigations were concerned with the correlation of traffic density and lead accumulation in plant tissues [190, 191]. Here again the preferential accumulation in mosses became evident. Explanted tissues of cauliflower, carrot, potato, and lettuce showed inhibition in cell proliferation by mercury, lead, copper, and zinc and detailed toxic levels for each species were ascertained[192].

A non-competitive inhibition of photosynthetic $^{14}CO_2$ fixation in isolated spinach chloroplasts starts with a concentration as low as 2 μM Pb $(NO_3)_2$. Concomitantly the ATP-synthesis is inhibited[193]. Lead salts (1.6—2.4 mM) do not affect photosystem I activity in isolated chloroplasts; however, photosystem II is inhibited and there are indications that the site of inhibition is between the primary electron donor of photosystem II and the site of water oxidation[194].

6. The action of air polluting gases on membranes

Recently much evidence accumulated that the complex symptoms of damage may originate in the damaging action of air polluting gases on membranes[43, 44, 46, 47]. The sulfhydryl groups are involved in the permeability of membranes and the oxidizing effect of ozone is supposed to be related to changes in the membrane permeability[158]. Ozone, acting on model membrane systems prepared with egg lecithin, causes production of peroxides and malonaldehyde if no glutathione is present. In its presence no peroxide is created and malonaldehyde is reduced[162]. Also the lipids of isolated chloroplasts react with ozone even though they do not produce peroxide; glutathione reduces formation of malonaldehyde too[162]. Thus the effect of ozone on membranes not only involves interference with lipid biosynthesis but it exerts a damaging effect on the membrane's lipids themselves. Inhibition of CO_2 fixation by 2-pyridine methanesulfonate is supposedly connected with its damaging effect on the inner or outer chloroplast membranes[179]. This is supported by the fact that transient changes in H+ flux and chloride transport

across cell membranes are strongly inhibited by 0.5 M HPMS[180]. However, since identity in the mode of action of SO_3^{--} and HPMS seems to be most unlikely[3], possible damage of membranes by SO_3^{--} itself deserves further investigations.

As to heavy metal poisoning, special "binding sites" seem to exist within the cell. Binding of the metals on the cytoplasmic membrane causes a leakage through the diffusion barrier and thereby an outflow of potassium, followed by anions such as phosphate. Finally, metal ions penetrate into the interior of the cells and as a consequence of these processes photosynthesis is affected[195].

References

[1] Georgii, H. W.: Umschau 63: 757–762, 1963
[2] Tebbens, B. D.: In: Air pollution, Vol. I. Ed. by A. C. Stern. Academic Press, New York, London pp. 23–46, 1968
[3] Schlipköter, H. W.: Jahrbuch 1964 Ministerpr. d. Landes Nordrhein-Westfalen, 475–503, 1964
[4] Katz, M.: In: Die Verunreinigung der Luft; Ursachen, Wirkungen, Gegenmaßnahmen. Verlag Chemie Weinheim pp. 91–151, 1964
[5] Taylor, O. C., F. M. Eaton: Plant Physiol. 41: 132–135, 1966
[6] Chatot, G., M. Castegnaro, J. L. Roche, R. Fontanges: Chromatographia 3: 507–514, 1970
[7] Garber, K.: Luftverunreinigung und ihre Wirkungen. Gebr. Bornträger, Berlin 1967
[8] Corn, M.: In: Air pollution, Vol. I. Ed. by A. C. Stern. Academic Press, New York, London pp. 47–94, 1968
[9] Korte, F., W. Klein, B. Drefahl: Naturwiss. Rundschau 23: 445–457, 1970
[10] Haagen-Smit, A. J., C. G. Wayne: In: Air pollution, Vol. I. Ed. by A. C. Stern. Academic Press, New York, London pp. 149–186, 1968
[11] Kongreßbericht Reinhaltung der Luft (1969), Arbeits- u. Sozialminister des Landes Nordrhein-Westfalen. Verlag für Wirtschaft u. Verwaltung, Essen 1969
[12] Koeppe, D. E., J. Miller: Science 167: 1376–1378, 1970
[13] Dugger, W. M., I. P. Ting: Ann. Rev. Plant Physiol. 21: 215–234, 1970
[14] Mudd, J. B.: Arch. biochem. biophys. 102: 59–65, 1963

[15] Mudd, J. B.: J. biol. Chem. 241: 4077–4080, 1966

[16] Mudd, J. B., R. Leavitt, W. H. Kersey: J. biol. Chem. 241: 4081–4085, 1966

[17] Todd, G. W.: Physiol. Plant. 11: 457–463, 1958

[18] Ordin, L., A. Altmann: Physiol. Plant 18: 790–797, 1965

[19] Ordin, L., M. A. Hall: Plant Physiol. 42: 205–212, 1967

[20] Mudd, J. B.: Arch. Envir. Health 10: 201–216, 1965

[21] Dugger, W. M., I. P. Ting: Phytopath. 58: 1102–1107, 1968

[22] Hall, M. A., R. L. Brown, L. Ordin: Phytochemistry 10: 1233–1238, 1971

[23] Ordin, L.: Plant Physiol. 37: 603–608, 1962

[24] Ordin, L., B. P. Skoe: Plant Physiol. 39: 751–755, 1964

[25] Dugger, W. M., J. Koukol, R. L. Palmer: J. Air Poll. Control Assoc. 16: 467–471, 1966

[26] Lee, T. T.: Plant Physiol. 42: 691–696, 1967

[27] McDowall, F. D. H.: Can. J. Bot. 43: 419–427, 1965

[28] Freebairn, H. T.: Science 126: 303–304, 1957

[29] Mudd, J. B., W. M. Dugger: Arch. biochem. biophys. 102: 52–58, 1963

[30] Koukol, J., W. M. Dugger, N. O. Belser: Plant Physiol. 38: xii, 1963

[31] Koukol, J., W. M. Dugger, R. L. Palmer: Plant Physiol. 42: 1419–1422, 1967

[32] Dugger, W. M., O. C. Taylor: Plant Physiol. 36: xlix, 1961

[33] Dugger, W. M., O. C. Taylor, W. H. Klein, W. Shropshire: Nature 198: 75–76, 1963

[34] Newton, J. W.: J. biol. Chem. 237: 3282–3286, 1962

[35] Asahi, T., S. Masaki: J. Biochem. (Tokyo) 60: 90–92, 1966

[36] Dugger, W. M., J. Koukol, W. D. Reed, R. L. Palmer: Plant Physiol. 38: 468–472, 1963

[37] Tomlinson, H., S. Rich: Phytopath. 57: 972–974, 1967

[38] Mader, P. P., G. Cann, L. Palmer: Plant Physiol. 30: 318–323, 1955

[39] Koukol, J., W. M. Dugger: Plant Physiol. 42: 1023–1024, 1967

[40] Howell, R. K.: Phytopath. 60: 1626–1629, 1970

[41] Goldstein, B. D., Ch. Lodi, Ch. Collinson, O. J. Balchum: Arch. Environ. Health 18: 631–635, 1969

[42] Felmeister, A., M. Amanat, N. D. Weiner: Environ. Sci. Technol. 2: 40–43, 1968

[43] Lee, T. T.: Plant Physiol. 43: 133–139, 1968

[44] Wedding, R. T., L. C. Erickson: Am. J. Bot. 42: 570–575, 1955

[45] Dugger, W. M., J. Koukol, R. L. Palmer: Plant Physiol. 40: xx, 1965

[46] Thomson, W. W., W. M. Dugger, R. L. Palmer: Bot. Gaz. 126: 66–72, 1965

[47] Thomson, W. W., W. M. Dugger, R. L. Palmer: Can. J. Bot. 44: 1677–1682, 1966

[48] Erickson, L. C., R. T. Wedding: Amer. J. Bot. 43: 32–36, 1956

[49] Miller, P. R., J. P. Parmeter, O. C. Taylor, E. A. Cardiff: Phytopath. 53: 1072–1076, 1963

[50] Thomas, M. D.: In: Die Verunreinigung der Luft; Ursachen, Wirkungen, Gegenmaßnahmen. Verlag Chemie Weinheim pp. 229–277, 1964

[51] Brandt, C. S., W. W. Heck: In: Air pollution, Vol. I. Ed. by A. C. Stern. Academic Press, New York, London pp. 401–445, 1968

[52] Thomas, M. D., E. W. Alther: In: Handb. d. exp. Pharmakol., Bd. XX/I. Ed. by A. Heffter, fortgef. W. Heubner. S. 231–306. Springer-Verlag, Berlin pp. 231–306, 1966

[53] Benedict, H. M., J. M. Ross, R. H. Wade: I. Air Poll. Control Assoc. 15: 253–255, 1965

[54] Knabe, W.: Staub 30: 384–385, 1970

[55] Hill, A. A., M. R. Pack, L. G. Transtrum, W. S. Winters: Plant Physiol. 34: 11–16, 1959

[56] Applegate, H. G., D. F. Adams: Bot. Gaz. 121: 223–227, 1960

[57] Jacobson, J. S., L. H. Weinstein, D. C. McCune, A. E. Hitchcock: J. Air Pollution Control Assoc. 16: 412–417, 1966

[58] Thomas, M. D.: Ann. Rev. Plant Physiol. 2: 293–322, 1951

[59] Warburg, O., W. Christian: Biochem. Zschr. 310: 384–421, 1942

[60] Ordin, L., B. P. Skoe: Plant Physiol. 38: 416–421, 1963

[61] Yang, S. F., G. W. Miller: Biochem. J. 88: 509–516, 1963

[62] Arrigoni, O., E. Marré: Giorn. Biochim. 4: 1–9, 1955

[63] Yang, S. F., G. W. Miller: Biochem. J. 88: 517–522, 1963

[64] Yang, S. F., G. W. Miller: Biochem. J. 88: 505–509, 1963

[65] Ordin, L., B. Propsy: Plant Physiol. 37: lxviii, 1962

[66] Adams, D. F., M. T. Emerson: Plant Physiol. 36: 261–265, 1961

[67] Chang, Ch. W., C. R. Thompson: Physiol. Plant 19: 911–918, 1966

[68] Chang, Ch. W.: Physiol. Plant 23: 536–543, 1970

[69] Chang, Ch. W.: Plant Physiol. 43: 669–674, 1968

[70] Terofik, S., P. K. Stumpf: J. biol. Chem. 192: 519–526, 527–533, 1951

[71] Müller, D.: In: Handbuch der Pflanzenphysiol. XII/I. Ed. by H. Ruhland. Springer-Verlag, Berlin pp. 543–571, 1960

[72] Ross, C. W., H. H. Wiebe, G. W. Miller: Plant Physiol. 37: 305–309, 1962

[73] McNulty, I. B., J. M. Lords: Science 132: 1553–1554, 1960

[74] McNulty, I. B., D. W. Newman: Plant Physiol. 36: 358–388, 1961

[75] Bredemann, C., H. Radeloff: Phytopath. Zeitschr. 5: 179–194, 1933

[76] Materna, J.: Arch. Forstwes. 15: 691–692, 1966

[77] Thomas, M. D., R. H. Hendricks, G. R. Hill: Plant Physiol. 19: 212–220, 1944

[78] Weigl, J., H. Ziegler: Planta 58: 435–447, 1962

[79] Thompson, J. R., D. M. Pace: Canad. J. Biochem. Physiol. 40: 207–217, 1962

[80] Arrigoni, O.: Ital. J. Biochem. VIII: 181, 1959

[81] Asahi, T., T. Minamikawa: J. Biochem. (Tokyo) 48: 548–556, 1960

[82] Steward, F. C., J. F. Thompson, F. K. Millar, M. D. Thomas, R. H. Hendricks: Plant Physiol. 26: 123–135, 1951

[83] Wilson, L. G., R. S. Bandurski, T. Asahi: J. biol. Chem. 236: 1822–1829, 1961

[84] Asahi, T., R. S. Bandurski, L. G Wilson: J. biol. Chem. 236: 1830–1835, 1961

[85] Schmidt, A., A. Trebst: Biochim. biophys. acta 180: 529–535, 1969

[86] Asahi, T.: Biochem. biophys. acta 82: 58–66, 1964

[87] Asahi, T.: J. Biochem. (Tokyo) 48: 772–773, 1960

[88] Asada, K., G. Tamura, R. S. Bandurski: Biochem. biophys. Res. Comm. 30: 554–559, 1968

[89] Mothes, K., W. Specht: Planta 22: 800–805, 1934

[90] Harrison, B. F., M. D. Thomas, G. R. Hill: Plant Physiol. 44: 245–257, 1944

[91] Biddulph, S. F.: Amer. J. Bot. 43: 143–148, 1956

[92] Spálený, J., M. Kutácek, K. Oplištilová: Int. J. Air Water Poll 9: 525–530, 1965

[93] Thomas, M. D., R. H. Hendricks, T. R. Collier, G. R. Hill: Plant Physiol. 18: 345–371, 1943

[94] Thomas, M. D., G. R. Hill: Plant Physiol. 12: 285–307, 1937

[95] Bersin, Th.: Advances Enzymol. X: 223–311, 1950

[96] Wieler, A.: Untersuchungen über die Einwirkung schwefliger Säure auf die Pflanze. Gebr. Bornträger, Berlin 1905

[97] Dässler, H. G.: Wiss. Z. Techn. Univ. Dresden 11: 567–569, 1962

[98] Thomas, M. D., R. H. Hendricks, G. R. Hill: Ind. Eng. Chem. 42: 2231–2235, 1950

[99] Osmond, C. B., P. N. Avadhani: Plant Physiol. 45: 228–230, 1970

[100] Pfleiderer, B., D. Jeckel, Th. Wieland: Biochem. Zschr. 328: 187–194, 1956

[101] Pfleiderer, B., E. Hohnholz: Biochem. Zschr. 331: 245–253, 1959

[102] Grewe, R.: Zeitschr. f. physiol. Chem. 242: 89–96, 1936

[103] Holzer, M., K. Beaucamp: Biochim. biophys. acta 46: 225–243, 1961

[104] Vogl, M., S. Börtitz, H. Polster: Arch. Forstwes. 13: 1031–1043, 1964

[105] Keller, H., J. Müller: Veröff. d. Forstbot. Inst. d. Bay. Forstl. Forschungsanst. P. Parey, Hamburg, Berlin 1958

[106] Pearson, L., E. Skye: Science 148: 1600–1602, 1965

[107] Noack, K., O. Wehner, H. Griessmeyer: Zschr. angew. Chem. 42: 123–126, 1929

[108] Griessmeyer, H.: Planta 11: 331–358, 1930

[109] Wieler, A.: Jahresber. Wiss. Bot. (Leipzig) 78: 483–543, 1933

[110] Thümmler, R,: Protoplasma 36: 254–315, 1941

[111] Dörries, W.: Zeitschr. f. Pflanzenkrankh. Pflanzenschutz 42: 257–273, 1932

[112] Kändler, U., H. Ullrich: Naturwiss. 51: 518, 1964

[113] Hawksworth, D. L.: Intern. J. Environm. Studies 1: 281–296, 1971

[114] Vogl, M.: Biol. Zbl. 83: 587–594, 1964

[115] Materna, J.: Naturwiss. 48: 723–724, 1961

[116] Dässler, H. G.: Flora 154: 376–382, 1964

[117] Härtel, O., E. Papesch: Ber. dtsch. Bot. Ges. 68: 133–142, 1955

[118] Todd, G. W., B. Probst: Physiol. Plant 16: 57–65, 1963

[119] Heiling, A.: Phytopath. Zeitschr. 5: 435–492, 1933

[120] Hiller, R. G., J. A. Bassham: Biochim. biophys. acta 109: 607–610, 1965

[121] Vines, H. M., R. T. Wedding: Plant Physiol. 35: 820–825, 1960

[122] Krall, A. R., N. E. Tolbert: Plant Physiol. 32: 321–325, 1957

[123] Chappelle, E. W., A. R. Krall: Biochim. biophys. acta 49: 578–580, 1961

[124] Stokinger, H. E., D. L. Coffin: In: Air pollution, Vol. I. Ed. by A. C. Stern. Academic Press, New York, London pp. 446–546, 1968

[125] Plesnicar, M., W. D. Bonner, B. Storey: Plant Physiol. 42: 366–370, 1967

[126] Ducet, G., A. J. Rosenberg: Ann. Rev. Plant Physiol. 13: 171–200, 1962

[127a] Vennesland, B., C. Jetschmann: Arch. biochem. biophys. 144: 428–437, 1971

[127b] Burris, R. H.: Ann. Rev. Plant Physiol. 17: 155–179, 1966

[128] Clayton, G. D., T. S. Platt: Amer. Ind. Hyg. Assoc. J. 28: 151–160, 1967

[129] Pratt, H. K., J. D. Groeschl: Ann. Rev. Plant Physiol. 20: 328–541, 1969

[130] Brennan, E., I. A. Leone, R. H. Daines: Science 143: 818–820, 1964

[131] Dörr, R.: Naturwiss. 52: 166, 1955

[132] Blochinger, A.: Protoplasma 54: 35–100, 1961

[133] Kannan, S.: Plant Physiol. 44: 517–521, 1969

134 Darley, E. F.: J. Air Poll. Control Assoc. 16: 145–150, 1966
135 Czaja, A. T.: Staub 22: 228–232, 1962
136 Ryrie, I. J., K. J. Scott: Plant Physiol. 43: 687–692, 1968
137 Heber, U. W., K. A. Santarius: Biochim. biophys. acta 109: 390 408, 1965
138 Bonsignore, A., I. Lorenzoni, A. Cancedda, A. de Flora: Biochem. biophys. Res. Comm. 39: 142–148, 1970
139 Lendzian, K., H. Ziegler: Planta 94: 27–36, 1970
140 Menser, H. A., H. E. Heggestad: Science 153: 424–425, 1966
141 Gilbert, O. L.: New Phytol. 69: 629–634, 1970
142 Menser, H. A., H. E. Heggestad, O. E. Street: Phytopath. 53: 1304–1308, 1964
143 Menser, H. A.: Plant Physiol. 39: 564–567, 1964
144 Rohmeder, E., W. Merz, A. v. Schönborn: Forstwiss. Zentralbl. 81: 321–332, 1962
145 Zahn, R.: Zeitschr. Pflanzenkrankh., Pfl.-schutz 70: 81–95, 1963
146 Lee, T. T.: Can. J. Bot. 44: 487–498, 1966
147 Leone, I. A., E. Brennan, R. H. Daines: J. Air Poll. Contr. Assoc. 16: 191–196, 1966
148 Dugger, W. M., O. C. Taylor, E. Cardiff, C. R. Thompson: Plant Physiol. 37: lxx, 1962
149 Glater, R. B., R. A. Solberg, F. M. Scott: Amer. J. Bot. 49: 954–970, 1962
150 Seidmann, G., J. H. Ibrahim: Plant Physiol. 37: lxix, 1962
151 Heck, W. W., J. A. Dunning: J. Air Poll. Contr. Assoc. 17: 112–114, 1967
152 Freebairn, H. T.: J. Air Poll. Contr. Assoc. 10: 314–317, 1960
153 Ordin, L., O. C. Taylor, B. E. Probst, E. A. Cardiff: Int. J. Air Water Pollut. 6: 223–227, 1962
154 Siegel, S. M.: Plant Physiol. 37: 261–266, 1962
155 Jones, J. L.: Science 140: 1317–1318, 1963
156 McNulty, I. B., D. W. Newman: Plant Physiol. 32: 121–124, 1957
157 Newell, R. E.: Scientific American 224: 32–42, 1971
158 Tomlinson, H., S. Rich: Phytopathology 60: 1842–1843, 1970
159 Chang, Ch. W.: Phytochemistry 10: 2863–2868, 1971
160 Chang, Ch. W.: Biochem. biophys. Res. Comm. 44: 1429–1435, 1971
161 Chang, Ch. W.: Phytochemistry 11: 1347–1350, 1972
162 Mudd, J. B., T. T. McManus, A. Ongun, T. E. McCullogh: Plant Physiol. 48: 335–339, 1971
163 Ordin, L., M. J. Garber, J. I. Kindinger: Physiol. Plant 26: 17–23, 1972
164 Barnes, R. L.: Canad. J. Bot. 50: 215–219, 1972
165 Dass, H. S., G. M. Weaver: Canad. J. Plant Sci. 48: 569–574, 1968
166 Curtis, C. R., R. K. Howell: Phytopath. 61: 1306–1307, 1971
167 Cobb, F. W., E. Zavarin, J. Bergot: Phytochemistry 11: 1815–1818, 1972
168 Evans, L. S., P. R. Miller: Amer. J. Bot. 59: 297–304, 1972
169 Evans, L. S., P. R. Miller: Canad. J. Bot. 50: 1067–1071, 1972
170 Peters, R. A., M. Shorthouse: Phytochemistry 11: 1337–1338, 1972
171 Vickerey, B., M. C. Vickerey: Phytochemistry 11: 1905–1909, 1972
172 Comeau, G., F. Le Blanc: Canad. J. Bot. 50: 687–907, 1972
173 Ziegler, I.: Planta 103: 155–163, 1972
174 Libera, W., H. Ziegler, I. Ziegler: Planta (in press) 1972
175 Buchanan, B. B., P. P. Kalberer, D. I. Arnon: J. Biol. Chem. 246: 5952–5959, 1971
176 Ziegler, I.: Phytochemistry (in press)
177 Hatch, M. D., C. R. Slack: Ann. Rev. Plant Physiol. 21: 141–162, 1970
178 Hill, D. J.: New Phytol. 70: 831–836, 1971
179 Murray, D. R., J. W. Bradbeer: Phytochemistry 10: 1999–2003, 1971
180 Lüttge, U., C. B. Osmond, E. Ball, E. Brinckmann, G. Kinze: Plant and Cell Physiol. 13: 505–514, 1972
181 Jäger, H. J., E. Pahlich: Oecologia 9: 135–140, 1972
182 Pahlich, E.: Planta 104: 78–88, 1972
183 Yeh, R., D. Hemphill, H. M. Sell: Canad. J. Biochem. 49: 162–165, 1971
184 Meudt, W. J.: Phytochemistry 10: 2103–2109, 1971
185 Hayatsu, H., R. C. Miller: Biochem. biophys. Res. Comm. 46: 120–124, 1972
186 Shapiro, R., B. Braverman: Biochem. biophys. Res. Comm. 47: 544–550, 1972
187 Delwiche, C. C.: Ann. N.Y. Acad. Sci. 174: 116–121, 1970
188 McConnell, J. C., M. B. McElroy, S. C. Wofsy: Nature 233: 187–188, 1971
189 Bidwell, R. R. S., D. E. Fraser: Canad. J. Bot. 50: 1435–1439, 1972
190 Lee, J. A.: Nature 238: 165–166, 1972
191 Briggs, D.: Nature 238: 166–167, 1972
192 Barker, G. W.: Canad. J. Bot. 50: 973–976, 1972
193 Hampp, R., H. Ziegler, I. Ziegler: Biochemie u. Physiol. der Pflanzen (in press)
194 Miles, C. D., J. R. Brandle, D. J. Daniel, O. Chu-Der, P. D. Schnare, D. J. Uhlik: Plant Physiol. 49: 820–825, 1972
195 Kamp-Nielsen, L.: Physiol. Plant 24: 556–561, 1971

Characterization of Acid Phosphatase and its Relation with Insecticide Metabolism in Bruchus chinensis (Linnaeus)

S. Rashid, R. A. Roohi, S. N. H. Naqvi

Institut für physiologische Chemie, 355 Marburg/Lahn, Deutschhausstraße 1 – 2, West Germany

Summary. Acid phosphatase has been characterized biochemically in the homogenate of Bruchus chinensis (L.) and optimum factors have been reported. Optimum pH was found to be 4.0, K_m value 1.25×10^{-3}M, and an optimum temperature of 41 ° C for a 30 minute incubation period. Enzyme activity decreased by in vivo treatment of the chlorinated (Ovex and Petkolin) insecticides while it increased by organophosphate (Zolone) insecticide. The oxygen rate was simultaneously measured to find any correlation between metabolic reactions and enzyme behavior.

The rate of respiration was found to increase with increase of insecticide concentration at 1 hour in the case of Petkolin and Zolone (more toxic) and even at 3-hour intervals in the case of Ovex (less toxic) insecticide. At 24-hour intervals it was found to decrease with increasing concentration in the case of Petkolin and Zolone while not in the case of Ovex.

It was concluded that metabolic reactions have some correlation with the toxicity of the insecticides while changes in the phosphatase level are under different genetic controls and thus depend on some other biochemical reaction. However, there is some correlation between this biochemical reaction and insecticide metabolism.

Zusammenfassung. Säurephosphatase wurde biochemisch im Homogenat von Bruchus chinensis (L.) bestimmt und optimale Faktoren werden berichtet. Der pH-Wert wurde als optimal bei 4.0, der K_m-Wert bei 1.25×10^{-3}M, die optimale Temperatur bei 41° C und 30-minütiger Inkubationszeit erkannt. Die Aktivität der Enzyme ließ bei in vivo Behandlung mit den chlorierten Insektiziden (Ovex und Petkolin) nach, während sie sich bei dem Organophosphat (Zolone) steigerte. Der Sauerstoffgehalt wurde gleichzeitig gemessen, um jegliche Korrelation zwischen Metabolismusreaktionen und Enzymverhalten herauszufinden. Es wurde gefunden, daß Metabolismusreaktionen in Korrelation mit der Toxität der Insektizide stehen, während Veränderungen im Phosphatasegehalt unter anderer genetischer Kontrolle stehen und dadurch von einer anderen biochemischen Reaktion abhängen. Es besteht jedoch eine Korrelation zwischen dieser biochemischen Reaktion und dem Insektizid-Metabolismus.

Introduction

The use of insecticides is increasing in fields and stores day by day thus creating problems of residues and resistance. Much work has been done on established and well-known insecticides but very little work has been done on insecticides presently used, i.e. Zolone, Petkolin (a new chlorinated insecticide; Ashrafi and Naqvi[6]) and Ovex. Moreover, very little attention has been paid to the effects of insecticide spraying on the *bio-ecosystem*. Resistance is one of the products of this disturbance. Therefore, in the present paper an attempt has been made to understand the basic cause of *resistance* by studying *phosphomonoesterase levels* in treated and normal insects because various authors have held the conversion of aliesterase into phosphatase as a cause of resistance.

The role of various enzymes has been studied in the metabolism of insecticides by various workers. Among them, some [1,2,5,7–10,12,20,28–33,37–40,42–51,54,56,60,61] have worked on the *esterases*, while others[16,19, 21,27,53,57,58] have worked on *dehydrochlorinase*. Antiesterase activity and the poisoning effect of insecticides have been reported by Brady and Sternburg[11] and Lord and Potter[25]. The role of microsomal oxidation has been reported by El Bashir and Oppenoorth[14], Oppenoorth and House[52] and others. A detailed review on the problem has been published by Terriere[59] and Georghiou[15]. In a personal discussion, Lord reported the role of transferases in this mechanism[26].

Since Ashrafi et al.[5], Belden[9], Naqvi et al.[37–39], Matsumura and Brown[29], Ogita[42–44], Ogita and Kasai[45–47] and Rockstien and Inashima[54] have reported the effect of insecticides on phosphatases in fast-flying insects, it was decided to study this effect in a slow moving stored grain pest, *Bruchus chinensis* (L.). The rate of

O_2 consumption was simultaneously studied to investigate any correlation between these two phenomena.

Materials and Methods

Insecticide Application

For determining the LD_{50} of *Petkolin**, *Ovex* (chlorinated) and *Zolone* (Organophosphate) in the case of *Bruchus chinensis* (L.), solutions were prepared in acetone, (v/v for Petkolin and Ovex; w/v for Zolone). On the basis of the probit mortality curves (Fig. 1) 0.01% concentration of Petkolin and Zolone and 1.0% concentration of Ovex were prepared and 1.2 µl dose was topically applied over the prothorax of each *Bruchus* by means of a calibrated micrometer syringe. For control (untreated) and check (treated with 1.2 µl/Bruchus, acetone) observations also, 30 insects were taken as in the case of treated samples. The enzyme quantity was determined after different intervals of treatment i.e. 1 hour, 3 hours, and 24 hours.

Assay Procedure

Assay procedure for acid phosphatase was the same as described by Naqvi et al.[34] with some differences, specific to the enzyme found in *Bruchus* and characterized before application of insecticide. For each insecticide, 30 *Bruchus*/ml were taken for the preparation of homogenate in double-distilled deionized water and ground in Teflon Pyrex tissue grinder for 3 minutes. It was filtered through a 2-mm-thick glass wool layer under moderate suction pressure. The rest of the procedure was the same as described by Naqvi et al.[34]. The readings were recorded on a Beckman DB620 Spectrophotometer, and O.D converted into micro-

---

* P. S. Ashrafi and Naqvi (1969).

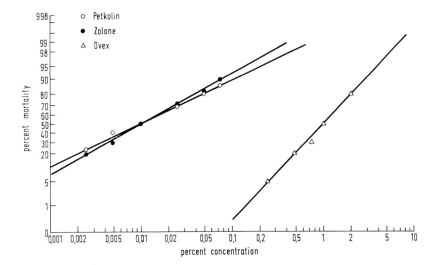

Fig. 1

moles of p-nitrophenol which were divided by tissue weight to get the number of micromoles released per milligram fresh tissue weight. The readings were compared by keeping the value of normal as 100%.

Measurement of the Rate *of Metabolic Gas Exchange* by Manometer

Effect of different concentrations (0.0025%, 0.01%, 0.05% of Petkolin and Zolone; 0.5%, 1.0%, 2.0% of Ovex) of insecticides on the *rate of respiration* or O_2 consumption of *Bruchus* was determined after 1, 3, and 24 hours of treatment (the intervals at which enzyme activity was also estimated) by using Warburg's apparatus. For this purpose 20 *Bruchus* were treated (1.2 µl/*Bruchus*) with each concentration of insecticide. Normal (untreated) and positive control (treated with acetone) were kept with each experiment. A small strip of filter paper (1 sq. cm) soaked in 0.1 ml of 10% KOH solution was placed in the central cavity of each flask, to absorb the CO_2 given off by the insects during respiration, so that it would not influence the gas phase. The flasks were fixed with the

specific manometer meant for each concentration, normal and positive control (the total volume of the flask and manometer already calibrated). The flasks were kept half immersed in water and the temperature was maintained at 37° C. When the temperature of the water reached 37° C, the columns of all manometers (filled with Brodie's solution) were adjusted at 150 mm. The upper arm, which remained open until the temperature of water bath reached 37° C was closed. The readings of the open arm (side arm) were recorded after every five minutes (in order to avoid maximum error) for 30 minutes. Experiments were done in triplicate with all the concentrations of the insecticides at different intervals and the mean value is reported in Tables 1—3.

Calculations

Amount of oxygen consumed during respiration was calculated as follows (Norman Richard 1963):
Amount of O_2 exchanged = h × k
where h = change in pressure measured on the open arm of the manometer
k = Flask constant

$$K = \frac{V_g \dfrac{273}{T} + V_f a}{P_0}$$

Table 1   Effect of Petkolin on the rate of oxygen consumption in µl/insect

% Concentration	% Mortality	Rate of respiration after		
		1 Hour	3 Hours	24 Hours
0.0025	20	49.83 ± 0.12	55.70 ± 0.96	46.01 ± 0.87
0.0100	50	77.25 ± 0.56	93.24 ± 1.44	45.81 ± 1.62
0.0500	80	96.47 ± 1.04	64.34 ± 0.88	18.00 ± 1.23
Control	–	25.14 ± 0.22	27.41 ± 0.49	24.70 ± 2.10
Check	–	28.79 ± 0.68	28.50 ± 0.93	27.17 ± 1.02

Table 2   Effect of Zolone on the rate of oxygen consumption in µl/insect

% Concentration	% Mortality	Rate of respiration after		
		1 Hour	3 Hours	24 Hours
0.0025	20	34.50 ± 0.27	52.24 ± 1.25	23.36 ± 0.86
0.0100	50	47.83 ± 0.68	71.60 ± 1.54	20.75 ± 0.59
0.0500	80	98.10 ± 2.13	59.50 ± 0.96	9.66 ± 0.85
Control	–	25.00 ± 1.07	23.44 ± 0.77	27.40 ± 0.89
Check	–	28.10 ± 0.76	28.70 ± 0.99	25.90 ± 0.63

Table 3   Effect of Ovex on the rate of oxygen consumption in µl/insect

% Concentration	% Mortality	Rate of respiration after		
		1 Hour	3 Hours	24 Hours
0.5	20	32.38 ± 1.31	44.14 ± 0.38	34.45 ± 1.45
1.0	50	41.30 ± 3.01	57.81 ± 1.43	36.68 ± 0.98
2.0	80	85.53 ± 0.84	69.28 ± 1.03	22.97 ± 0.64
Control	–	24.90 ± 1.06	25.70 ± 0.34	28.50 ± 1.98
Check	–	29.90 ± 0.46	27.40 ± 1.33	25.40 ± 0.52

V_g = Volume (in µl) of the gas space in the particular flask and manometer.

T = Absolute temperature of the system.

V_f = Volume (in µl) of the fluid in the system.

a = Solubility of the gas involved in the fluid where the gas is at atmospheric pressure.

P_0 = Manometer is filled with the Brodie's solution which has the density of 1.033, P_0 = 10.000

e.g.:

V_g = 15.65

V_f = 1.5 ml (volume of the fluid)

a = o (solubility of CO_2)

T = 273 + 37 = 310

P_0 = 10.000

V_g = 15.65 – 1.5 ml = 14.15

$$K = \frac{14.15 \times \dfrac{273}{310} + 1500 \times o}{10.000}$$

= 1.246 µl

Amount of oxygen exchanged = h × K/minute/insect

h = 10.000 µl

K = 1.246

T = 30 minutes

No. of Bruchus = 20

$$= \frac{10.000 \times 1.246}{30 \times 20}$$

= 25.14 µl/Bruchus

Results and Discussion

Effect of Varying pH

Solutions of critical acid buffer with different pH were used by keeping all other factors constant. The maximum activity was found at pH 4.0 (Fig.2) which is comparable with 4.2 for silkworm[13], silkworm testes[17], and Bagrada picta[36]; 4.4 for stable fly[3] and desert locust[35] but different from 4.8—5.0 for housefly[2] and 4.6—5.3 for Aedes aegypti[22] (L.).

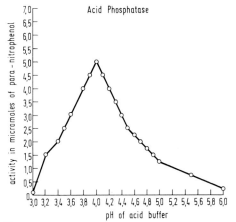

Fig. 2

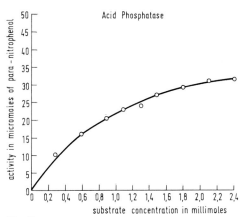

Fig. 3

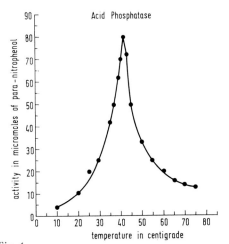

Fig. 4

Effect of Varying Substrate Concentration

Various substrate concentrations were used to find the optimum substrate concentration. A linear response was found up to 0.0009 M, after which the enzyme entered into zero order reaction (Fig. 3). The K_m value was calculated by plotting double reciprocal graph (Lineweaver and Burk 1934) and was found to be 1.25×10^{-3} M (Fig. 4). This value agrees to some extent with the K_m value 1.74×10^{-3}M for *Oncopeltus*[4], 9.7×10^{-4} M for stable fly[3], 9.39×10^{-4} M for pyrophosphatase of boll weevil[23], 9.5×10^{-4}M for *Bagrada picta*[34] and 3.0×10^{-4} M for intestinal acid phosphatase of the desert locust[35].

Effect of Temperature Variation

The phosphatase activity was determined at different temperature from $10°$ C to $75°$ C for 30 minute incubation period. The maximum activity was found at $41°$ C as shown in Fig. 5. This is quite near to $40°$ C reported for housefly[55], for stable fly[3], for large milkweed bug[4] and desert locust[35]; $37.5 °$C for silkworm[18] and $42 °$C for *Bagrada picta*[36]. Temperature coefficient (Q_{10}) value was found 2.512 between 20 and $30°$ C (Fig. 6). This value is also close to the values for *Aedes*[22], for stable fly[3], testes of silkmoth[17], desert locust[35] and *Bagrada picta*[36].

Effect of Varying Incubation Period

The reaction mixtures were incubated for various intervals at $41°$ C. Zero order reaction was maintained up to 30 minutes incubation (Fig. 7). It is comparable with the results reported for large milkweed bug[4], Bagrada[36], silkworm[13] but differs with that reported for silkmoth testes[18].

Effect of Insecticides on the Enzyme

Effect of Petkolin, Ovex (chlorinated) and Zolone (organophosphate) was studied after different intervals (1, 3, and 24 hours).

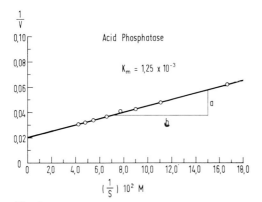

Fig. 5

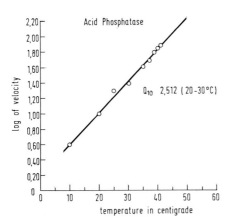

Fig. 6

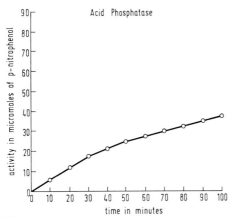

Fig. 7

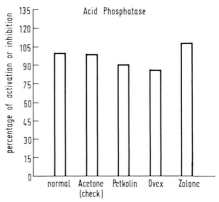

Fig. 8   enzyme activity after 1 hour

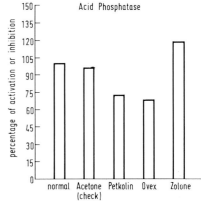

Fig. 9   enzyme activity after 3 hours

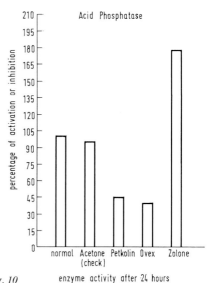

Fig. 10   enzyme activity after 24 hours

Petkolin inhibits the enzyme activity less than *Ovex* (Fig. 8—10) at its respective LD_{50} but its toxicity for *Bruchus* is 100 times more than Ovex. After 1 hour, about 10%, after 3 hours, 28%, and after 24 hours, 55% enzyme activity was inhibited as compared to normal while 8%, 23% and 50% respectively as compared to check. A similar effect of Petkolin has been reported by Ashrafi et al.[5] and Naqvi et al.[39].

Ovex also inhibited the enzyme but at a very high dose as compared to Petkolin and Zolone because it is less toxic to insects. The inhibition was 14%, 31%, and 60% respectively after 1, 3, and 24-hour intervals as compared to the normal, while 12%, 27%, and 54% as compared to check. The inhibition by chlorinated insecticides is in accordance with the findings of Ashrafi et al.[5], Onikenko[48], and Naqvi et al.[37-39]. This effect is perhaps similar to the cholinesterase inhibition and poisoning reported by Brady and Sternburg[11].

Zolone, instead of inhibition, showed an *increase in the enzyme activity*, with an increase in the reaction time. This increase was 9%, 19%, and 78% after 1, 3, and 24-hour intervals in comparison to control (normal), while in comparison to positive control it was 11%, 23%, and 83% respectively. This increase in the activity has been reported by Ashrafi et al.[5], Afsharpour and O'Brien[1], Matsumura and Brown[29], and Naqvi et al.[37, 39]. The decrease in phosphatase activity may be explained as poisoning of aliesterase and thus lesser conversion of aliesterase into phosphatase. While increase in the case of Zolone (organophosphate) is perhaps due to a genetic change which results in more conversion of aliesterase into phosphatase and thus creating the problem of resistance.

Effect of Insecticides on Respiratory Rate or O₂ Consumption

Rate of respiration or O₂ consumption was manometrically measured to investigate whether some correlation exists between the phosphatase reaction with insecticide and the general metabolic rate. The quantity of oxygen utilized by the normal and treated insects is shown in Tables 1—3, after 1, 3, and 24 hours of treatment. It was observed that the respiratory rate increased with the increase in concentration up to 1 hour in all the insecticides and up to 3 hours in the case of Ovex (less toxic). This was perhaps due to higher *microsomal oxidation* and rapid degradation of the insecticides by certain oxidative enzymes. In the case of Petkolin and Zolone at 3-hour intervals, there was decrease in the O₂ consumption perhaps due to the major poisoning of the enzymes resulting in general decrease in the metabolic activities. After 24 hours a decrease in O₂ consumption was observed in the case of Petkolin and Zolone (more toxic insecticides) and the reason is perhaps the same as mentioned above. This is further supported by the readings in the case of Ovex and it is more probable, that toxicity and the poisoning effect of the insecticide play an important role. There seems to be some correlation between toxicity and O₂ consumption when we compare the data in Fig. 1 and Tables 1—3.

However, it is evident from a comparison of Fig. 8—10 and Tables 1—3 that the *phosphatase activation* or *inhibition* has no correlation with the *rate of respiration* or O₂ consumption. The enzyme was inhibited by chlorinated insecticides and catalyzed by organophosphate insecticides, with the increase of time while it is not so in the case of respiration. This behavior of the enzyme has been reported by other researchers as well[48, 29]. Whether this is a step towards resistance or merely an ionic reaction is difficult to say definitely. However, on the basis of previous observations of different researchers it may be said that there is poisoning of the enzyme by chlorinated insecticides and conversion of aliesterase into phosphatase by mutant gene allele in the case of organophosphate insecticides. It is also possible that the phosphate present in the organophosphate insecticide acts as a substrate for the enzyme in addition to the substrate used during estimation of the enzyme and thus resulting in an apparently high activity of phosphatase. However, it is very difficult to conclude anything in this respect until a detailed investigation is made on the quantity of aliesterase, phosphatase and the oxidative enzymes in insects treated with the three groups of insecticides (both susceptible and resistant). The role of oxidative enzymes seems to be quite clear but that of phosphomonoesterases is not clear and requires further investigation.

Thus, whatever the cause of the *resistance*, one should be very particular in using the insecticides indiscriminately which disturb the natural balance in the bio-ecosystem. In developed countries these aspects are kept in mind but in the developing countries, little attention is being paid to these factors. Different agencies spray insecticides (land and aerial) without keeping these factors in mind and thus polluting the atmosphere, killing the predators and causing problems like resistance. We suggest that more work be done in these countries on such problems.

Acknowledgement

The authors are indebted to Dr. S. H. Ashrafi, Dr. M. Abdullah, Mr. A. H. Quraishi and Mr. R. A. Khan of PCSIR Karachi, for cooperation during the work.

References

[1] Afsharpour, F., R. D. O'Brien: Colum chromatography of insect esterases. J. Insect Physiol. 9: 521–529, 1963

[2] Alexander, B. H., R. J. Barker, F. H. Babers: The phosphatase activity of susceptible and resistant houseflies and German roaches. J. econ. Ent. 50: 211–213, 1958

[3] Ashrafi, S. H., F. W. Fisk: Acid phosphatase in the stable fly, Stomoxys calcitrans (L.). Ann. Entomol. Soc. Am. 54: 598–603, 1961

[4] Ashrafi, S. H., F. W. Fisk: Acid phosphatase activity of the alimentary tract of the large milkweed bug, Oncopeltus fasciatus (Dallas) (Hemiptera-Lygeidae). Pakistan J. scient. ind. Res. 5: 24–26, 1962

[5] Ashrafi, S. H., S. N. H. Naqvi, S. A. Qureshi, S. A. Muzaffar: Comparative effect of insecticides on the activity of high energy producing enzymes. Pakistan J. Sci. Res. 20: 170–173, 1968

[6] Ashrafi, S. H., S. N. H. Naqvi: The use of chlorinated petrol, Petkolin, as a Pesticide. PANS 15: 549–552, 1969

[7] Asperen, K. van, F. J. Oppenoorth: Organophosphate resistance and esterase activity in houseflies. Ent. exptl. Appl. 2: 48–57, 1959

[8] van Asperen, K., F. J. Oppenoorth: The interaction between organophosphorus insecticides and esterase in homogenates of organophosphorus resistant strain of houseflies. Ent. exptl. Appl. 3: 68–83, 1960

[9] Belden, D. A., jr.: Histochemical localisation of acid and alkaline phosphatases in male Musca domestica (L.). Dissertation Abstract, Washington State University, Pullman, Washington 1958

[10] Bigley, W. S., F. W. Plapp: Cholinesterase and aliesterase activity in organophosphorus susceptible and resistant houseflies. Ann. Entomol. Soc. Am. 53: 360–364, 1960

[11] Brady, U. E., J. Sternburg: Studies *in vivo* cholinesterase inhibition and poisoning symptoms in houseflies. J. Insect Physiol. 13: 369–379, 1967

[12] Dauterman, W. C., F. Matsumura: Effect of malathion analogues upon resistant and susceptible Culex tarsalis mosquitoes. Science 138: 694–695, 1962

[13] Denuce, J. M.: Recherches sur le systeme phosphatasique des glandes sericigens chez le ver a soie (Bombyx mori L.). Experientia 8: 64-68, 1952

[14] El Bashir, S., F. J. Oppenoorth: Microsomal oxidation of organophosphorus insecticides. Nature 233: 210–211, 1969

[15] Georghio, G. P.: Genetics of resistance to insecticides in houseflies and mosquitoes. Exptl. Parasitol. 26: 224–255, 1969

16 Girgolo, A., F. J. Oppenoorth: The importance of DDT-dehydrochlorinase for the effect of resistance gene "kdr" in housefly *Musca domestica* L. Genetica 37: 159–170, 1966

17 Gilbert, L. I., C. J. Huddleston: Testicular acid phosphatase in giant silkmoth. J. Insect Pwysiol. 11: 177–194, 1965

18 Huddleston, C. J., L. I. Gilbert: Acid phosphatase activity in testes of cynthia silkworm. Amer. Zool. 1: 221, 1961

19 Khan, M. Q. A., L. C. Terriere: DDT-dehydrochlorinase activity in housefly strains resistant to various groups of insecticides. J. econ. Ent. 61: 732–736, 1968

20 Kasai, T., Z. Ogita: Studies on malathion resistance and esterase activity in green rice leafhopper. SABCO J. 1: 130–140, 1965

21 Kimura, T., A. W. A. Brown: DDT-dehydrochlorinase in *Aedes aegypti* L. J. econ. Ent. 57: 710–716, 1964

22 Lambremont, E. N.: A study of acid phosphatase of the mosquito, *Aedes aegypti* L. J. Insect Physiol. 3: 325–334, 1959

23 Lambremont, E. N., R. M. Schrader: Enzyme of bollweevil, II. Inorganic pyrophosphatase. J. Insect Physiol. 10: 37–52, 1964

24 Lineweaver, H., D. Burk: Determination of enzyme dissociation constants. J. Amer. Chem. Soc. 56: 658–666, 1934

25 Lord, K. A., C. Potter: Differences in esterase from insect species; Toxicity of organophosphorus compounds and *in vivo* antiesterase activity. J. Sci. Food Agric. 5: 490–498, 1954

26 Lord, K. A.: Role of transferases in the degradation of insecticides. (personal communication) 1970

27 Lovell, J. B., C. W. Kearns: Inheritence of DDT-dehydrochlorinase in housefly. J. econ. Ent. 52: 931–935, 1959

28 Main, A. R., P. E. Braid: Hydrolysis of malathion by alesterase *in vitro* and *in vivo*. Biochem. J. 84: 255–263, 1962

29 Matsumura, F., A. W. A. Brown: Biochemical study of malathion tolerant strain of *Aedes aegypti*. Mosquito News 21: 192–194, 1961

30 Matsumura, F., A. W. A. Brown: Studies on carboxyesterase in malathion resistance *Culex tarsalis*. J. econ. Entomol. 56: 381–388, 1963

31 Matsumura, F., W. C. Dauterman: Effect of malathion analogues on a malathion resistant housefly strain, which possesses a detoxification enzyme, carboxyesterase. Nature 202: 1536–1538, 1964

32 Matsumura, F., C. J. Hogendijk: The enzymatic degradation of malathion in organophosphate

resistant and susceptible strain of *Musca domestica* L. Ent. exptl. Appl. 7: 179–193, 1964

33 Menzel, D. B., R. Craig, W. M. Hoskins: Electrophoretic properties of esterases from susceptible and resistant strains of houseflies (*Musca domestica* L.). J. Insect Physiol. 9: 479–493, 1963

34 Naqvi, S. N. H., S. H. Ashrafi, M. A. H. Qadri: Quantitative determination of acid and alkaline phosphatase in different tissues of the alimentary canal of desert locust, *Schistocerca gregaria* (Forskal). Pakistan J. scient .ind. Res. 10: 204–207, 1967

35 Naqvi, S. N. H., S. H. Ashrafi, M. A. H. Qadri: Acid phosphatase activity in the digestive system of desert locust, *Schistocerca gregaria* (Forskal). Austral. J. biol. Sci. 21: 1047–1052, 1968

36 Naqvi, S. N. H., F. Yasmeen, S. H. Ashrafi: Acid phosphatase activity in the homogenate of *Bagrada picta* (Fabr.). Scientific Researches 6: 1969

37 Naqvi, S. N. H., S. A. Muzaffar, S. A. Qureshi: Studies of in vivo effect of varying doses of insecticides on the phosphomonoesterases of desert locust, *S. gregaria*. Folia Biologica 17: 1969

38 Naqvi, S .N. H., S. A. Muzaffar, S. H. Ashrafi: Detoxification of DDT and its relation with inhibition of phosphomonoesterases in desert locust, *S. gregaria*. Z. Pflanzenkrankh. 77: 577–581, 1970a

39 Naqvi, S. N. H., S. Rashid, S. H. Ashrafi: Inhibition of phosphomonoesterases by insecticides in *Poekilocerus* egg. Pakistan J. Zool. 2: 149–158, 1970b

40 Naqvi, S. N. H., M. Y. Zubairi, S. Rashid: *In vivo* inhibition of phosphomonoestrases by dimetilan and its metabolites in desert locust. Pakistan J. Biochem. 1971 (in press)

41 van Norman Richard, W.: Experimental Biology. Prentice Hall, Englewood Cliffs. New Jersey 130–142, 1963

42 Ogita, Z.: Genetical relationship between aliesterase activity and insecticide resistance in *D. melanogaster*. Botyu Kagaku 26: 93–97, 1961

43 Ogita, Z.: Genetico-biochemical analysis on the enzyme activities in the housefly by agar-gel electrophoresis. Jap. J. Genetics 37: 518–521, 1962

44 Ogita, Z.: Genetico-biochemical analysis on the esterase activity in the housefly. Jap. J. Genetics 38: 200, 1963

45 Ogita, Z., T. Kasai: Biochemical differences between insecticide resistant and susceptible flies. Information Cir. Insecticide Resistance WHO 46: 32–33, 1964

[46] Ogita, Z.: Genetico-biochemical analysis of specific esterases in housefly. *M. domestica*. Jap. J. Genetics 40: 173–184, 1965

[47] Ogita, Z.: Genetic control of multiple forms of the acid phosphomonoesterase in *M. domestica*. Jap. J. Genetics 40: 185–197, 1965

[48] Onikenko, F. O.: The effect of Heptachlor on the phosphorylase activity of some albino rat organs. Fiziol. Zh. Akad. Nauk. Ubr. R.S.R. 10: 125–127, 1964

[49] Oppenoorth, F. J.: Genetics of resistance to organophosphorus compounds and low aliesterase activity in housefly. Ent. exptl. Appl. 2: 304–319, 1959

[50] Oppenoorth, F. J.: Some cases of resistance caused by alteration of enzymes. International Congr. Entomol. pp. 240–242, 1964

[51] Oppenoorth, F. J., Asperen, K. van: Allelic genes in the housefly producing modified enzymes that cause organophosphate resistance. Science 132: 298–299, 1960

[52] Oppenoorth, F. J., N. W. H. House: DDT resistance in the housefly caused by microsomal degradation. Ent. exptl. Appl. 11: 81–93, 1968

[53] Perry, A. S., W. M. Hoskins: The detoxification of DDT by resistant houseflies and inhibition of this process by Piperonyl cyclonene. Science 111: 600–601, 1950

[54] Rockstien, M., M. D. Inashima: Enzymes in insects. – Alkaline phosphatase. Bull. Bklyn. Entomol. Soc. 48: 20–23, 1953

[55] Rockstien, M.: Phosphatases of housefly, Musca domestica (L.). Bull. Bklyn. Entomol. Soc. 51: 8–17, 1956

[56] Stegwee, D.: Esterase inhibition and organophosphorus poisoning in the housefly. Nature 184: 1553–1554, 1959

[57] Sternburg, J., E. B. Winson, C. W. Kearns: Enzymatic dehydrochlorination of DDT by resistant flies. J. econ. Entomol. 46: 513–515, 1953

[58] Sternburg, J., C. W. Kearns, H. H. Moorefield: DDT-dehydrochlorinase, an enzyme found in DDT resistant flies. J. Agric. Food. Chem. 2: 1125–1130, 1954

[59] Terriere, L. C.: The oxidation of Pesticides. The comparative approach. In "The Enzymatic Oxidations". (E. Hodgson editor) pp. 175–196, 1968

[60] Voss, G., F. Matsumura: Biochemical Studies on a modified and normal cholinesterase found in Leverkusen strains of two spotted spider mite, *Tetranychus telarius*. Can. J. Biochem. 43: 63–72, 1965

[61] Velthius, H. H. W., Asperen, K. van: Occurrence and inheritance of esterases in *Musca domestica* (L.). Ent. exptl. Appl. 6: 79–87, 1963

Studies on the Toxicology of Nitrites

N. Gruener, H. I. Shuval

Environmental Health Laboratory, Department of Medical Ecology,
Hebrew University, Hadassah Medical School, Jerusalem, Israel

Summary. Toxicological data about the influence of nitrites on rats are given. There are indications concerning the increase of methemoglobin levels, and the pathological changes in lungs and heart are discussed. Besides normal rats, the same experimental investigations were made with pregnant rats and their newborn. The transfer of nitrites through the placenta causes methemoglobinemia. Upon chronic administration of sodium nitrites in water, changes in the brain electrical activity were detected.

Zusammenfassung. Über toxikologische Daten des Einflusses von Nitriten auf Ratten wird berichtet. Es werden Angaben über den Anstieg der Methämoglobin-Konzentrationen gegeben und die pathologischen Veränderungen in Lunge und Herz diskutiert. Neben normalen Ratten werden auch gravide Ratten und Neugeborene den gleichen experimentellen Untersuchungen unterworfen. Durch plazentaren Übertritt von Nitriten wird Methämoglobinämie hervorgerufen. Nach chronischer Verabreichung von Natronnitrit im Wasser entstehen Veränderungen der Hirnströme.

Introduction

There is growing concern over the possible consequence resulting from *human exposure to nitrates and nitrites from various environmental sources*. Nitrate levels in ground and surface water are rising in many areas due to pollution from organic wastes or nitrogenous chemical fertilizers. Nitrates in drinking water can be reduced to nitrites by stomach and intestinal micro-flora and have led to the development of thousands of cases of methemoglobinemia, particularly in young infants [3, 7, 9]. Nitrates and nitrites found as natural constituents in certain vegetables such as spinach have also given rise to methemoglobinemia in infants in some cases fatal [6, 12]. These same chemicals are widely used as food preservatives in many "corned" meat products, most countries allowing the use of nitrite salts in concentrations of 200 ppm. There is also some indication that oxides of nitrogen in polluted air can lead to raised levels of methemoglobin [14].

In order to assist in the evaluation of the toxicological impact of such environmental exposure, a series of experimental studies was initiated. Of particular concern were the chronic and sub-acute effects of exposure to sodium nitrite. This compound was chosen since nitrate salts can be converted to nitrite, which is the more toxic form, by nitrate reducing bacteria within the intestinal tract or during processing and storage of certain foods. Sodium nitrite is also used extensively, directly as a food additive.

Chronic Exposure of Rats to Sodium Nitrite in Drinking Water

Previous chronic studies by others have not shown significant increased *methemoglobin (MetHb) levels* or other pathological findings in rats consuming water containing nitrites as high as 2000 mg/l [4]. Our studies of the kinetics of MetHb disappearance after the administration nitrites showed that the MetHb level is reduced by 50 % every 90 minutes. We have also shown that rats, being nocturnal animals, consume 80 % of their water at night and that by 10 am they may no longer show significant MetHb levels even when consuming water high in nitrites. Our measurements showed a peak in water consumption and MetHb level in the middle of the night. In designing our chronic studies we adjusted the "day" and "night" hours in the animal rooms so that blood examinations would be taken at about the rats "midnight", so as to detect the expected maximum daily MetHb concentration.

Five groups of male, 3-month-old, Hebrew University, Sabra, albino rats were used with 8 per group. The first group, "A", receiving normal tap water, was the control. The other four groups received drinking water containing sodium nitrite at the following concentrations: "B" — 100, "C" — 1000, "D" — 2000 and "E" — 3000 mg/l.

Rats were weighed monthly and samples of tail blood were taken for determination of MetHb and Hb. MetHb determinations were made by the sensitive micromethod requiring 0.2 ml of blood developed for these studies [8]. After 24 months the animals were killed and examined for gross pathology as well as for histological differences.

In summary the findings of this study which can be considered only as a pretest for the full scale chronic toxicity experiments in progress are as follows:

1) There were no significant differences in growth and development or mortality between the controls and the experimental group.

2) There were no significant differences in Hb levels between the controls and experimental group.

3) The MetHb levels of experimental groups C, D, and E were significantly raised throughout the study and averaged respectively 5 %, 10 %, and 20 %. The MetHb level of "B" — receiving 100 mg/l of sodium nitrite was slightly raised above the controls for the first 60 days of the study only. Afterwards the level of "B" and "A", the controls, were identical.

4) Examination for blood glucose, pyruvate and lactate did not show any differences between the controls and experimental group.

5) Pathology:

At the end of 24 months exposure to nitrites the five groups were killed. The animals were anesthetized with ether, bled and their internal organs were inspected, weighed, and fixed in 10 % neutral formaline.

Histological examinations were done on tissues from the heart, lungs, kidneys, liver, spleen, pancreas, adrenals, and some brains.

The last three organs showed no pathological features in any of the rats. The liver and spleen were frequently congested, while the kidneys sometimes showed focal inflammatory and degenerative changes. The main *pathological changes* in the experimental group were noted *in the lungs and heart.*

In the lungs, the bronchi were frequently dilated, their walls infiltrated with lymphocytes, while the mucosa and muscle were often atrophied. Frankly purulent bronchial exsudate also occurred. Interstitial round cells and fibrosis were sometimes encountered, emphysema was the rule. These changes while present in one or two rats of the control group and in group "B" (100 mg/l NO_2) were found with increasing frequency and severity in the last three groups treated with higher NO_2 doses (Fig. 1 and 2).

In the heart, there were small foci of cells and fibrosis in some animals, while a mode diffuse interstitial cellularity with pronounced degenerative foci was frequent in the highest nitrite groups only. The oil red O stain showed no increased lipid deposits in the hearts of the experimental animals.

The *intramural coronary arteries* provided the surprising feature of this study. In most of the control animals the blood vessels showed some degree of thickening and often even a marked hypertrophy and narrowing. In the experimental groups and especially in group E, who received about 250-350 mg/kg of $NaNO_2$ for 2 years the coronaries were thin and dilated, their appearance not what is usually seen in animals of advanced age (Fig. 3 and 4).

In each group, tissue was taken from the heart for electronmicroscopic examinations. Full details of the pathological findings are to be reported upon elsewhere.

The Effects of the Administration of Sodium Nitrite in Drinking Water on Pregnant Rats and their Newborn

Two experimental groups were used, each containing twelve pregnant albino "sabra" rats. Group II was given 2000 mg/l sodium nitrite and Group III 3000 mg/l in their drinking water. The control, Group I, of seven pregnant rats, received tap water without added nitrite.

The pregnant rats that received nitrites showed an increase in *methemoglobinemia*. Group II had a mean of 15.5% Met-Hb. The Group III mean was 24.0% and the mean of the Controls was 1.1%.

The pregnant rats that received sodium nitrite suffered from *anemia* in a direct

relationship to the concentration of the compound in their drinking water. The Hb results are presented in Table 1.

In view of the marked *hematological effect of nitrite* on the pregnant rats, we measured red cell fragility with hypotonic solutions. Erythrocytes from pregnant rats that received nitrite showed less fragility to hypotonicity than those from the control rats.

In an experiment where the blood of rats was mixed with 0.0 to 0.9% sodium chloride solutions (i. e. 0 to 100% normal saline), we obtained typical sigmoidal fragility curves. In red cells, 50% hemolysis corresponded (from the curves) with 25.4% normal saline for Group III with 30.7% normal saline for Group II and 37.8% for Group I, the controls. Thus red blood cells from the treated groups were more resistant to hypotonicity than those from the control group.

It is possible that nitrite has some contact effect on the erythrocyte in which the "weaker" cells only are affected, the "stronger", more resistant cells remaining unaffected. Incubation of erythrocytes, in vitro, with sodium nitrite does not have any effect on the reaction of hypotonic osmotic pressure.

MetHb is reduced in the erythrocyte to Hb with the aid of an enzyme system utilizing DPNH for this purpose:

$$MetHb + DPNH \rightarrow Hb + DPN$$

In future experiments we will test the possibility that the increase in MetHb changes the metabolic picture in the red

Table 1 Distribution of hemoglobin determinations in pregnant rats chronically exposed to sodium nitrite in drinking water

gm% Hb	Group I %	Group II %	Group III %
10 >	0	45.2	52.4
10.1–12.0	31.5	27.2	26.0
12.1–14.0	44.6	21.6	13.0
14.1 +	23.9	6.0	8.6

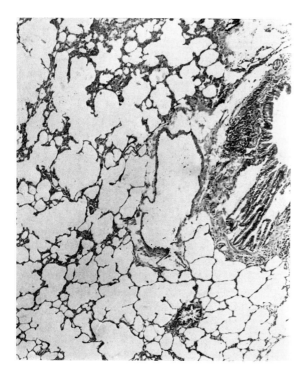

Fig. 1 Lung, control, 2-year-old rat. (×150)

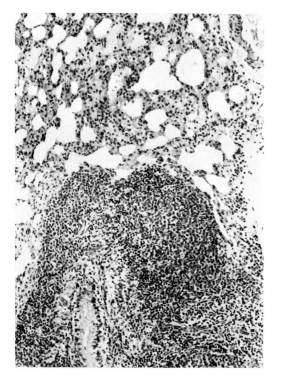

Fig. 2 Lung, rat, after two years of drinking water containing 1000 ppm NaNO₂ (×150)

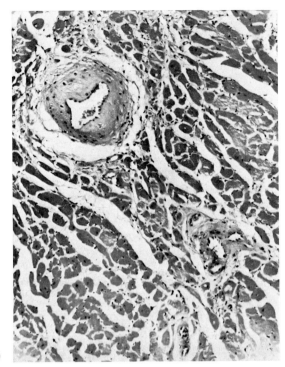

Fig. 3 Heart, control, 2-year-old rat (×105)

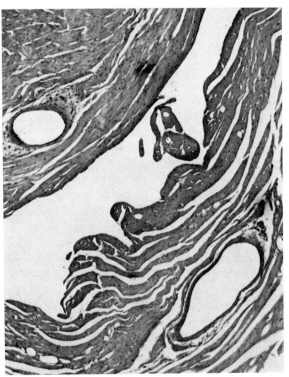

Fig. 4 Heart, rat, after two years of drinking water containing 1000 ppm NaNO₂ (×63)

blood cell. It is assumed that the DPNH/ DPN ratio is small during the MetHb increase, due to the rapid consumption of DPNH.

The *reduction of the MetHb* is rapid and uptake of DPNH outbalances the rate of regeneration of DPNH in the glycolytic pathway. By way of checking the DPNH/ DPN ratio we assayed the *pyruvate/lac-tate ratio* in blood from pregnant rats which, together with lactic dehydrogenase constitutes the main system in the ery-throcyte using DPNH/DPN.

Lactate + DPN → Pyruvate + DNPH

Competition for DPNH between the py-ruvate/lactate and the MetHb reductase systems may increase the pyruvate/lactate ratio. The results are presented in Table 2. The increase in the ratio was attributable to increase in pyruvate (up to 4 times more than in the control) as opposed to the small lactate increase. This points to possible increased glucose metabolism to provide extra DPHN supplies to the cells. There was a pronounced *effect on morta-lity,* among newborn rats of dams re-ceiving 2000 mg/l (Group II) and 3000 mg/l (Group III) in their water, particu-larly in the three-week period up till weaning.

The average litter in the control group con-tained 10 fetuses, 9.5 in Group II and 8.5 fetuses in Group III. The mortality with-in the first three weeks was 6% in the control, as opposed to 30% in Group II and 53% in Group III. Birthweights were similar with 5.5 gm for each group. How-ever, after birth, newborn rats in Groups II and III lagged behind the controls in their *growth rates.* For example, after 1 week the mean weight was 16.5 gm in the control group, 12.0 gm in Group II and 9.5 gm in Group III. After 21 days (at the end of the period of giving nitrites to the dams), 51.5 gm mean weight in the control group, 29.5 gm in Group II and 18.5 gm in Group III.

Table 2   Pyruvate/lactate ratio in pregnant rats chronically exposed to sodium nitrite in drinking water

Group	pyruvate/lactate ratio
I (controls)	0.013
II (2000 mg/l)	0.038
III (3000 mg/l)	0.042

Apart from the weights, a characteristic difference observed in the experimental groups was that the *fur thinned* and lost its luster. After separation from their dams and being put on water, there was an improvement in growth in the experi-mental groups. At the age of 32 days, the mean weight for the control group was 100 gm, for Group II 67 gm and for Group III 39 gm. At 62 days the control group attained a mean weight of 213 gm, Group I, 181 gm and Group II, 172 gm.

During the whole period from birth to weaning, the newborn showed no abnor-mally high MetHb. The mean hemoglobin of the newborn from the experimental groups was low — about 20% of the con-trol group.

Induced Methemoglobinemia by Transplacental Transport of Nitrites in Rats

Since clinical methemoglobinemia from nitrates in water apparently appears only in babies, one possible solution proposed for areas with high nitrate water is to supply the infants with bottles of low nitrate content water from alternate sources. This measure will not exclude the risk of the exposure to nitrates in the prenatal stage, i.e. the transfer of nitra-tes or nitrites through the placenta to the fetus, where methemoglobinemia may be induced or other toxic effects may occur. Pregnant women can also consume nitri-tes in "corned" meat products.

The possibility of this occurring was tes-ted on pregnant rats. Nitrites were given

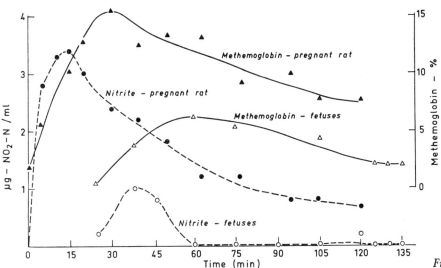

KINETICS OF NITRITE AND METHEMOGLOBIN IN BLOOD OF A PREGNANT
RAT AND THE FETUSES   (30 mg/kg NaNO₂ per os)

Fig. 5

in drinking water or by injection to the pregnant rats and subsequently nitrites and methemoglobinemia were tested in the fetal blood.

Suckling "sabra" rats whose dams got nitrites in the drinking water only after giving birth showed no rise in MetHb levels. By contrast, their dams showed high MetHb levels. This demonstrates that nitrites are apparently not transferred in appreciable amounts to the young rats via the milk.

The transfer of nitrites to the fetus in situ and the production of MetHb was tested in the following experiment. Pregnant "sabra" rats were used in this experiment. Each pregnant rat was weighed and anesthetized with ether. Suitable quantities of blood were collected from the tail at regular intervals throughout the experiment. After opening the abdomen the fetuses were removed serially at 10—15 minute intervals from alternate sides of the womb over a two hour period, the umbilical cords being cauterized. The

fetuses were washed in saline solution at 37° C, and then decapitated. Blood was collected and the initial MetHb and nitrite levels were determined. From 2.5—50 mg/kg of sodium nitrite were given orally or injected intraperitoneally to the pregnant rat and the kinetics of nitrites and MetHb in the dam as well as in the fetuses were measured. Micro-methods developed by our group for determining MetHb [8] and nitrites [15] in blood enabled us to carry out these experiments with the small amount of blood available from each fetus.

The characteristic picture obtained is shown in Fig. 5. After a 30 mg/kg dose of NaNO₂ per os to the pregnant rat, nitrites rose in the fetal blood though with a lag of about 20 minutes behind the dam. The rise in nitrite was followed by a rise in MetHb. The kinetic picture was similar between the dam and the fetuses.

Different concentrations of nitrites caused similar kinetic pictures differing only in their timing and methemoglobin levels. Table 3 shows that the threshold of the effect was at a sodium nitrite dose of 2.5 mg/kg.

The increase in effect was steep with increased sodium nitrite dosage. The possibility that the placenta was damaged during the experiment and that this led to increase permeability was excluded when sodium nitrite was given to normal pregnant rats after labor had started. The first fetus showed a normal MetHb level of 1.6% but those who were born after the chemical had been given showed levels of 10.1% MetHb and 0.3 µg/ml of sodium nitrite (as N) in their body. All births were unassisted.

These experiments demonstrate the *transfer of nitrites through the placenta* and the production of MetHb. The low levels of methemoglobin found 12 hours after birth from dams chronically exposed to nitrites in their water throughout the period of gestation can be explained by the rapid recovery from methemoglobinemia even in the new-born rats. In mature rats, the time needed for the reduction of half of the MetHb level is about 90 minutes. The results underline the possible risk of intrauterine methemoglobinemia when water or foods containing nitrates or nitrites are consumed during pregnancy.

Behavioral Studies in Mice Chronically Exposed to Sodium Nitrites in Water

In an effort to develop sensitive tests to detect possible effects of subclinical methemoglobinemia, behavioral studies with mice were undertaken (I). Groups of 57-black-6 J male mice were given nitrites in their drinking water at doses aimed at producing MetHb levels varying from slightly above normal (1%) to 15% which can be considered in the sub-clinical range. $NaNO_2$ doses in water were 100,

Table 3   Blood nitrite and methemoglobin levels after injection of different doses of sodium nitrite

$NaNO_2$ dose mg/kg	Peaks of MetHb level as per cent of total Hb		Peak of nitrite level in blood as $NaNO_2$ µg/ml	
	mother	fetus	mother	fetus
0	1.0	0.9	–	–
2.5	3.4	1.9	3.9	traces
5.0	5.0	2.7	6.9	traces
10.0	11.9	5.1	8.9	0.4
15.0	17.0	7.9	10.8	1.2
20.0	33.2	13.3	21.7	5.9
25.0	40.4	19.2	25.6	6.9
30.0	60.2	27.2	32.5	9.4

1000, 1500, 2000 mg/l. Controls received tap water. The level of motor activity was determined in a special barrier activity box designed for psychological studies with mice[11].

Analysis of the results shows a decided and significant *reduction of overall motor activity* in the groups receiving the highest levels of nitrites. There is a significant inverse relation between MetHb level and motor activity with a coefficient of correlation of– 65. An effort to counteract the methemoglobinemia was made by giving the group receiving the highest level of nitrites (2000 mg/l) vitamin C. The effect was to reduce the MetHb levels close to normal but the activity level of the group so treated remained low and about equal to the equivalent group receiving no antidote.

Brain Electrical Activity Changes in Rats Resulting from Chronic Exposure to Sodium Nitrite in Water

Studies were initiated to determine whether there were any detectable brain electrical activity changes in rats exposed to varying levels of sodium nitrite in their drinking water[13]. It was the specific objective of these studies to develop sensitive neurophysiological parameters to measure the effects of nitrite consumption at levels leading to sub-clinical methemoglobinemia.

Electrodes were implanted on the cortex of 4 groups of 3 month old male, albino "sabra", rats. Group A, the controls, received tap water; Group B, 100 mg/l; Group C, 300 mg/l; and D, 2000 mg/l of $NaNO_2$. For controls EEG recordings were made on each rat several times before starting the regime of nitrites in their drinking water. After exposure to nitrites had begun, regular EEG recordings were made about every 3 days for a period of 2 months. After 2 months, nitrites were removed from the drinking water. EEG recordings were taken at intervals for another 4 month period.

Analysis of the results shows that there was an *increase in the frequency of the EEG background waves* in the experimental Group D over the controls while in Groups B and C frequency of the EEG background was slightly reduced. Rats in all experimental groups showed paroxysmal outbursts not appearing in the Control Group or in their own pre-nitrite EEG recordings. On ending of nitrite intake the electrical outbursts disappeared only in Group B which received 100 mg/l in their water, which nevertheless continued to show significant brain electrical activity changes. Groups C and D showed continuation of the same type of EEG outbursts over the 4 month period. The findings of this study are to be reported upon in full elsewhere.

Discussion

Previous work on the toxicology of nitrates and nitrites has not given rise to any serious concern [4, 10] and both compounds are widely used as legally approved food additives. Two-hundred ppm of either nitrate or nitrite are allowed in sausages and other "corned" meat products, despite the fact that nitrite is many fold more toxic than nitrate.

Nitrates are also widely found as a natural component of many foods [12]. In our own studies [6] we detected as much as 4850 ppm of NO_3 and 233 ppm of NO_2 in a local variety of spinach. Samples of beets, cauliflower, cabbage, rhubarb, and radishes were all found to contain over 1000 ppm of NO_3. Under certain conditions of preparation and storage a portion of the nitrates can be reduced to nitrites by bacterial flora [6, 12]. Numerous cases of *methemoglobinemia*, some fatal, have been reported on in connection with infants who ate spinach.

Phillips [12] calculated that in a typical Canadian meal, adults may consume some 313 mg of nitrate from various fresh foods or about 4.5 mg/kg. Under normal circumstances this is not considered detrimental. However all, or a major portion of the nitrates might be converted to nitrites prior to eating, or by stomach flora after eating.

A 20-kg child who consumes a 100 gram portion of processed meat treated with 200 ppm of nitrite would be exposed to a dose of some 1 mg/kg. The same child would be exposed to 10 mg/kg of nitrites if he drank 2 liters of water having a nitrate concentration of 100 ppm, and all of it was converted to nitrite by intestinal microflora. Likewise the dose might reach 15 mg/kg if 100 grams of spinach, having 5000 ppm of nitrate, were consumed. From this it appears that nitrate or nitrite doses of 1—10 mg/kg may occur in humans and in extreme cases the exposure may be several times greater.

Our findings of distinct pathology in the heart and lungs of rats chronically exposed to 2000 and 3000 mg/l of $NaNO_2$ in their drinking water for a two year period, providing a sodium nitrite dose of about 250—350 mg/kg is striking although difficult to extrapolate to humans, both because the pathology was detected only at massive dose levels and since there is always the possibility of species-specific toxic effects.

The lack of the usual aging changes in the coronary arteries is noteworthy and may possibly be associated with known pharmacological properties of nitrogen compounds as dilating agents. Organic nitrogen compounds are considered as a possible cause of heart damage among explosives workers[2].

The high mortality rate among newborn rats of dams who received large doses of sodium nitrite during the gestation period and up to weaning may be associated with some form of prenatal damage to the fetus which did not lead, however, to abortions or a lower live litter size.

However, it also might be suggested that dams treated with such high doses of sodium nitrite and suffering from severe methemoglobinemia would not function normally and the high mortality of the suckling rats was not necessarily related to any direct toxic effect of nitrites on them.

Nevertheless, we have clearly demonstrated that nitrites fed to the dam are transferred to the fetus through the placenta leading o the development of methemoglobinemia in the fetus. The fact that this effect was detectible when a dose of 2.5 mg/kg of sodium nitrite was administered to the dam is a cause for some concern since such a dose is not much larger than might be consumed by humans under certain conditions. Here again we have no evidence that this same mechanism occurs in humans. A study of MetHb levels in infants cord blood, borne of mothers consuming drinking water high in nitrates has been initiated to help clarify this question.

The results of the behavioral studies in mice chronically exposed to sodium nitrite in their water seem to indicate that nitrites reduce motor activity.

The finding of regular paroxysmal outbursts in rats not appearing in the control group or in their own pre-nitrite EEG recordings for all levels of sodium nitrite administered was unexpected. Even more unexpected was the finding that in all treated groups, brain electrical activity changes did not disappear after the exposure to nitrites ceased. This appears to indicate some irreversible brain electrical changes resulting from chronic exposure to drinking water containing as little as 100 ppm of sodium nitrite.

No histopathology was done on these animals, but controls with the same electrodes showed no changes in brain electrical activity.

The mechanism of the toxic effect is unclear at this time and our investigations of the phenomenon are continuing.

In conclusion, we must caution that it is too early to extrapolate from these preliminary experimental findings with laboratory animals to humans, but some of the results point to the possibility that sodium nitrite, which is so widely used as a food additive is not as safe as has generally been considered to date. In addition, these findings should be carefully considered in evaluating the possible health impact of increasing concentrations of nitrates in water supplies as well as nitrates found naturally in many vegetables, since nitrates can under certain circumstances be converted to nitrites which are potentially more hazardous.

Acknowledgements

Many persons participated in the studies which were carried out as part of a team effort. The contributions of our key collaborators, Mrs. S. Cohen, Mr. K. Behroozi, and Dr. H. Shecter are particularly noteworthy.

The pathological work-up was made by Dr. R. Yarom of the Department of Pathology. The behavioral studies with mice were supervised by Dr. R. Guttman of the Department of Psychology and carried out by Mr. K. Behroozi. The EEG studies were supervised and interpreted by Dr.

S. Robinson of the Talbieh Psychiatric Hospital and carried out skillfully by Mr. K. Behroozi. Mrs. H. Eshed provided valuable assistance in the laboratory work.

References

[1] Behroozi, K., R. Guttman, N. Gruener, H. I. Shuval: Changes in the motor activity of mice given sodium nitrite in drinking water. Proceedings Annual Symposium on Environmental Physiology, Beersheva, Dec. 1971 (in press)

[2] Carmichael, P., J. Lieben: Sudden death in explosive workers. Arch. Environ. Health 7: 424–439, 1963

[3] Comly, H. H.: Cyanosis in infants caused by nitrates in well water. J. Amer. Med. Ass. 129: 112–116, 1945

[4] Druckrey, H., D. Steinhoff, H. Beuthner, H. Schneider, P. Klarner: Screening of nitrite for chronic toxicity in rats. Arzneimittel Forsch. 13: 320–323, 1963

[5] Evelyn, K. A., H. T. Malloy: Microdetermination of oxyhemoglobin methemoglobin and sulfhemoglobin in a single sample of blood. J. Biol. Chem. 126: 655–662, 1938

[6] Eisenberg, A., E. Wisenberg, H. I. Shuval: The public health significance of nitrate and nitrite in food products, Research Report Ministry of Health, Jerusalem (in Hebrew). 1970

[7] Gruener, N., H. I. Shuval: Health aspects of nitrates in drinking water. In: Developments in Water Quality Research. Ed. by H. I. Shuval. Ann Arbor-Humphrey Science Publishers, Ann Arbor pp. 89–106, 1970

[8] Hegesh, E., N. Gruener, S. Cohen, R. Bochkovsky, H. I. Shuval: A sensitive micromethod for the determination of methemoglobin in blood. Clin Chem. Acta. 30: 679–682, 1970

[9] Knotek, Z., P. Schmidt: Pathogenesis, incidence and possibilities of preventing alimentary nitrate methemoglobinemia in infants. Pediatrics 34: 78–83, 1964

[10] Lehman, A. J.: Nitrates and Nitrites in meat products. Assoc. of Food and Drug Officials, U.S. 22: 136–138, 1958

[11] Lieblich, I., R. Guttman: The relation between motor activity and risk of death in audiogenic seizure of DBA mice. Life Sciences 4: 2295–2299, 1965

[12] Phillips, W. E. J.: Nitrate content of foods, public health implications. Canadian Inst. of Food Tech. Jour. 1: 98–103, 1968

[13] Robinson, S., K. Behroozi, N. Gruener, H. I. Shuval: Changes in the electrical activity of the brain of rats fed with a sodium nitrite solution. Proceedings, Annual Symposium on Environmental Physiology. Bersheba, Dec. 1971 (in press)

[14] Schmidt, P.: The influence of an atmosphere contaminated with sulphur dioxide and nitrous gas on the health of children. Zeitschr. Ges. Hyg. & Grenz. 13: 34–38, 1967

[15] Schechter, H., N. Gruener, H. I. Shuval: A micromethod for determining nitrites in blood. Anal, chim. Acta 60: 93, 1972

Cadmium Content in Sea Water, Bottom Sediment, Fish, Lichen, and Elk in Finland

T. Jaakkola, H. Takahashi, J. K. Miettinen

Department of Radiochemistry, University of Helsinki, Finland

Summary. A survey is given of the Cd-levels in different samples throughout Finland. The highest Cd-level in water measured just outside the sewer of the zinc refinery at Kokkola was about 50 times higher than the presumable natural level, but still within the international permissible level in drinking water. The handling of finely powdered zincsulfide during 7 years in open air has increased the Cd-level five-fold in 1 kilometers' distance.

Zusammenfassung. Dieser Beitrag gibt einen Überblick über Kadmium-konzentrationen in einer Reihe von Umweltproben aus verschiedenen Gegenden in Finnland. Die Kadmiumkonzentration, welche in Quellwasser nahe der Abfallschlammhalde der Zinkraffinerie *Kokkola* gemessen wurde, war etwa 50mal höher als der vermutliche natürliche Wert; aber immer noch innerhalb des international zugelassenen Wertes für Trinkwasser. Durch Umgang mit fein pulverisiertem Zinksulfiterz während 7 Jahren im Freien wurde die Kadmiumkonzentration im Abstand von einem Kilometer um das 5fache erhöht.

The great *toxicity of Cd* to all forms of life is well known. For most aquatic organisms, cadmium is the second in order of toxicity, of the heavy metals, mercury being first[1]. The great toxicity of Cd for laboratory animals and man has been known for many years.

In *rats*, Cd, administered in doses capable of producing tissue concentrations comparable to those found in modern man, causes *arterial hypertension, sclerosis* of small arteries in kidneys, heart and other organs, *proteinuria, testicular atrophy,* and *neoplasms*[2-7]. In *man*, Cd is suspected of causing *pulmonary emphysema, proteinuria* and *prostatic carcinoma*[2]. It is absent at birth, but accumulates, especially in the kidney, with increasing age[8]. Perry et al.[9] found in the human kidney values (as ppm of ash) varying from 750 (Ruanda Urundi Africa) through 2250 (USA) to 5300 (Japan) of

Cd. Schroeder and Balassa[8] reported similar high values in Japanese samples. They calculated that the diet could contain 4 to 100 µg/day of Cd depending on whether sea food and kidneys were eaten. In Japan, rice could also contribute significantly to the total intake of Cd[8].

In Japan Cd has been recently identified by Kobayashi[10] as the main causative agent of a chronic disease which has afflicted the residents at *Jintsu river,* in Toyama prefecture, since the second world war. The patients have severe osteomalacia and suffer from intense pain in their bones. Kobayashi[10] reported the following Cd values (ppm of ash) found in the organs of a patient who died: kidney 4,900, liver 7,050, rib 11,500. The disease was called *"itai-itai-byo"* (ouch-ouch disease) because of the shrieks of the patients suffering from chronic pain. In February 1968 the death toll was 119 persons, mainly middleaged women who had borne several children. Only 5 were men. In addition, 94 non-lethal typical cases and an estimated 1000 latent cases were reported[11]. In the advanced stage of the sickness, the bones become brittle and fractures occur easily[10]. A zinc mine at the upper course of the Jintsu River was identified as the origin of the cadmium. Production of zinc and lead was increased rapidly during the war without proper treatment of the wastes, which were discharged into the river. Fine ore particles still containing heavy metals were deposited on rice fields downstream causing not only high cadmium levels in the rice, but also damage to the crop. In addition, many farmers used river water for drinking and food making, and river fish for food.

Kobayashi reports 0.5—1.1 percent cadmium, 0.02 percent lead, and 0.3 percent zinc in the bone ashes of one patient. Polished rice from the district contained 1.19 ppm of Cd in dry matter and 125 ppm in ash. Later the disease was identified on Tsushima Island outside a mine. The well water used by one patient on Tsushima contained 0.225 ppm of Cd, 13.37 ppm of Zn and 0.41 ppm of Pb[10]. Kobayashi[10] confirmed, by feeding experiments with rats, that cadmium replaces calcium, a fact known for some time[12].

In the USA, Carroll[13] observed a positive *correlation* between the *Cd content* of air and the incidence of *hypertension* and *arteriosclerosis* in 28 North-American cities.

In Scandinavia and Finland, a marked increase in the death rates caused by *cardiovascular diseases* has occurred in recent years. Between 1955—64 this increase was 8 in Sweden, Finland 14, Denmark 17, and Norway 25 percent, while the increase of deaths occurring from *ischemic heart disease* was even sharper: in Finland 33, Sweden 36, Denmark 41, and Norway 73 percent[14]. Masironi[15] speculates that increased industrialization in Norway has brought about higher pollution with *Cd* which, being *accumulated by shellfish* and other marine animals, is eventually ingested by humans. This hypothesis is based on the facts that both the consumption of seafood, which is generally notoriously rich in *Cd*[13], and the increase of cardiovascular diseases are exceptionally high in Norway. An increase of the *Cd level* in fish and shellfish in Norway has not been reported, however. Masironi concludes that the experimental, clinical and statistical evidence indicates that *Cd* may be an etiological factor in some forms of cardiovascular disease, but further investigations are needed to ascertain whether *Cd* really plays a role and to establish what mechanism is involved[15].

In Sweden, occurrence of *Cd in the environment* has been studied recently by two groups of investigators: Rühling and Tyler in Lund[16-19] and Ljunggren et al. in Stockholm and Lysekil[20]. The former authors analyzed several heavy metals in the moss *Hylocomium splendens* from northern and southern Scandinavia, finding, for *Cd*, 0.18 ppm (in dry material) in northern and 0.99 ppm in southern samples[18]. Tyler reports values for seven heavy metals in the cypress moss (*Hypnum cupressiforme*) in the neighborhood of the industrial city of Norrköping in southern Sweden. In an area of a few square kilometers at the north-east corner of the city (the windward portion) values as high as 3 to 4 ppm were found, and in an area of about 100 km² values around 1 ppm were found. Ljunggren et al[20] report the cadmium levels in plankton, plants, insects, crustaceans, and fish from a fresh water system beginning at a non-polluted lake and running through urban and industrialized regions. The *Cd* levels of those organisms increased in the respective order. Homogenized dried samples originated in the non-polluted lake had the lowest values, 0.6—1.5 ppm, while the highest values 0.6—17 ppm, were found in samples from the polluted river. Pike from polluted fresh water gave 0.17 ppm of fresh weight. Six pike from the Baltic outside the city of Oskarshamn, which is known to spread pollution, gave $\leqq 0.020$ but one pike gave $\leqq 0.1$ ppm Cd per fresh weight.

Landner[21] in a preliminary experiment kept the aquarium fish guppy in water containing 0.1 ppm Cd. The cadmium level of the fish rose from 2 to 10 ppm in 7 weeks. When the fish were placed into clean water, their Cd level decreased with a half-time of about 18 days.

The purpose of the present study is twofold: First, to get a broad picture of the *cadmium levels in environmental samples in Finland,* in comparison with other countries, both in coastal waters and inland, and second, to check the local environmental levels in the vicinity of the city of Kokkola, where a large zinc refinery with an annual capacity of 90,000 tons of zinc became operational at the end of 1969. The locations from which the first group of samples were taken are presented in Fig. 1; locations of the Kokkola samples are presented in Fig. 2. As hydrospheric samples, water, bottom sediments, and organs of pike were analyzed. The *flesh of pike* is probably *representative* of the sea food that man is likely to consume. As terrestrial sample, reindeer lichen *(Cladonia alpestris)* was chosen, because it grows very slowly (a 10-cm-high plant may be 50 years old) and retains all heavy metals very efficiently, as is well known from fallout studies. One analysis of elk kidney was also performed to have an idea of its im-

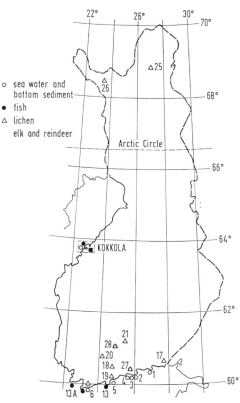

Fig. 1 Map of Finland showing the locations where samples were collected for cadmium analyses

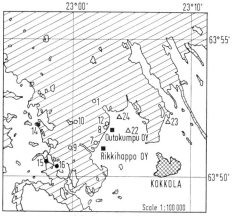

Fig. 2 Map of sampling locations in the vicinity of Kokkola. For location of this area see Fig. 1

portance as a dietary source of *Cd*. About 8,000 elk are consumed annually in Finland, which has about 4.7 million inhabitants.

Methods Used

Collection of Samples

Sea water samples were taken, from a boat, at a depth of about 0.5 m and placed into a 5-liter polyethylene container which had been rinsed 3 times with the local sea water before sampling. The samples had not been filtered but were homogenized by shaking before the aliquot for analysis was taken.
Bottom sediment samples were taken by a Birge-Ekman sampler accepting an area of 20 x 20 cms. Collection of about a 1 cm thick surface layer of the sediment was attempted.
The intact *fish* were sent fresh to the laboratory where the organs were separated, dried, and homogenized.
Lichen samples consisted of the clean upper portion containing about half green "flower" and half grayish older cells.

Direct Analysis of Wet-Ashed Solution

The samples of bottom sediment, fish, lichen, and tissues of elk and reindeer were dried overnight at 110° C. The bottom sediment samples were leached with 6 N HCl and the precipitate was removed by filtering. The digestion of the fish, lichen, elk, and reindeer samples was carried out by HNO_3 — $HClO_4$ wet-ashing. All sample solutions were diluted with water to a known volume.
The Cd levels of solutions of these samples, with the exception of the fish flesh, were determined directly by *atomic absorption spectrophotometry*. The atomic absorption spectrophotometer used was the Perkin-Elmer Model 303. The detection limit of this instrument for Cd in

water solution was 0.005 ppm. It was found that the total amount of solid material in the sample higher than 0.5% interferred with the determination of Cd. This matrix effect is due to the light scattering, which is especially potent at low wavelengths and when a high concentration of Ca is present[22].

Concentrative Extraction Using the Radioisotope [109]Cd for Checking the Yield

Sea Water. The *Cd content of sea water* samples is so low that it is necessary to concentrate the samples and to separate cadmium from the alkali metals and alkaline earths in order to allow atomic absorption analysis.
To determine the Cd content of sea water samples, the following method was developed:

1) 1000 ml of sea water is poured into an evaporating dish. Radioactive carrier-free [109]Cd, in solution, is added to check the yield of the procedure. (The half-time of the [109]Cd isotope is 453 days; it decays completely by electron capture and γ-radiation of 88 KeV by the daughter [109]Ag is emitted). The sample is then evaporated to dryness.

2) The residue is dissolved with 10 ml of concentrated HCl and 40 ml of distilled water is added. The sample then is filtered and the solution is adjusted to pH3 using NH_3.

3) The solution is then transferred to a separatory funnel. 10 ml of 5% aqueous solution of ammonium pyrrolidine dithiocarbamate (APDC) and 20 ml methyl-iso-butyl ketone (MIBK) are added[23, 24]. The mixture is thoroughly shaken for five minutes and the layers are allowed to separate. The water phase is discarded.

4) The Cd is transferred back into the water phase by shaking twice for five

minutes, each time with 5 ml of 1 N HCl. The combined water phase is transferred to a 10-ml centrifuge tube. The residues of MIBK are separated by centrifugation and discarded with the organic phase.

5) The solution is aspirated by the flame of the atomic absorption spectrophotometer. The Cd content of the sample is determined using standard solutions of Cd, which have been extracted in the same manner as the sample.

6) An aliquot of the 10 ml-water solution

obtained in step 4 is taken for yield determination. The radioactivity of this aliquot is determined using a NaI(Tl)-detector and a multichannel analyzer. By comparing the radioactivity found with the amount added, the yield of the procedure can be calculated.

Using the method described, a hundred-fold concentration of sea water samples is obtained. The yield of the method was, on the average, 92% (ranging from 79% to 100%). The precision of the method

Table 1   Cadmium content of sea water samples in 1970. Nos. 1–6, Gulf of Finland, Nos. 7–12, Gulf of Bothnia. Sampling depth is in all cases 0.5 m.

No.	Region	Cordinates	Depth of sampling place (m)	Sampling date (day, month)	Cd content in ppb[1]		
					1	2	average
1_W	Kotka, Mussalo	60° 25′ N, 26° 53′ E	1	1.12	0.2	–	0.2
2_W	Loviisa, Hamnholm	60° 13′ N, 26° 14′ E	40	2.8	0.4	0.4	0.4
3_W	Porvoo, Pellinki	60° 14′ N, 25° 51′ E	10	2.8	0.5	0.5	0.5
4_W	Porvoo, Wessö	60° 17′ N, 25° 41′ E	2[2]	5.6	2.0	2.3	2.2
5_W	Helsinki, Tervasaari	60° 11′ N, 24° 58′ E	3	18.8	1.0	0.4	0.7
6_W	Tvärminne, Lappohjan selkä	59° 51′ N, 23° 16′ E	1	22.7	0.1	0.1	0.1
7_W	Kokkola, Rikkihappo Oy[3]	63° 51′ N, 23° 02′ E	1	3.8	0.3	–	0.3
8_W	Kokkola, Outokumpu Oy[3]	63° 52′ N, 23° 03′ E	1	3.8	10.2	–	10.2
9_W	Kokkola, shipping channel	63° 51′ N, 23° 00′ E	10	3.11	0.6	0.6	0.6
10_W	Kokkola, shipping channel	63° 52′ N, 23° 00′ E	14.5	3.11	0.5	0.5	0.5
11_W	Kokkola, shipping channel	63° 53′ N, 23° 02′ E	13	3.11	0.5	0.5	0.5
12_W	Kokkola, Outokumpu Oy[3]	63° 52′ N, 23° 03′ E	1.5	3.11	5.1	5.1	5.1

[1] ppb = 10^{-6} g/l.
[2] High turbidity.
[3] Sample taken at about 40 m distance from the mouth of the sewer.

Table 2   Cadmium content in bottom sediments (1970). Nos. 1–6, Gulf of Finland, Nos. 7–12, Gulf of Bothnia

No.	Location	Sampling depth (m)	Sampling date	ppm Cd/dry wt	Ash %	ppm Cd/organic dry wt (calc.)
1_M	Kotka, Mussalo	1	1.12	0.47	86.0	3.4
6_{M1}	Tvärminne, Byviken	5	23.7	1.88	80.3	9.6
6_{M2}	Tvärminne, Lappohjan selkä	32	23.7	1.16	85.3	7.9
7_M	Kokkola, Rikkihappo Oy[1]	1	3.8	0.17	99.8	85.0
8_M	Kokkola, Outokumpu Oy[1]	1	3.8	1.37	98.0	69.0
9_M	Kokkola, shipping channel	10	3.11	0.92	93.7	15.0
10_M	Kokkola, shipping channel	14.5	3.11	1.23	93.8	20.0
11_M	Kokkola, shipping channel	13	3.11	0.29	99.0	29.0
12_M	Kokkola, Outokumpu Oy[1]	1.5	3.11	1.18	99.1	131.0

[1] Sample taken at about 40 m distance from the mouth of the sewer.

Table 3   Cadmium content in organs of marine fish (pike, Esox lucius L.) at the shores of Finland.

No	Tissue	Weight of fish (kg)	Location	Sampling date	ppm Cd/ dry wt	ppm Cd/ fresh wt
6_{F1}	flesh	0.5	Gulf of Finland, Tammisaari	1. 11. 70	0.011	0.002
	liver				0.014	0.003
	kidney				0.108	0.028
					0.760	0.153
6_{F2}	flesh	≈ 0.5	Gulf of Finland, Tvärminne	1. 11. 70	0.014	0.003
	liver				0.173	0.055
	kidney				1.150	0.232
$13A_F$	flesh	0.55	Gulf of Finland, Hiittinen	12. 6. 68	0.013	0.003
					0.021	0.004
13_F	flesh	0.6	Gulf of Finland, Porkkala	12. 6. 68	0.016	0.003
					0.019	0.004
14_F	flesh	0.9	Gulf of Bothnia, Kokkola, Långholmen	6. 8. 70	0.018	0.004
	liver				0.148	0.034
	kidney				1.519	0.169
15_F	flesh	0.65	Gulf of Bothnia, Kokkola	23. 11. 70	0.024	0.005
	liver				0.390	0.096
	kidney				1.550	0.339
15_{F2}	flesh	0.64	Gulf of Bothnia, Kokkola	23. 11. 70	0.018	0.004
	liver				0.280	0.070
	kidney				1.520	0.308
	flesh	2.48	Gulf of Bothnia, Kokkola	23. 11. 70	0.062	0.013
16_F	liver				0.409	0.113
	kidney				1.480	0.295

was good, according to the parallel determinations presented in Tables 1 and 3. *Fish Flesh.* To determine the *Cd level of fish flesh* samples, the radioactive ^{109}Cd solution was added at the beginning of the wet-ashing. The solution obtained by wet-ashing was adjusted to pH$_3$ and the determination of cadmium proceeded as described in connection with the sea water samples (step 3).

Results

The results of *water* analyses are presented in Table 1. The samples can be divided into 3 groups: the non-polluted ones (Nos. 1 and 6) having 0.1 to 0.2 ppb of Cd, the heavily polluted ones from outside the zinc refinery (Nos. 8 and 12) having 5 to 10 ppb, and the less polluted ones having 0.4 to 0.7 ppb. Sample No. 4 was an exception, as it was very turbid

and contained richly organic mud particles.

The results of the *bottom sediment* analyses are presented in Table 2. All results are between 0.17 and 1.9 ppm Cd/dry weight. As the sediments varied greatly, from nearly pure sand (Nos. 7, 11, 12) to those rich in organic matter (Nos. 1 and 6), the cadmium level was also determined per unit weight of organic dry matter. The "non-polluted" areas gave values between 4 and 10 ppm, the coastal area outside Kokkola gave 70 to 130 ppm and sediments taken about 2 kms. from the shore gave 15 to 30 ppm.

The results of the *fish organ* analyses are presented in Table 3. In most cases, the liver contains about 10 times more Cd than the muscle flesh of the same fish, and the kidney contains about 30 to 80 times more.

The muscle flesh and the liver of the older fish No. 16 contained 2 to 3 times more cadmium than the corresponding organs in the younger fish from the same area. There is no difference in the kidney values, however. The Kokkola values are roughly two times higher than the other ones showing the slight, initial effects of pollution.

The results of the *lichen* analyses are presented in Table 4. They can be roughly divided into 3 categories. The samples from Lapland contained about 0.1 to 0.2 ppm of cadmium those taken near the zinc refinery in Kokkola contained about 1 ppm, and the reminder contained, for the most part, 0.3 to 0.4 ppm.

The results of the *elk and reindeer* analyses are presented in Table 5. The blood and meat of the elk contained about 2 ppm of Cd per unit of dry matter, while the kidney contained about 50 ppm. Elk kidney is thus about 20 times greater as a source of Cd than elk meat, and 200 times greater than pike flesh from the same region.

Discussion

Comparison of the Cd levels in the *coastal waters of Finland* with the one value reported in *well water in Tsushima* (225 ppb) shows that even the highest value found in Finnish coastal sea water, 10 ppb in front of the sewer in Kokkola, falls far below the aforementioned Japanese well water value (which had been

fatal). The present values found in pike muscle, mostly 0.002 to 0.004 ppm in fresh material, are lower than the one value given by Ljunggren et al: ≤ 0.015 ppm. The present lichen values established for Lapland, 0.1 to 0.2 ppm, are the same as Rühling and Tyler found in the moss *Hylocomium splendens:* 0.18 ppm,

Table 4 Cadmium content in lichen *(Cladonia alpestnis)* in Finland

No.	Location	Sampling date	Cd content ppm/ dry wt.
6$_{L1}$	Tammisaari, Koverhar	21. 12. 70	0.41
6$_{L2}$	Tammissaari, Tvärminne	21. 12. 70	0.35
17$_1$	Virolahti, Klamila	30. 4. 67	0.32
17$_2$	Virolahti, Klamila	6. 11. 67	0.32
17$_3$	Virolahti, Klamila	22. 6. 68	0.29
18	Tuusula	2. 10. 64	0.21
			0.30
19$_1$	Helsinki, Kontula	22. 4. 67	0.46
19$_2$	Helsinki, Kontula	15. 10. 67	0.41
20$_1$	Loppi, Räyskälä	28. 11. 67	0.29
20$_2$	Loppi, Räyskälä	9. 10. 68	0.13
21	Asikkala, Salonsaari	20. 9. 70	0.18
22	Kokkola, Ykspihlaja	3. 8. 70	1.10
23	Kokkola, Kalvholm	23. 11. 70	0.32
			0.35
24	Kokkola, Sannanranta	23. 11. 70	0.94
			1.06
25$_1$	Lapland, Inari	26. 6. 67	0.09
25$_2$	Lapland, Inari	23. 7. 67	0.10
25$_3$	Lapland, Kaamanen	3. 9. 69	0.43
25$_4$	Lapland, Kaamanen	3. 9. 70	0.17
25$_5$	Lapland, Kaamanen	4. 9. 70	0.13
25$_6$	Lapland, Muddusjärvi	6. 9. 70	0.13
26$_1$	Lapland, Hetta	22. 8. 69	0.38
26$_2$	Lapland, Hetta	9. 9. 70	0.05
26$_3$	Lapland, Hetta	9. 9. 70	0.13

Table 5 Cadmium content in organs of elk and reindeer

No.	Sample	Animal age, sex	Location	Sampling date	Cd content ppm/ dry wt	ppm/ fresh wt.	Dry wt./ fresh wt. (%)
27	meat of elk	2–3 y, ♀	Porvoo	24. 10. 65	2.35	0.59	25.2
27	liver of elk	2–3 y, ♀	Porvoo	24. 10. 65	5.50	1.50	27.2
27	kidney of elk	2–3 y, ♀	Porvoo	24. 10. 65	51.9	8.05	15.5
28	blood of elk	1.5 y, ♂	Tuulos	28. 11. 65	2.35	0.51	21.5
25	backbone of reindeer	4.5 y, ♀	Lapland, Inari	11. 12. 65	6.90	5.10	73.8

and our highest overall lichen value, 1 ppm, was also the same as that found in southern Sweden: 0.99 ppm. Handling of zincsulfide has increased the Cd level five-fold at about 1 kilometers' distance from the factories (samples Nos. 22 and 24), but no increase is yet noticeable at a distance of about 4 kilometers (No. 23). On the whole, there does not seem to be an alarming level of Cd in the neighborhood of the *Kokkola zinc works*. Although these zinc works had been operating only for a few months when the samples were taken, zincsulfide had been loaded into ships in this place during 7 years already. Most of the increased level of Cd evidently is derived from this source.

Acknowledgements

Our thanks are due to Mr. M. A. Rainio, Water Administration, Kokkola District (Kokkolan maanviljelysinsinööripiiri), for delivering the Kokkola samples; to Mr. I. Vöry, for collecting the lichen samples from Lapland in 1970 and to Miss Aila Nuutinen for technical assistance in the analytical work.
This study was partially financed by a grant from Nordforsk, which is hereby gratefully acknowledged.

References

1 Keckes, S., J. K. Miettinen: Mercury as Marine Pollutant. FAO Techn. Conf. on Marine Pollution. Rome, Dec. 1970, published in: Marine Pollution and Sealife, ed. M. Ruivo, Fishing News Ltd., London, pp. 216–288 (1972)

2 Patty, F.: Industrial Hygiene and Toxicology, Vol. II. Ed. by D. Fassett and D. Irish. Interscience, New York 1963

3 Schroeder, H. A.: Amer. J. Physiol. 207: 62, 1964

4 Schroeder, H. A., J. J. Balassa: Amer. J. Physiol. 209: 433, 1965

5 Schroeder, H. A. et al.: J. Nutr. 86: 51, 1965

6 Schroeder, H. A. et al.: Arch. Environm. Health 13: 788, 1966

7 Schroeder, H. A.: Circulation 35: 570, 1967

8 Schroeder, H. A., J. J. Balassa: J. Chron. Dis. 14: 236–258, 1961

9 Perry, M. M. et al.: J. Chron. Dis. 14: 259–271, 1961

10 Kobayashi, J.: Relation between the "Itai-Itai" Disease and the Pollution of River Water by Cadmium from a Mine, Paper presented at the 5th Intern. Water Poll. Conf., San Francisco, July 1970

11 Shukan Asahi: Feb. 1968

12 Nicaud, P., A. Lafitte, A. Gros, J. P. Gautier: Bull. et Mém. Soc., méd. hop. Paris 19: 204, 1942

13 Carroll, R. E.: J. Amer. Med. Ass. 198: 177, 1966

14 World Health Org. (WHO), Cardiovascular Diseases, Annual Statistics 1955–1964 by sex and age, Epidem. Vital Statist. Rep. 20: 9–10, 1967

15 Masironi, R.: "Trace elements in cardiovascular diseases" in Uses of Activation Analysis in Studies of Mineral Element Metabolism in Man. IAEA Techn. Rept. No. 122, Vienna 1970

16 Rühling, Å., G. Tyler: Bot. Notiser 122: 248–259, 1969

17 Rühling, Å., G. Tyler: Oikos 21: No. 1, 1970

18 Rühling, Å., G. Tyler: Regional variations in fallout of heavy metals in Scandinavia, Report No. 10, Lund's University, Dept. of Ecological Botany. Feb. 1970 (in Swedish)

19 Tyler, G.: Heavy metal contaminants at the region around the city of Norrköping, Sweden. – A report from the Department of ecological botany, Univ. of Lund. Sept. 1969

20 Ljunggren, K., B. Sjöstrand, A. G. Johnesl, M. Olsson, G. Otterlind, T. Westermark: Activation analysis of mercury and other environmental pollutants in water and aquatic ecosystems, IAEA, symposium on the Use of Nuclear Techniques in the Measurement and Control of Environmental Pollution. Salzburg, Austria, Oct. 1970

21 Landner, L.: "Cadmium in aquatic systems", IVL internal report 91 / 21.3.69

22 Billings, G. K.: Atomic Absorption Newsletter 4: 357, 1965

23 Mulford, C. E.: Atomic Absorption Newsletter 5: 88, 1966

24 Mansell, R. E., H. W. Emmel: Atomic Absorption Newsletter 4: 365, 1965

Research in the Gesellschaft für Strahlen- und Umweltforschung on the Evaluation of the Risks Involved in Irradiation

W. Pohlit

Gesellschaft für Strahlen- und Umweltforschung MBH, Munich, West Germany

In this report, of course, not all the activities of the Gesellschaft für Strahlen- und Umweltforschung (GSF) can be considered. However, a general survey about the problems in the *estimation of risks* due to *irradiation with ionizing radiation* is given using examples form the work of GSF. Some of the basic assumptions and formulations used here may be of importance in other fields of environmental research too. Therefore an exchange of scientific ideas as well as suitable solutions of problems between these different fields should be intensified.

The term "risk" is in use in various scientific as well as in non-scientific areas with quite different meanings. Therefore before entering a discussion of radiation risk a clear *definition* will be given here for a *risk factor*, η, which should be explained first in a simple but useful example.

Assume, for example, that different types of airplanes are produced — a fleet of exactly hundred planes of each type. Then these airplanes are used in a number of flights of equal length under equal weather conditions. The number of planes in original state N, will then decrease as a function of the number of flights n, of the fleet due to different aircraft accidents. This decrease in number, dN, would be proportional to the number of planes in the fleet N, and would be proportional to the increase of number of flights dn. Therefore, one can write the equation

$$dN = -\eta \cdot N \cdot dn \qquad (I)$$

where η is a constant.

The factor η can be explained best by using numerical examples: Assume, for example, N = 100 and η = 0.03, then after the first flight of the fleet (dn = 1) it follows

$$dN = -\eta \cdot N \cdot dn = -0.03 \cdot 100 \cdot 1$$
$$= -3.$$

This means that dN = 3 planes are lost.

If one would have used a larger value for η, for example, η = 0.1 then the change in the plane number under the same conditions could be dN = 10. One can see from these examples that the factor η expresses, what we would call "risk" in the common language. Therefore η sould be named the "*risk factor*" and its precise definition would be (from equation I):

The risk factor η is the quotient of dN and $N \cdot dn$, where dN is the change in the number of objects, N, from their original state due to the measurable amount or number of stress dn.

$$\eta = -dN/N \cdot dn \qquad (II)$$

Before using this defined risk factor in the field of radiation reactions, some other general facts should be explained.

The differential equation (I) can easily be integrated and results in

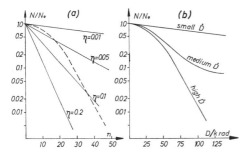

Fig. 1a Examples for the decrease of the number of objects N in original state as a function of the number of stress, n (N_0 is N at n = o;   = risk factor). The dotted line is a demonstration for = f (n)

Fig. 1b Examples for the decrease on the number of biological objects, N, as a function of absorbed dose, D, (N_0 is N at D = 0) (1)

$$N = N_0 \cdot e^{-\eta \cdot n} \tag{III}$$

where N_0 is the number of objects for n = o. In Fig. 1a examples for this relation are drawn in semilogarithmic coordinates.

Different values of η may occur from various reasons, for example, in the case of aircraft accidents:

a) due to different reactions considered (small defect, total crash)

b) due to different types of planes (new, old, badly constructed)

c) due to different conditions during flight (quiet, stormy, hurricane).

Therefore η generally is different for different reactions a), different objects b), and different environmental conditions c):

$$\eta = f (a, b, c) \tag{IV}$$

Especially from b) one can see that in most cases η would not be a constant for increasing numbers of flights; the planes get older by the time and η changes. Therefore the curve for N/N_0 in a semilogarithmic plot gets steeper with time or and shows a "shoulder" as indicated in Fig. 1a by the dotted line. Airlines

take this into account by repeated repair to keep the machines "new" or by excluding old machines from use.

All these basic statistical considerations generally can be applied also for radiation effects in living objects with the absorbed dose as the *stress quantity*. A certain effect is considered by which the object is changed from its original state into a "changed state". All environmental conditions should be kept constant, then the number of objects in the original state is given by:

$$N = N_0 \cdot e^{-\eta \cdot D} \tag{V}$$

where D is the absorbed dose and the risk factor in this case has the dimension (absorbed dose)$^{-1}$. Correspondingly the number of objects in the changed state, N', will be:

$$N' = N_0 (1 — e^{-\eta \cdot D}). \tag{VI}$$

In many cases from reasons as discussed above, η is not a constant but changes with absorbed dose. Therefore in many cases "shoulder curves" are obtained for *biological radiation reactions* (Equation V) or sigmoid curves in the case of (Equation VI). Up to now one is not able to analyze quantitatively such shoulder curves for radiation reactions. As a consequence, extrapolation to the region of small absorbed doses is not possible. This is. however, of extreme interest because experiments cannot be performed with sufficient accuracy there. Therefore some groups in the GSF perform radiobiological experiments with different types of living cells to clarify the mechanism leading to such *dose-effect curves* as shown in Fig. 1b. *Mathematical and cybernetic models* for the radiation reactions in living cells and the biological counter reactions have been developed and are under experimental examination.

Quite different types of *radiation reactions* have been investigated to clarify, which one would be of highest importance for radiation protection in the different stages of development of the organism:

1) *Induction of mutations*
2) *production of tumors*
3) *shortening of life span.*

For example, the production of *lethal mutations* has been studied by irradiation in different spermatogenic stages in mice. As can be seen from Fig. 2, this radiation reaction is dependent on the state of cell development. These differences are expressed especially at lower absorbed doses. As a result one cannot decide from an experiment with a relatively large absorbed dose, which reaction or which type of cell will be the most sensitive. At least one must take this into consideration in the estimation of radiation risk factors. The most sensitive stage in the development of mice seems to be in the interval of change from spermatogonia to spermatozytes. Another high sensitivity or large risk factor can be observed from irradiation near mating of female animals after mating. If a proportionality of lethal mutations in late spermatogo-

nia with absorbed dose is assumed, an increase of 0.01 in N/N_0 over the spontaneous rate would be caused by an absorbed dose of about 4 rad. The corresponding absorbed dose for such an increase of lethality during mating would be even smaller.

These data are one of the scientific reasons for the exclusion of female workers during pregnancy from control areas by legal regulations. The relatively low sensitivity of a developed fetus and of grown-up animals to radiation, on the other hand, reflects the high degree of security in the maximum permissible absorbed doses for radiation workers (5 rad per year) and for the population (0.5 rad per year).

The reasons for this different *radiation sensitivity* through spermatogenetic and embryogenetic phases are unknown up to now. It is not even possible to decide, if many different loci for lethality are present in the living cell which are responsible for the different sensitivity, or the same locus of lethality is reached by different pathways. Also it must be stated that the transfer of these data from mice to man remains uncertain.

In contrast to these radiation effects during the early stages of the development of an organism, late *radiation damage* can be expected especially in the field of incorporation of radionuclides into the living organism. In these cases, the absorbed dose rate can be quite low so that somatic radiation effects can be eliminated or be repaired by the organism but oncogenetic effects predominate.

Such reactions have been under investigation also in the GSF, especially using the short living radionuclide ^{224}Ra, which has a half life of only 3.6 days. These investigations have been performed mainly to clarify the differences between reactions of *alpha particles* and *beta particles*, between short exposure and con-

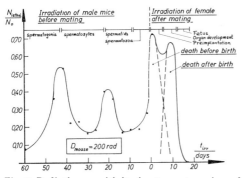

Fig. 2 Radiation sensitivity in spermatogenic and embryogenic stages of mice. Drawn is the quotient of lethal embryos, N_{lethal}, and normally expected embryos, N_0, as a function of time of irradiation, t_{irr}, before (2) and after time of mating ($t_{ii} = 0$) (3)

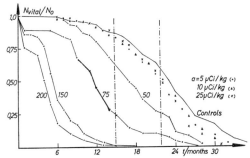

Fig. 3 Number of vital animal, N_{vital}, relative to the number of control animals, N_o, as a function of time after treatment, t, with different administered specific activity a = A/m, where A is the activity and m the mass of the animal, of ^{224}Ra (4)

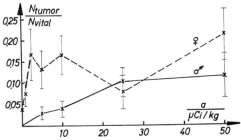

Fig. 4 Number of animals with tumors, N_{tumor}, relative to the number of control animals, N_o, as a function of administered specific activity, a, of ^{224}Ra (5)

tinuous exposure, between homogenous and inhomogenous distribution of radiation absorbed dose in the organism.

As can be seen from Fig. 3, an application of 5 to 25 μCi per kg of body weight reduces the life expectancy of mice just significantly. In this case of ^{224}Ra the total absorbed dose in the skeleton is highest and amounts to 140 to 700 rad respectively. For these low specific activities and also for 50 μCi/kg the life time seems to be shortened mainly due to the development of tumors in the animals. But at specific activities higher than 75 μCi/kg other additional effects seem to be responsible for shortening of the life time.

The *tumor incidence* as a function of ad-

ministered specific activity is given in Fig. 4 for male and female mice. As can be seen the number of male tumor animals N_{tumor}, relative to the number of vital male animals at that time, N_{vital}, increases up to a value of 0.1 which remains about constant for higher specific activities. This corresponds to the curve of type b in Fig. 1b. In female animals a higher number of tumors is induced and the increase is much steeper. The normal tumor incidence in unirradiated mice is doubled for about 2 μCi/kg which corresponds to an absorbed dose of 56 rad in the skeleton.

The two vertical lines in Fig. 3 indicate the positions between which the maximum of tumor incidence has been observed at all different specific activities. This shows that the tumor induction is a special late effect of radiation injury which needs several months to be manifested.

The question arises if this latency time is a characteristic one for tumor development in mammalian cells in general and therefore can be applied also to man, or if it is related to aging processes in the organism. Also the type of storage of radiation damage over such long time periods remains unknown until now, but is one of the topics of research in the GSF.

After having discussed some of the research work to determine basic information about the risk factors as used in Equation V, it should be stated that by reducing the absorbed dose D also a reduction of radiation damage can be achieved, of course. For *practical radiation protection* this is of outstanding importance and is an important factor for radiation risk in a broader sense. Therefore the evaluation of factors which reduce the absorbed dose received by man is also one of the topics of research in the GSF.

As indicated in Fig. 5 the absorbed dose D received by the object is the result of a reduction by the transmitting medium from an absorbed dose D* which would be present in the object without the medium. Therefore extensive physical measurements have been made in the GSF to determine optimal data for the *shielding* of all types of *radiation sources;* only one typical example should be given here:

If x-rays are used for medical diagnosis, the radiation absorbed dose received by the examining doctor partially is due to radiation scattered from the walls of the room. Due to high absorbed dose rates in the vicinity of the walls, large distances must be provided in medical rooms as can be seen from Fig. 6. About 2 m distance are necessary to reduce this scattered radiation down to the mean room level. Very often an unnecessary high absorbed dose rate is accepted if this distance is not available. The scattering of photons can largely be influenced by a thin layer of specially developed tiles and the radiation absorbed dose rate in the body of the practicing doctor can be reduced by a factor up to four in such case as shown in Fig. 6.

Other research in the GSF is related to the questions of K_{source} in Fig. 5. From *atomic power reactors,* waste with high radioactivity of several Megacurie per year is produced which has to be stored under extreme safe conditions. The GSF therefore operates an unused old salt mine, capable to receive all the *radioactive waste* of the FRG during the next decades. The validity of an extremely small value of K_{source}, reducing the potentially high absorbed dose D** to a negligible dose D*, has been checked by research work of the "Institut für Tieflagerung" and the "Institut für Radiohydrometrie", both operated by the GSF. These investigations have included the

measurements of mechanical strength of the salt caves under extreme conditions of heat and irradiation, the heat transfer from the radioactive material to the surrounding salt and the distribution and movement of ground water around the salt mine under the earth.

All factors together in Fig. 5, the technical and the biological, finally are responsible for the injury from radiation sources of different kinds. Only the complete knowledge about all of these factors will enable us in future to use the benefit of these sources in technical and medical applications without harmful influence to living organisms.

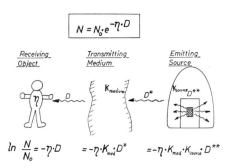

$$N = N_0 \cdot e^{-\eta \cdot D}$$

$$\ln \frac{N}{N_o} = -\eta \cdot D \quad = -\eta \cdot K_{med} \cdot D^* \quad = -\eta \cdot K_{med} \cdot K_{source} \cdot D^{**}$$

Fig. 5 *Scheme of influencing factors in radiation protection*

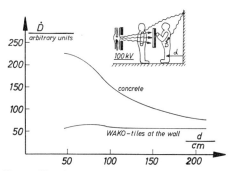

Fig. 6 *Absorbed dose rate in examining doctor, D, as a function of distance from the wall, d, for concrete wall and wall covered with WAKO tiles (6)*

References

[1] Pohlit, W.: Recovery and repair of radiation damage in yeast cells. Int. Congress Radiation Research Evian, 1970

[2] Ehling, U. H.: Comparison of radiation- and chemically-induced dominant lethal mutations in male mice. Mutation Research 11: 35–44, 1971

[3] Kriegel, H.: Über die teratogene Wirkung ionisierender Strahlung. Zentr. Veterinärmed., Beih. 11: 1969

[4] Hug, O., W. Gössner, W. A. Müller, A. Luz, B. Hindringer: Production of osteosarcoma in mice and rats by incorporation of Ra 224. In: Radiation-induced cancer. IAEA, Vienna, 393–409, 1969

[5] Gössner, W., B. Hindringer, O. Hug, A. Luz, W. A. Müller: Early and late effects of incorporated ^{224}Ra in mice. Int. Congress of French Soc. for Radioprotection Grenoble, 1971

[6] Göttel, K., A. Kraft: Rückstreuung von Röntgenstrahlung. In: Jahresbericht 1970, Institut für Strahlenschutz. GSF – S – 125

Research in the Gesellschaft für Strahlen- und Umweltforschung on the Evaluation of the Risks Involved in Environmental Chemicals

W. Klein

Institut für ökologische Chemie der GSF, St. Augustin 1, West Germany

Any approach to the evaluation of impact of environmental chemicals on environmental quality has in the first step to go via experimental investigations of model compounds which are representative and important in practice. There will be a deep gap in the knowledge of all influences of all environmental chemicals on the quality of our environment in the future, too, and therefore we should try to gain at least competent knowledge of the influences of the chosen *model compounds* on the chemical and toxicological qualities of the environment. The priorities for model compounds should be listed by using the following 6 parameters; each of these parameters has to be focussed differently for each different chemical:

1) *Production level*
2) *Use pattern*
3) *Persistence*
4) *Conversion under biotic* and *abiotic conditions*
5) *Dispersion tendency*
6) *Predictable biologic effects* based on structure — activity-relationship

The basis for any measurements or estimation of the quantitative anthropogenic load on the environment by chemicals is the knowledge of production level, use pattern, and persistence of the respective chemical. If additionally its dispersion tendency were also known, these figures would allow the concentrations of the chemical in the environment to be calculated. Today, however, we have still to determine the effective load by analyzing representative samples.

In cooperation with the Institute for Irradiation Protection, the Department for Physical Technology, the Department for Nuclear Biology, and groups outside the *Gesellschaft für Strahlen- und Umweltforschung*, the Institute for Environmental Chemistry tries to measure changes of the chemical environment caused by *inorganic chemicals,* such as mercury, cadmium, lead, copper, cobalt, beryllium, and other trace metals, by analyzing representative water-, soil-, air- and biologic samples using atomic absorption and activation analysis. Representative investigations on the occurrence of organic chemicals in the environment are done by the Institute for Environmental Chemistry mainly on the example of *PCBs, hexachloro-cyclohexane isomers,* and other *halogenated compounds.*

The Institute for Environmental Chemistry has a number of research projects concerned with the fate of potentially hazardous environmental chemicals under biotic and abiotic conditions. This activity is due to the fact that it is becoming more and more urgent to know the *fate* of these chemicals, as a basis for measuring *long-term risks to man* and his animate environment as well as for the *development of new technologies.* As model compounds for investigations into the fate of environmental chemicals we have selected DDT, PCBs, lindane, cyclodiene insecticides, and other chlorinated aromatic chemicals. We have chosen these compounds since, on the one hand, more is known about their break-down and behavior under environmental conditions than for other environmental chemicals; and, on the other hand, *terminal breakdown products*, neither metabolites nor chemical conversion (e.g. atmospheric) products are known, which naturally includes the lack of knowledge of their toxicological consequences.

The objectives of our investigations in this context are the identification of *significant conversion products* in the environment for their toxicological evaluation, and also *balance studies* which will give information on the long-term impact of these compounds on the environment.

I would like to briefly discuss two results of these investigations using the example of *aldrin* (Fig. 1).

During the last three years we have demonstrated that a number of halogenated pesticides are *metabolized by higher plants* under green-house conditions. In order to measure residues of these metabolites under practical conditions, we have set up an open field trial with aldrin-[14]C. In cooperation with institutions outside the GSF, several plant species (potatoes, wheat, sugar beets, and maize) were grown under practical conditions in the U.K., USA, Spain, and Germany. The different locations were chosen for the measurement of the *influence of climatic conditions* on total residues and conversion ratios.

Figure 1 shows high residues of conversion products more polar than dieldrin.

Fate of Aldrin—14C in Maize After Soil Application

Under Practical Conditions

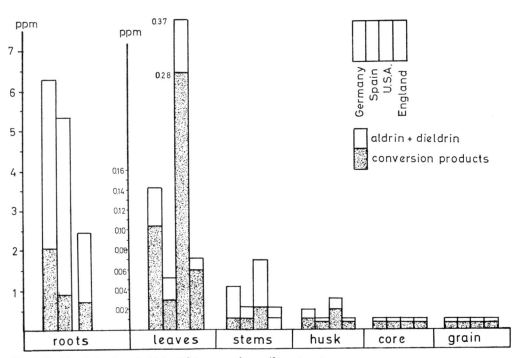

Fig. 1 Residue behavior of aldrin-14C in corn after soil treatment

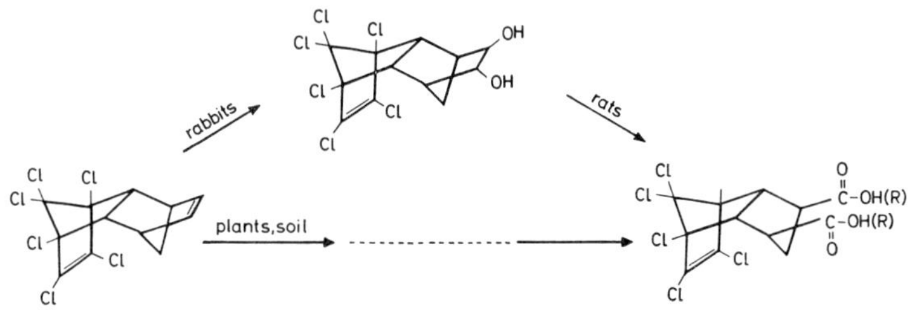

Fig. 2 Metabolism of Aldrin-Dieldrin-^{14}C

The major metabolite of aldrin in higher plants — a water-soluble metabolite — has recently been identified as dihydro-chlorden-dicarboxylic acid, shown in Fig. 2.

Since the same compound has also been identified as a mammalian metabolite of aldrin-dieldrin its toxicological evaluation ought to be feasible.

The *atmospheric* concentration of *dieldrin,* its conversion by sunlight to photo-dieldrin and its dechlorination upon UV-irradiation have been known for some years. Recently we found that upon *irradiation of dieldrin* in the presence of pollutants such as nitrogen-oxides or ozone, nitrated and hydroxylated derivatives as well as a known mammalian metabolite are formed. These substitutions do not occur in gaseous phase reactions, but it is known that environmental chemicals in the atmosphere are partly absorbed by particles. Since it is at present impossible to simulate atmospheric conditions including atmospheric particles, we do not know whether these derivatives might occur in the environment.

When discussing environmental chemicals absorbed by particles, a research project of the Department for Biophysical Irradiation Research has to be mentioned. This project is concerned with the transport and deposition or *resorption* respectively of *aerosols* with a particle size below 0.5 nm in the human lungs.

With this project we began to discuss toxicological consequences of environmental chemicals.

A research group which is exclusively concerned with the *toxicological evaluation* of *environmental chemicals* for men and animals is being built up in our organization at present and will, for a start, study the global occurrence of PCBs and mercury compounds.

Departments of the Institute for Biology, with sophisticated experience in the evaluation of biological irradiation effects, are now also using their know-how for the detection of the effects of environmental chemicals.

The Department for General and Experimental Pathology is, in this context, intensifying histochemical investigations of precursors of *liver cancer* upon the action of chemical carcinogens.

Mutation experiments of the Department for Genetics have the objective of testing the human *genetic risk* from environmental mutagens and of elucidating the mechanism of mutations.

The experimental methodology used is the *specific locus method* and the *dominant lethal method* in mice. So far, alkylsulfonates and mitomycin C have been investigated. From the results obtained I would only like to mention that mitomycin induces, when applied in ppm-concentrations, dominant lethal mutations in specific states of spermatogenesis in mice and also effects on reproduction in fermale mice.

The Department for Irradiation Biology and Biophysics is investigating *cytogenetic effects* of chemicals in *human peripheral lymphocytes*. Women with ovarian carcinoma were treated with 1—18 gm endoxane for a period of 6–8 weeks. On the average more than 15% of the cells studied in the blood samples had chromosomal defects. After cyclamate treatment lymphocytes of a group of patients showed significantly elevated chromosomal defects (3.3%) compared to control groups. No significant chromosomal defects could be detected after intake of lead, either in in vitro experiments, in occupationally exposed workers or in persons with acute lead poisoning.

The Department for Nuclear Biology is studying the effects of DDT on *mineral metabolism* of mammals. Facilities are being developed now, which can be used as routine tests for the detection of the effects of other environmental chemicals on mineral and bone metabolism respectively. Furthermore, this department prepares experiments on the toxic effects of chemicals in the embryonal and fetal period and on the diaplacental penetration of these compounds.

We expect that by progressive cooperation of the research groups in the GSF with scientists and organizations outside the GSF we will succeed in better evaluating the risks involved in environmental chemicals with exact experimental data.

Environmental Quality:
When Does Growth Become too Expensive?

S. F. Singer

Department of the Interior, Washington, D. C. 20240, USA

Summary. It is clear that the present costs of maintaining environmental quality are huge but not yet excessive when measured in terms of other public expenditures. It is also clear that we cannot continue in the same manner since the costs are rising faster than either the population or the GNP. Various technological fixes are possible which will alleviate the present situation and postpone the inevitable confrontation. These include reuse and recycling to overcome some of the problems of solid wastes (particularly junk automobiles), the problems of industrial wastes, mining wastes, and agricultural wastes. Reuse of water by chemical-physical purification, or by desalting, will become increasingly important, coupled with better and more economical ways of using water.

Another mandatory requirement is a more adequate spreading out of the existing and especially of the future population. We simply cannot let our present cities expand without limit; nor are satellite cities located within the megalopolis belts a real long-term solution. They just diffuse the problem a little bit. What we need are brand new cities in areas of the country that have not hitherto been settled. We have many areas where the geology and the climate are highly desirable; with transportation available and with communication established; we only require water to support an economic base of industry and population.

Ultimately, however, the general upward trend in the cost to maintain environmental quality must set a limit to a population in a city or in a region which can be established fairly accurately, and to the country as a whole, with the optimum range established less accurately since it depends on the distribution of the population.

Zusammenfassung. Es steht fest, daß die heutigen Unkosten zur Erhaltung der Umweltqualität hoch, jedoch nicht übertrieben hoch sind, wenn man sie mit denen anderer öffentlicher Ausgaben vergleicht. Weiterhin steht fest, daß es so nicht weitergehen kann, denn die Kosten nehmen schneller zu als die Bevölkerung oder das Bruttosozialprodukt. Es besteht die Möglichkeit, verschiedene technologische Bestimmungen zu erlassen. Dadurch würde die heutige Situation gemildert und die unvermeidbare Gegenüberstellung hinausgezögert werden. Diese Bestimmungen umfassen Wiedergebrauch und Wiederverarbeitung, um die Probleme der festen Abfallstoffe (bes. Auto-

teile) und Probleme der Industrie-, Bergbau- und landwirtschaftlichen Abfälle, zum Teil zu bewältigen. Der Wiedergebrauch von Wasser durch che-
misch-physikalische Reinigung oder durch Entsalzen im Zusammenhang mit
günstigerem und wirtschaftlicherem Wassergebrauch gewinnt immer mehr
an Bedeutung. Eine weitere verbindliche Forderung ist ein angemessenes
Anwachsen der heutigen und hauptsächlich der zukünftigen Bevölkerung.
Wir können unsere Städte einfach nicht bis ins Grenzenlose ausdehnen, und
Satellitenstädte innerhalb der Umgebung von Großstädten sind auf die
Dauer keine Lösung. Sie würden das Problem nur ein wenig verteilen. Was
wir brauchen, sind ganz neue Städte auf bisher unbesiedelten Landgebieten.
Es gibt viele Gegenden, in denen Boden und Klima günstig sind, wo es
gute Transportmöglichkeiten und Verbindungen gibt. Man brauchte nur
Wasser zur Versorgung einer wirtschaftlichen Grundlage für Industrie und
Bevölkerung. Schließlich müßte jedoch der allgemeine Aufwärtstrend der
Ausgaben zur Erhaltung der Umweltqualität der Stadtbevölkerung oder
den Bewohnern eines sorgfältig ausgestatteten Gebietes und dem gesamten
Land mit optimaler Ausdehnung, das aufgrund seiner Abhängigkeit von
der Bevölkerungsverteilung weniger sorgfältig ausgestattet ist, eine Grenze
setzen.

Environmental quality is not a luxury; it is an absolute necessity of life. Pollution increases the costs of various economic operations. In a manufacturing plant, if the intake water is not pure, then money must be spent in cleaning it before it can be used. In an electric power plant, if the intake water is too warm, then it will not provide adequate cooling and the thermodynamic efficiency of the power plant goes down. *Air pollution* causes economic losses, including sickness and loss of productivity, as well as direct damages to crops, equipment, and buildings.

I would like to organize my talk as follows: First, I would like to discuss the current and planned costs for various pollution control programs and put these in some meaningful relationship.

Second, I want to mention the need for technological fixes to overcome some major pollution problems. This need becomes quite obvious when you start to examine the cost figures. But not all problems can be solved by technology and, in many cases, technology only alleviates the immediate problem and postpones the inevitable.

Third, I would like to discuss the need for more basic changes of the attitudes of people, corporations, and governments: 1) avoiding *waste by* inherent reliability and *long life of products;* 2) improving the fundamental efficiency of products rather then merely and regularly changing their appearance; 3) the need for *recycling* and *reuse,* especially of water and metals; and 4) the need for real dispersion of a population to remove the high concentration in urban areas and megalopoles.

Finally, I want to discuss the trends which appear to be inevitable; namely, that environmental quality goes down as the level and affluence of the population rises. In fact there appears to be a fairly direct correlation between pollution and

the GNP. And, since the costs of achieving and maintaining environmental quality go up faster than the GNP, it follows
that a larger and larger fraction has to
be spent in order to maintain environmental quality. It is difficult to say at
what point such costs become unacceptable, whether at one percent, five percent,
or even twenty-five percent. But wherever
the fraction becomes unacceptable, it is
at that point that we have passed the
optimum level of population.

This is the general plan of my presentation today. Let me turn now to look
at the cost figures for pollution control
of land, air, and water*.

Land Pollution

Municipal Wastes

A recent survey shows that it costs $ 4.5
billion a year to dispose of 360 million
tons of *solid wastes* created annually in
the United States (or 10 pounds per person per day). Even though only half this
amount is regularly carted away, the
collection and disposal problem has become critical. Current estimates show
that public and private agencies in the
United States collect an average of 5.3
pounds per person per day of solid wastes,
or more than 190 million tons per year.
By 1980, this figure will have risen to
340 million tons per year and costs, too,
will have doubled[1]. In addition, about

* It should be realized at this point that the various
pollution problems are interrelated and that, in
solving one kind of pollution, one may only shift
it into a different part of the environment. For
example, the burning of solid waste produces air
pollution. Chemicals released into the atmosphere
will eventually wash down in rain and pollute
rivers. The ultimate sink, however, appears to be
the ocean which covers three-quarters of the
Earth's surface and, through its general circulation, carries water, waste materials, and pollutants
over the whole globe. Nevertheless, it is useful
to discuss each form of pollution separately, but
keeping in mind the interrelationships.

7 million passenger cars, trucks, and buses
are junked annually in the United States.
Waste disposal ranks right after education and road construction cost to municipalities. A new HEW report indicates
that it will take an additional $ 4.2 billion
over the next five years just to develop a
satisfactory system[2].

Mining and agriculture are the largest
waste producers, approximately ten times
the amount of municipal refuse.

Mining Wastes

Solid wastes from mining and processing
are accumulating at the rate of 1.5 billion
tons per year from mine waste, mill tail-
ings, washing plant rejects, processing
plant wastes, and smelter slags and rejects. This figure will rise to 2 billion
tons by 1980 and will go higher as lower
quality ores are developed[3]. There are
additional special pollution problems associated with mining; for example, the
problem of *acid mine drainage*, a widespread form of water pollution which is
produced by water seeping through abandoned and operating mines and leaching
acids into streams; the problem of brine
disposal in oil fields; and future problems
in the retorting of oil shale and in underground nuclear retorting.

Even without an increase in population
or in the GNP or in the demand for minerals — just satisfying the continuing
demand, but through lower quality ores,
will result in increased pollution.

Agricultural Wastes

These constitute the largest solid waste
problem in the United States, exceeding
even the mining wastes. They consist primarily of *animal manure* and other organic material which can be either burned
or oxidized more slowly by bacteria. In
addition, there are 2 billion tons of sedi-

ment which are washed down from farm-lands into rivers and streams every year[4]. Adding up all these numbers, we arrive at a total figure per person of 32 tons of solid waste per year. If all this material had to be disposed of, at about $ 25 per ton, this would be about $ 800 per person per year, or 20 percent of the GNP. The actual expenditures are probably less than 1 percent of GNP, but are rising.

There is a form of land pollution which is the complement to the types of pollution just discussed. I have in mind here the continual destruction of forest areas, grasslands, and even producing farms and orchards in order to make way for more housing, more highways, and more factories at the rate of 5 million acres per year.* This is directly related to an expanding population and, more particularly, to an expanding GNP. The cost to the Nation in terms of loss of recreational and resources is difficult to quantify. Like the *slag piles* that are not removed, the *strip mines* which are left

---

* The situation can be alleviated but not entirely eliminated by proper land use planning. It should be possible to arrange land uses so as to minimize needless destruction and so as to accomodate competing uses to the largest possible extent. One example of such accomodations is given in a piece of legislation recently introduced by the Department of the Interior dealing with estuarine and coastal zone management. The coastal zone is probably one of the most valuable, but also one of the most endangered, pieces of land in the United States. Here the population pressure is greatest and, at the same time, the amount of shoreline and estuarine land is limited. What has been proposed is a joint Federal-State endeavor in which the States prepare a long-range comprehensive plan for the management of the coastal zone, with guidance and support from the Federal Government. After this plan has been examined and accepted, the State then carries out its development in the coastal zone, for all purposes, in accordance with such a plan and uses the plan to determine land use assignments at the local government level. They make sure that zoning proposed by local governments is consistent with the plan.

unfilled, the silt which fills up lakes and estuaries — these are all costs which we pass on to the next generation.

For solid waste pollution, a partial answer lies in technological fixes and in recycling. One place where a technological fix is needed is in packaging which contributes about 46.5 million tons (about 13.3 %) to the Nation's annual refuse burden. This includes *cartons and paper wrappers* ($^2/_3$ by weight), about 25 billion glass bottles and jars, and some 50 billion metal cans[5].

Recycling of wastes is another necessity, with considerable economic potential. The ordinary household and commercial refuse collected annually contains more than 11 million tons of *iron and steel scrap*, almost 1 million tons of *nonferrous metals*, including such valuable materials as aluminum, copper, lead, zinc, and tin. In addition, 15 million tons of glass and lesser amounts of other worthwhile materials are recoverable from these two types of municipal wastes alone and the energy value of the waste is equivalent to the heating value of 60 million tons of coal. The total scrap and energy value that could be derived annually from these wastes approaches $ 1 billion. The potential scrap value of the iron and steel alone is nearly $ 200 million[6].

The principal problems in metal recovery are to remove copper from iron pieces and tin cans so as to get high-grade ferrous scrap; and to separate and recover the aluminum, copper, lead zinc, and tin values from the nonferrous fractions. Other efforts should be directed at reclamation of *cobalt, nickel, and chromium* from high temperature or super-alloy scrap, and the recovery of these metals from waste electroplating solutions which, incidentally, also eliminates a major waste pollution hazard.

Vast quantities of precious metals, such as gold, silver, and platinum are presently lost from *electronic equipment* in commercial and defense applications because of lack of effective recovery techniques. But electrolytic methods are being developed to transform contaminated aluminum into commercially pure metal.

A tremendous potential in recycling and recovery lies in the 7 million *automobiles junked* every year. A major problem is to eliminate copper which is a troublesome impurity. The eventual answer may lie in educating our manufacturers, and indeed all manufacturers, to pay great attention to the ultimate fate of the article they manufacture and to design it so that it can be easily recycled. Various positive and negative tax incentives could be used to accelerate the educational process.

Air Pollution

Air pollution has become mainly the problem of the cities, with the major contributors the *internal combustion engine, fossil fuel power plants,* industry, and the combustion of various materials — from leaves to refuse. The major pollutants are carbon monoxide (51%), sulfur oxide (18%), hydrocarbons (13%), nitrogen oxide (9%), and particles (9%). Of the 142 million tons of pollutants released into the atmosphere annually, about 60 percent comes from automobiles[7]. Carbon monoxide is produced almost entirely by automobiles. Sulfur oxides come mainly from the combustion of fossil fuels in *electric power plants* while hydrocarbons, nitrogen oxides, and lead come mainly from the gasoline-powered internal combustion engine.

There are some rather specialized problems. Hydrocarbons evaporate from *cleaning fluids,* and other volatiles leak into the atmosphere, e. g. nitrogen oxides from the excessive use of artificial fertilizers. Asbestos particles have been recognized as presenting a severe health hazard. Carbon dioxide is produced in vast quantities by the combustion of fossil fuels, but is not generally considered as a pollutant although it may have long-range effects on the climate. As far as we can tell now, even though the percentage of carbon dioxide in the atmosphere has risen by 15 percent and will continue to rise, the heating produced is small (less than 1° C) and may be overpowered and neutralized by the radiating effects of particulate material introduced into the atmosphere by pollution.

Another point of long-range concern is the *heat energy* released to the atmosphere by human industrial activities. In the United States, the average electric power generated in 1970 is 1.5×10^{12} KWH per year or 1.7×10^{11} watt. This works out to 70 milliwatts per m^2, about 0.1% of the value of solar heat contribution of about 100 watts per m^2. Total energy consumption in the United States in 1970 comes to 2.2×10^{12} watts, or nearly 1% of solar heat contribution. On a world-wide basis, energy consumption is $\sim 6.6 \times 10^{12}$ watts, area is 5.1×10^{14} m^2, giving an energy release of 1.3×10^{-2} watt/m^2, or 0.01% of solar heating.

Another specialized problem is the air pollution produced by *aircraft*. We are concerned here not only about the combustion products but also about the emitted water vapor which causes *high altitude contrails*. The artificial cirrus clouds which are thereby produced can have substantial effects on local weather and on world climate generally, if the frequency of occurrence is large enough. With the tremendous increase in high altitude jet travel, this is a distinct possibility.

Technological Fixes

The point about air pollution problems is that they are amenable to technological fixes, but at a cost. *Sulfur* can be removed from fossil fuels and, in some cases, the *sulfur extracted* has commercial value. In any case, sulfurous fuels can be and should be burned as far away as possible from populated regions so as to eliminate the synergistic effects of different types of air pollutants. Particles and most industrial air pollutants can be eliminated either by process design or by properly designed *pollution control devices* which are added at the end of the process. It is clear that we should provide positive incentives to industry to develop processes so that pollution will not be produced in the first place.

The pollution abatement costs are not excessive. For example, the additional annual cost of reducing human exposure to sulfur oxides and particulates by 60 to 75 percent in all metropolitan areas (containing 70 percent of the population) was estimated to be about three quarters of a billion dollars. Of this, manufacturing and electric power companies would likely bear about one-half the cost or $ 350 million — the equivalent of one-sixth of 1 percent of the value-added by industry to the production of goods and services in the United States in 1967. In contrast, another industry cost, labor costs rose by 5 percent or 30 times as much in 1967. Service and distributive firms and households would bear the balance of the abatement cost, averaging about $ 3 per person in cities.*

Other estimates, however, lead to higher costs. A major midwestern electric utility company prefers to import 6 million barrels of low sulfur residual oil (which produces 6.5 million BTU per barrel) at a cost of 45 c per million BTU; this oil would replace high sulfur local coal (which produces 26 million BTU per ton) at 30 c per million BTU.

* Jack W. Carlson, "What Price a Quality Environment", Journal of Soil and Water Conservation, May–June 1969.

If we were to apply this differential of 15 c per million BTU to the 400 million tons of coal used annually by utilities, the total national cost would be $ 1.5 billion/year, or $ 15 billion over a ten-year period which is a reasonable time scale for new technology development.

It is against such costs that one must judge the desirability of expending funds for new technical approaches, such as *sulfur removal* during the burning of coal in *fluidized beds*.

But the chief polluter, as mentioned earlier, is the internal combustion engine which accounts for nearly all the carbon monoxide, most of the hydrocarbons, nitrogen oxide, and nearly all of the lead. It is instructive to look at this example in more detail because it illustrates the great value of a national investment in technology which gets rid of pollution.

Today, the total *vehicle pollution* emission into the atmosphere is close to 90 million tons per year. By 1980 this will have been reduced to nearly 60, or about one-third less. However, from then on, in spite of ideal maintenance and with the use of the latest California standards, the emission load will reach 70 million tons by the year 2000 when the total registered vehicles will have climbed from 100 to 220 million[8].

But can we afford this ideal control program? Over a fifteen year period, say 1975 to 1990, the total costs to the public for emission control devices and maintenance has been estimated by one industry source as between $ 96 and $ 141 billion, without any essential improvement in air quality[9]. In fact, since it must be assumed that these devices will not work perfectly between inspections, the air quality will probably deteriorate. It is against costs like these that we have to weigh the importance of making major technological investments in low pollution engines or of devising other alternatives for mass transportation.

Water Pollution

Water pollution in some sense seems to be more fundamental than either land or air pollution: 1) Pollution persists and accumulates, especially in lakes; and 2) there are no technological fixes to the fundamentals of *waste production by human beings*. Water pollution itself, however, is due to the use of water to transport the wastes; it was the invention of the water closet over a hundred years ago in England that gave rise to the serious problems of river pollution which are just now being overcome by appropriate treatment plants.

In our discussion it is well to keep in mind that there are conservative pollutants, such as chlorides and other inorganic salts which dissolve in water, and non-conservative pollutants, such as heat, or organic waste materials which are oxidized naturally by biological action.

Domestic Wastes

What one would like to do is develop a method for determining the per capita cost to supply and clean up water to a certain degree of purity, as a function of population level and of concentration of population. Into this model we would feed the fact that there is a certain natural rate of disappearance of a non-conservative pollutant and that there is also a critical level of pollution above which many undesirable things take place. This critical level may be either a legal criterion incorporated into water quality standards, or it may be a physical-chemical level such as the *eutrophication* threshold for a lake; e.g., the critical concen-

tration levels for *nitrogen* and *phosphorus* above which we get prolific growth of algae.

Some simple considerations will illustrate how such a methodology can be constructed:

— 10^{12} gallons of water flow to the ocean in the continental U.S. each day.

— 18% into the Pacific, 21% into the Atlantic, 50% into the Gulf of Mexico, 11% into the Gulf of California.

— Oxidation 1 gallon of raw sewage requires 140 ppm of oxygen (or 0.0011 pounds of O_2 per gallon).

It has been estimated that each person contributes 135 gallons of sewage per day; this would demand 0.167 pounds of oxygen or $135 \times 140 = 19,000$ ppm of O_2 demand per person per day.

Water at 10° C contains about 10 ppm of O_2 when saturated. In order to bring the O_2 content to a minimum value of 5 ppm, 4000 gallons of water are required to self-purify the sewage per person per day.

These simple considerations show that there is not enough fresh water in the United States to self-purify the sewage of more than 250 million people, even if these people were perfectly distributed. As a matter of fact, the amount of waste water from various manufacturing industries came to 13 billion gallons per year in 1963 and is probably of the order of 50 billion gallons per day at this time. The biochemical *oxygen demanding material* in the industrial waste water came to 22 billion pounds per year in 1963 (or 60 million pounds per day), which is just about three times as much as that of the 120 million people served by sewers in the United States[10].

It is therefore clearly necessary to treat waste water discharged into fresh water streams and lakes. In the case of ocean outfalls, it may also be necessary to apply treatment, particularly if the dispersal is not adequate. In 1962, about 20 percent of the waste water in communities served by sewer systems was untreated; 30 percent received only primary treatment; and 50 percent some kind of secondary treatment[11]. This situation is now changing rapidly, largely because of the national water pollution control legislation and federal funds which have become available to spur the construction of treatment plants.

Primary treatment, which is basically mechanical removal of solids, costs between 3 and 4 cents per thousand gallons and removes about one-third of the BOD and only small amounts of the nitrogen and phosphorus. *Secondary treatment* is an accelerated and controlled form of natural oxidation by *bacteria.* When done in a properly designed plant, as much as 90 percent of the BOD can be removed, about half of the nitrogen, and a third of the phosphorus. The total cost (including primary treatment) is about 15—20 c per 1000 gallons[12]. However, nitrogen and especially *phosphorus* are very troublesome, particularly in lakes and estuaries where the flushing rate may be very low. These nutrients can give rise to the growth of *blue-green algae* and other noxious phytoplankton which can cause fish kills and deplete the lake of oxygen. Various types of tertiary treatment are now available which can purify water to any desired extent. In fact, purification can be carried to the stage where the water can be directly used for drinking purposes, at an additional cost of 15—20 c per 1000 gallons[13]. Currently, complete reuse of waste water is practiced in Windhoek, South Africa; other communities are cleaning up water to an extent where it is available for many domestic uses other than drinking.

One of the exciting possibilities coming out of the laboratory research program, and now in the pilot plant stage, is the complete elimination of secondary (biological) treatment; primary treatment can be followed directly by physical-chemical methods which remove nearly all of the phosphorus and nitrogen, and more than 99 percent of the BOD as well. It is clear that advanced waste treatment methods will be required where the population concentration is very high.

Thermal Pollution

A similar analysis can be performed for *thermal pollution.* One starts with a certain level of energy production, which for the United States in 1965 was about 5,000 kilowatt hours per capita. This production is assumed to have a doubling time which is conservatively given as ten years. Since the thermal efficiency of nuclear power plants (which will form the predominant form of power) is approximately one-third, then the energy which has to be rejected in the form of waste heat is now 10,000 kilowatt hours per capita and will double every ten years. This heat has to be carried away, at least partly, by cooling water.

In principle, rivers may be reused over and over for cooling waters since they dissipate the heat to the atmosphere as the water flows downstream. But the establishment of water temperature limits in order to protect water quality markedly reduces the number of available river sites and, in addition, requires the installation of supplemental or alternative cooling methods such as cooling ponds or cooling towers. This, again, increases costs and also loss of water by evaporation. It is becoming clear that an increasing number of power plants will have to be located along the costs where the waste heat can be efficiently discharged into the oceans.

Predictions indicate that in 1990 between 92 and 94 percent of the power generated will be produced thermally, with almost two-thirds of this produced by *nuclear plants* which require 40 percent more cooling water than *fossil-fueled plant*[14]. By 1980, electrical needs will require the use of 1/6 of the total available *fresh water run-off* in the entire Nation for cooling purposes. Now if we discount flood flows which usually occur about $1/3$ of the year and account for $2/3$ of the total run-off, it becomes apparent that the power industry will require about half the total run-off for the remaining $2/3$ of the year. More recent projections of the waste heat load by the year 2000, which take into account probable technical improvements, indicate that the requirement for cooling water will equal two-thirds of the total national daily run-off of 1200 billion gallons[15].

It is clear also that maintenance of environmental quality in the face of increasing demands for energy means an increase in costs, but not an excessive one. The increase in production cost normally would not exceed 0.3 mill per KWH, adds about $ 300 million to the total national cost, and of the order of 1 % to energy costs for consumers.

Agricultural Pollution

Even if municipal and industrial waste discharges can be controlled at a reasonable cost, the discharge from diffuse sources and especially from agricultural sources will continue to plague us. The chief problems come from the improper application of and excess application of *fertilizers*, from *pesticides* which wash into rivers or vaporize into the atmosphere, and from *organic wastes*, particularly from feedlots, which are allowed to wash into rivers; 10,000 cattle produce as much waste as a city of 164,000[16].

The answer here lies in stricter control and conservation methods. Full waste treatment may be necessary for feedlots; the application of fertilizers and pesticides will have to be strictly controlled and its excess or unnecessary uses prevented; and soil erosion will have to be reduced by proper conservation methods. This is an expensive but necessary program.

Water Conservation

There are a few essential steps which will alleviate the immediate situation and postpone the inevitable.

Chief among these is water conservation. In the United States, water is used in excess quantities and for many unnecessary purposes. Other countries which are water short are leading the way in experimentation to reduce water consumption.

Irrigation is the chief consumer of water; it will have to be made much more efficient than it presently is.*

In domestic use, great savings are possible by properly designed baths, showers, and toilet facilities. A number of locations throughout the world have installed so-called *vacuum toilets* in which toilet wastes are moved by pneumatic pressure rather than purely by water. This system also has advantages in that the sanitary wastes are highly concentrated and can be treated apart from other wastes which do not contain the large amounts of BOD as well as bacteria and viruses.

Another *remedy is reuse of water*. Waste water can be cleaned up to drinking water quality and reintroduced into the water supply system. Three or four cycles

---

* Examples are automatic sprinkling devices which respond to the actual water need of plants or which do not operate unless the meteorological conditions are proper. Another approach is trickle irrigation which supplies water, nutrients, and occasional pesticides directly to each plant rather than spreading it over the whole field.

are possible before chlorides build up to excessive values. (Parenthetically we might remark that waste water which is often completely untreated is reused naturally but in diluted form from rivers and lakes.)

Even without reusing waste water for drinking purposes, it is possible to set up a hierarchy of water supply and waste water uses in which the cleaned-up waste water is used for laundry, gardens, golf courses, etc.

Water Treatment Costs

The cost of remedial measures to abate pollution are high, of course. Studies performed by the Federal Water Pollution Control Administration for their report "The Cost of Clean Water" indicate the following with respect to certain waste sources for the five-year period from 1968 to 1973.

Municipal Wastes: Construction of new treatment facilities, replacement and expansion of existing facilities, removal of phosphates by the Great Lakes communities, operation and maintenance expenses, and construction of *sanitary sewer lines* range from $ 17.4 to $ 23.4 billion. (About $ 10 billion of this construction is eligible for federal grants on a cost-sharing basis.)

Industrial Wastes: Acceptable treatment of industrial wastes, including thermal effluents from power plants, will cost from $ 8.5 to $10.9 billion for both capital costs and for operation and maintenance. To achieve complete cleanup, the following programs will have to be carried out — necessarily on a longer time scale.

Overflows from Combined Storm and Sanitary Sewers: Total separation of lines handling storm water and *sanitary sewage* is estimated to amount to approximately $ 49 billion, $ 30 billion of which would be applied to public sewers. Effec-

tive alternate solutions to this problem, other than complete separation, might reduce the cost to perhaps $ 15 billion.

Erosion and Sedimentation: Total cost of controlling *erosion* is difficult to estimate since it is not entirely a pollution problem, but the range extends from a minimum of $ 300 million to perhaps as much as $ 10 billion.

Acid Mine Drainage: Estimated total costs range from $ 1.7 billion for a 40% reduction to as much as $ 6.6 billion for a 95% reduction. Actual costs will depend on the amount of neutralization necessary for each specific area to meet the water quality standards.

There are other pollution problems such as pollution from vessels, *oil spills, agricultural wastes,* and *radioactive wastes* where the costs cannot yet be well defind.

The Cost-Population Relationship

It is clear from preceding discussions that the physical environment has the ability to assimilate a limited amount of pollution. The atmosphere can disperse local pollution concentrations, and rainfall can clean the atmosphere — up to a certain point. Rivers, lakes, and the oceans can assimilate varying amounts of BOD and dissipate a certain amount of waste heat, before becoming overloaded. Therefore, pollution abatement becomes necessary only when the population concentration — and therefore the waste inputs — exceed certain limits. For man living in ecological balance with his environment, the cost of pollution control is zero.

If we view the human population/water pollution situation, we can see the following picture: As the level of population increases, the degree of treatment must be constantly increased in order to keep the level of pollutants below a critical level. However, the cost of treatment rises quite rapidly as we increases the degree of purification. We are therefore back to a kind of Malthusian equation: The cost of maintaining environmental quality increases at a faster rate than the population. Inevitably, therefore, at some point in the future, a major fraction of the GNP would have to be devoted to maintaining environmental quality. Everyone would agree that spending more than 50% of the national income on pollution control is unrealistic. On the other hand, 1% does not appear unreasonable. Perhaps the right level is of the order of a few percent which is very close to what is being contemplated now.

A partial answer to the Malthusian problem might be to spread out the population so that each river basin has its appropriate number of people. In a sense, this is an ecologically beneficial solution since the water supply also has to be in balance with the population level and density. Perhaps this is the way to aim, both for a short-term and for a longer-term solution.

In conclusion, there are a number of points, all of them important, which should at least be mentioned:

1) What should be our goal for environmental quality: How do we measure the benefits, including those which are difficult to quantify, so that we can gauge the extent of our national investment?

2) How to distribute the costs of maintaining a quality environment so that each sector of society pays its fair share? For example, groups that pay for abatement, such as municipalities and industries, do not favor a very high standard of quality; groups that benefit do not pay directly, such as fishermen and outdoor recreationists, are likely to demand very high standards.

3) Finally, the question of passing along environmental problems to future generations. To what extent should we invest our current income so that future gene-

rations which will have a higher income will be able to spend less? To what extent can we leave a legacy of pollution — do we have the moral right to do so?

References

[1] Black, R. J., A. J. Muhick, A. J. Klee, H. L. Hickman, R. D. Vaughan: An Interim Report: 1968 National Survey of Community Solid Waste Practices. 1968 annual meeting, Institute for Solid Wastes, AmericanPublic Works Association, Miami Beach, Florida, October 1968. Bureau of Solid Waste Management, U. S. Department of Health, Education, and Welfare, Washington, D.C., p. 47

[2] Ibid, p. 51

[3] Vogely, W. A.: The Economic Factors of Mineral Waste Utilization. Proceedings of the Symposium: "Mineral Waste Utilization", IIT Research Institute, Chicago, Illinois, p. 7, March 1968

[4] Wadleigh, C. H.: Wastes in Relation to Agriculture and Forestry. U.S. Department of Agriculture, Miscellaneous Publication No. 1065, Government Printing Office, Washington, D.C., p. 6, March 1968

[5] Darnay, O., W. E. Franklin: The Role of Packaging in Solid Waste Management, 1966 to 1976. U.S. Department of Health, Education, and Welfare, Public Health Service Publication No. 1855, Government Printing Office, Washington, D.C., 1969

[6] Rampacek, C.: Reclamation and Recycling Metals and Minerals Found in Municipal Incineration Residues. Proceedings of the Symposium: "Mineral Waste Utilization", IIT Research Institute, Chicago, Illinois, p. 124, March 1968

[7] The Sources of Air Pollution and Their Control. U.S. Department of Health, Education, and Welfare, Public Health Service Publication No. 1548, Government Printing Office, Washington, D.C., 1966

[8] Private communication from William P. Lear, Chairman of the Board, Lear Motors, Reno, Nevada, February 1970

[9] Ibid

[10] The Cost of Clean Water: Vol. 1: Summary. U.S. Department of the Interior, Federal Water Pollution Control Administration, Government Printing Office, Washington, D.C., p. 21, January 1968

[11] Glass, A. C., K. H. Jenkins: Statistical Summary of 1962 Inventory of Municipal Waste Facilities in the United States. Division of Water

Supply and Pollution Control, Department of Health, Education, and Welfare, Public Health Service, Government Printing Office, Washington, D.C., 1964

[12] Cleaning Our Environment: The Chemical Basis for Action. Subcommittee on Environmental Improvements, Committee on Chemistry and Public Affairs, American Chemical Society, Washington, D.C., p. 108, 1969

[13] Ibid, p. 125

[14] Industrial Waste Guide on Thermal Pollution. U.S. Department of the Interior, Federal Water Pollution Control Administration, p. 7, September 1968 (revised)

[15] Holcomb, R. W.: Power Generation: The Next 30 Years. Science, Vol. 167, p. 160, January 9, 1970

[16] Wadleigh, C. H.: Wastes in Relation to Agriculture and Forestry. U.S. Department of Agriculture, Miscellaneous Publication No. 1065, Government Printing Office, Washington, D.C., p. 41, March 1968

Economics of Fertilizer Use by U.S. Farmers — Productivity and the Environment

V. W. Davis

U. S. Department of Agriculture, Economic Research Service,
Farm Production Economics Division, Washington, D. C., USA

Summary. During the last two decades, economic returns from the use of fertilizer have been unusually favorable and U.S. farmers have increased their use of fertilizer almost $4^1/_2$ times. The higher use of fertilizer is one of the major reasons U.S. farmers are producing 40 percent more food and fiber on 11 percent fewer acres than in 1950.

Concern about fertilizer pollution encompasses three principal areas, eutrophication of water areas, high nitrate content in drinking water, and high nitrate accumulation in food plants. The extent of pollution from fertilizer is not clear.

Farmers view the use of fertilizer differently than society. To farmers, the current level of use fertilizer appears justified economically; but as an individual, they do not bear all the social costs. They fail to take environmental factors into account as a necessary part of their decisionmaking.

Alternatives for reducing pollution from fertilizer can be of two kinds: 1) technological changes in the production and application, and 2) direct restrictions on use. A policy of complete restriction on the use of fertilizer is untenable even though it would presumably improve the quality of the environment. A limited restriction on the use of fertilizer could be accomplished with negligible increases in costs and prices of farm products. For example, a limitation of 150 pounds of nitrogen per acre of corn in the Corn Belt would affect 24 percent of the acreage, reduce the quantity of nitrogen used by 7 percent, reduce corn production 2 percent, and have little effect on costs per bushel.

Zusammenfassung. Der Ertrag vom Gebrauch von Düngemitteln war in den letzten zwei Jahrzehnten ungewöhnlich günstig, und die Landwirte in den Vereinigten Staaten haben den Kunstdünger-Verbrauch fast auf das $4^1/_2$fache ausgedehnt. Der erweiterte Gebrauch von Düngemitteln ist einer der Hauptgründe dafür, daß die amerikanischen Landwirte heute, verglichen mit 1950, 40 Prozent mehr landwirtschaftliche Erzeugnisse auf 11 Prozent weniger Landfläche herstellen.

Besorgnis über Verunreinigung durch Düngemittel schließt drei Hauptgebiete ein, die Übersättigung von Gewässern mit Nährstoffen (eutrophication), den hohen Stickstoffgehalt in Trinkwasser, und die hohe Konzentration von

Stickstoff in Nahrungsmittelpflanzen. Das Ausmaß der Umweltbelastung durch Düngemittel ist nicht klar.

Landwirte beurteilen den Gebrauch von Dünger anders als die Gesellschaft. Für die Landwirte erscheint das heutige Ausmaß des Düngerverbrauches wirtschaftlich tragbar; sie müssen jedoch als Einzelne nicht die sozialen Kosten tragen. Sie unterlassen es, die Faktoren der Umweltbelastung als einen notwendigen Teil ihrer Entscheidungen in Betracht zu ziehen.

Es bestehen zwei Möglichkeiten, um die Verunreinigung durch Düngemittel herabzusetzen: 1) technologische Verbesserungen in der Herstellung und im Gebrauch, und 2) direkte Einschränkung des Verbrauches. Ein vollkommenes Verbot von Düngemitteln ist unhaltbar, obwohl es vermutlich die Umwelt verbessern würde. Eine teilweise Einschränkung des Düngemittelverbrauches könnte jedoch mit unbedeutender Erhöhung der Kosten und Preise für landwirtschaftliche Produkte erreicht werden. Zum Beispiel, eine Beschränkung auf 170 kg Stickstoff je Hektar Mais im Maisgürtel würde 24 Prozent der Fläche betreffen, den Verbrauch von Stickstoffdünger um 7 Prozent verringern, die Mais-Erzeugung um 2 Prozent verringern, und sich kaum auf den Preis von Mais auswirken.

Introduction

Since 1950, through the use of fertilizer, pesticides, improved varieties of crops and breeds of livestock, and other modern practices, farmers in the United States have increased the total farm production about 40 percent. They are now using 11 percent fewer acres of land for crops and less than half as much labor as they did 20 years ago. *Fertilizer* is an essential part of the complex package of total inputs responsible for this result.

However, there is increasing concern for the *"spill over"* effect (sometimes referred to as technical diseconomies or social costs)* of *agricultural chemicals* on the environment and on persons who have no choice in their use. Thus, conflicting objectives do exist, and we have choices and decisions to make that range from those affecting individual farmers to those concerning total society.

My main concern in this discussion is with the economic consequences of the use of fertilizer by U.S. farmers — the effect on productivity and the potential environment damage.** First, I will present some facts on the past and current use of fertilizer in the United States and an indication of future demand. These data illustrate indirectly the physical and economic productivity of fertilizer. Next comes a brief discussion of some of the environmental problems associated with the use of fertilizer, especially with nitrogen and phosphorus. Then, I would like

* A technical externality is one that is transferred from one decision unit to another by a technical or physical linkage (4, page 41). Externalities can also be pecuniary in which case they are transmitted by a market mechanism and do not generally lead to what is considered an inefficient resource allocation.

** Environmental pollution has been defined as "the unfavorable alteration of our surroundings through direct or indirect effects on the *chemical, physical,* and *biological characteristics* of our *air, land,* and *water,* influenced primarily by *man's actions"* (1).

to point out the conceptual difference between the optimal use of fertilizer for an individual farmer and that for society. This difference in viewpoint is the essence of current public concern about the alleged environmental hazards from the use of fertilizer. And finally, I will outline some alternatives for reducing environmental deterioration from fertilizer.

Fertilizer Use

The major share of fertilizer used in the United States — an estimated 11 million tons (elemental equivalent basis) in 1970 or 85 percent of the total — was used by farmers (Table 1). The remaining 15 percent was used for nonfarm purposes such as home lawns, airports, and sides and centerstrips of highways.

In 1970, farmers paid about $ 2 billion for fertilizer. This was approximately 5 percent of total farm production expenses ($ 40 billion). As compared with other production items, the total outlay for fertilizer ranked below most other costs.

Fertilizer is unique among farm inputs.

Table 1   Fertilizer use on corn, wheat, and all crops and percentage of harvested acres receiving any fertilizer, United States, selected years, 1950–1970

Year	Total fertilizer used	Average rate per acre receiving			Percentage of harvested acres receiving		
		N	P	K	N	P	K
	Thous. tons	Pounds			Per cent		
			All Crops				
1950	729	15	10	12	48	48	48
1954	1,413	27	11	17	60	60	60
1959	2,017	41	15	27	61	59	52
1964	2,851	58	18	29	85	78	72
1969*	4,428	109	27	51	92	87	82
1970*	4,952	112	31	60	94	90	85
			Wheat				
1950	238	8	13	12	22	22	22
1954	320	15	9	7	28	28	28
1959	431	26	10	12	33	31	16
1964	477	28	12	13	47	37	17
1969*	578	39	15	17	55	42	14
1970*	633	49	7	27	58	35	17
			Corn for Grain				
1950	2,617	15	15	15	NA	NA	NA
1954	4,041	27	15	22	NA	NA	NA
1959	5,278	42	18	37	14	14	11
1964	7,037	56	20	38	17	16	12
1969*	10,508	NA	NA	NA	NA	NA	NA
1970*	10,786	NA	NA	NA	NA	NA	NA

* Farm use estimated to be 85 per cent of total fertilizer used in the United States in 1969 and 1970. NA = Not available.

Sources: "Fertilizer Used on Crops and Pasture in the United States, 1954 Estimates", U.S. Dept. Agr., Stat. Bull. 216, Aug. 1957; "Commercial Fertilizer Used on Crops and Pasture in the United States, 1959 Estimates", U.S. Dept Agr., Stat. Bull. 348, July 1964; "Fertilizer Use in the United States by Crops and Areas", U.S. Dept. Agr., Stat. Bull. 408, Aug. 1964; "Fertilizer Use on Selected Crops in Selected States, 1969 and 1970" (published in *Crops Production,* Jan. 1970 and 1971); "Consumption of Commercial Fertilizers in the United States", SpCr 7 (5–70) and SpCr 7 (10–70).

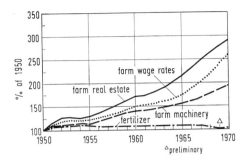

Fig. 1 Prices of selected farm inputs

While prices of nearly all things farmers buy have increased, some spectacularly over the last 20 years, fertilizers prices have remained virtually unchanged in current dollars (Fig. 1, Table 2). In terms of real values, prices of fertilizer have declined substantially because of improved fertilizer manufacturing technology and increasing competition within the industry. Consequently, use of plant nutrients in 1970 increased to a level four and a half times that of 1950 (Fig. 2, Table 3).

An outstanding example of declining price can be seen best in the record of *anhydrous ammonia* (82 percent elemental nitrogen), by far the major source of fertilizer nitrogen in the United States. In the early 1950's, the average price paid by farmers for ammonia was $ 180 a ton. By 1970, the price had fallen to $ 75! These are prices paid by farmers and therefore include transportation costs. Price levels for the other basic fertilizer ingredients also have declined or remained fairly stable during the last 20 years. Most farmers now use some fertilizer. Thus, it is difficult to realize that only a few years ago a relatively large number of farmers in the Corn Belt — Ohio, In-

Table 2   Prices of selected farm inputs, 1950 to 1970 (1950 = 100)

Year	Farm wage rates	Farm machinery	Fertilizer	Farm real estate
1950	100	100	100	100
1951	111	108	106	115
1952	118	111	108	126
1953	121	112	109	128
1954	120	113	110	126
1955	121	113	108	131
1956	126	118	106	137
1957	131	123	106	146
1958	135	129	106	152
1959	144	134	106	163
1960	148	138	106	171
1961	151	141	107	172
1962	155	144	106	182
1963	159	146	106	189
1964	163	149	105	202
1965	171	154	106	214
1966	185	160	106	231
1967	199	167	106	246
1968	216	175	103	262
1969	238	184	99	275
1970*	253	193	101	286

* Preliminary

Table 3   Quantities of selected farm inputs, 1950 to 1970 (1950 = 100)

Year	Labor	Land used for crops	Mechanical power and machinery	Fertilizer	All other inputs
1950	100	100	100	100	100
1951	101	101	107	119	106
1952	96	100	112	133	106
1953	92	100	113	147	107
1954	88	100	114	156	109
1955	85	100	115	163	113
1956	80	97	115	161	119
1957	73	94	116	171	118
1958	70	93	115	176	125
1959	68	95	117	203	131
1960	65	93	121	204	134
1961	62	90	117	216	138
1962	59	88	116	234	143
1963	57	89	121	265	148
1964	54	89	119	292	152
1965	51	89	122	308	153
1966	49	88	128	350	160
1967	48	91	133	394	165
1968	46	90	133	429	169
1969	45	89	133	435	173
1970*	44	89	136	458	173

* Preliminary

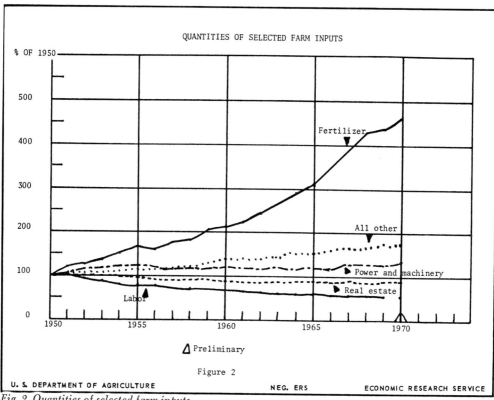

Fig. 2 Quantities of selected farm inputs

diana, Illinois, Iowa, and Missouri — were not using fertilizer at all. They relied completely on crop rotations and livestock manure for plant nutrients.

To illustrate the dramatic change in the use of fertilizer in the United States, let us examine the use of fertilizer in corn production. Currently, almost half of all agricultural fertilizer is used on corn. In 1950, an average application of 15 pounds of nitrogen per acre was used on 48 percent of the corn acreage (Table 1). By 1970, this had risen to 112 pounds per acre on 94 percent of the corn acreage. Total tons of fertilizer (elemental equivalent basis) used on corn increased from less than a million tons in 1950 to almost 5 million tons in 1970. During this period, the total used on all crops increased from 2.6 million tons to 10.8 million tons.

The economic return from the use of fertilizer has been unusually favorable during the last two decades. Although most farmers have probably reached the point of diminishing economic returns from the use of fertilizer, many are still not using the optimum amount for greatest profits. Ibach estimated that in 1960 to 1964, farmers received a marginal return of $ 2.50 per dollar spent for fertilizer[3]. With increased application rates of fertilizer in recent years, the added value of crops per marginal unit of fertilizer has declined. Based on Ibach's computations for 1960—64, the marginal return for the 1970 use of fertilizer would be approximately $ 1.50 per dollar spent for fertilizer. This would suggest that some farmers still lack knowledge and/or are uncertain concerning the profitability of fertilizer use, as there is still potential for increased use.

The opportunities for further expansion in use of fertilizer are certainly less than in the last two decades; we can expect farmers to continue to increase their use of fertilizer, although less rapidly than before. Opportunities for expansion will vary depending on the crop and current levels of fertilizer use. For example, with over 94 percent of the corn acreage fertilized in 1970, opportunities for expansion must come mainly from economic incentives to apply larger quantities per acre or to increase the acres planted to corn.

Environmental Problems

Phosphorus and nitrogen are two elements essential for plant growth. They are of particular concern in connection with pollution. These nutrients can move into streams, lakes, and reservoirs in the drainage water from land, and nitrate nitrogen can move into ground water. *Nitrogen* and *phosphorus* may come from several sources; for example, farm land, barnyards, and feedlots, municipal and rural sewage, and industrial wastes [6]. The source of chemicals washed from farmland can be the natural *geological "weathering"* of the mineral and organic components of the soil, as well as those applied by man in the form of fertilizer.

Concern about fertilizer pollution encompasses three principal areas, *eutrophication* of water areas, high nitrate content in drinking water, and *high nitrate* accumulation in food plants.

The process of eutrophication occurs in streams, lakes, ponds, and reservoirs. This is considered to be one of the major pollution problems today. Nutrient enrichment of waters, primarily with nitrogen and phosphorus, can result in excessive *algal bloom* and undesirable growth of *aquatic plants* rendering these waters less

suitable, and occasionally unsuitable, for drinking, recreation, and other beneficial uses. Eutrophication accelerated by man must be differentiated from natural aging processes occurring in lakes, because different corrective measures are needed. This problem affects both rural and urban areas.

The second area, that of high nitrate content in drinking water, (either surface or groundwater) is undesirable. In sufficient concentrations it may poison by causing oxidation of the hemoglobin in the blood *(methemoglobinemia)* in infants or in ruminant livestock. More precise criteria on the effects of nitrates in water on humans and animals are needed.

The third area of concern relates to high nitrate accumulations in plants. Certain vegetables, such as *spinach* and *beets*, normally accumulate large amounts of *nitrates*. This accumulation can be increased through use of fertilizer, and is characteristic only of certain plant species. Under certain conditions, nitrates in food can be harmful to *infants*. High nitrate concentrates also occur in some forage crops, particularly if the crops are subjected to drought or other stress. Animals grazing such forage can develop *nitrate toxicity*. High nitrate forage used as silage may also release noxious nitrogen gases on fermentation.

The extent of pollution from fertilizer is not clear. Some ecologists point an accusing finger at fertilizer and suggest the need for greater reliance on nitrogen-fixing legumes, composting of plant residues, and animal wastes. At the same time, scientists are busy searching for answers to the fate of chemicals in our soil, water, and air.

During the last 20 years, tremendous growth in the use of fertilizer in the United States may have increased the potential from that source for the pollution of the environment. Without restrictions on its use, we can expect farmers to continue to increase the application of fertilizer and this may further increase the potential for pollution. But we are not certain whether the potential is a reality or not, or if it is, whether it is widespread or restricted to certain situations or soil conditions. We do not know how much of the nitrogen and phosphorus in our water supply comes from fertilizer, and how much comes from other sources including the geological decomposition of the soil and the parent rock.

Optimum Use-Farmers vs. Society

President Nixon, in his August 1970 message to Congress on "Environmental Quality" stated that the basic causes of our environmental troubles are complex and deeply rooted[2]. He identified six causes but two seem especially applicable to a discussion of the effects of agricultural fertilizer on the quality of the environment. They are: 1) "the failure to take environmental factors into account as a normal and necessary part of our planning and decisionmaking", and 2) "the failure of our economy to provide full accounting for the social costs of environmental pollution"[2]. I would like to elaborate on these two causes of pollution by outlining the conceptual difference between the optimum use of fertilizer for an individual farmer and the optimum use for society.

First, the viewpoint of society. When we consider ways of restricting farmers' use of *fertilizer*, we are implicitly saying that the social costs (hazards to health and environment) of using current quantities of fertilizer have exceeded the social benefits (larger supplies of food and fiber). We are implying that we no longer have an optimum use of fertilizer from the standpoint of society. We are saying that it is no longer acceptable to society for

farmers to use fertilizer without regard for the ultimate effect on the environment, and on the health of society.

Second, the viewpoint of the farmer. He sees the fertilizer issue differently. To him, the current level of use of fertilizer appears justified and economic, but as an individual he does not bear all of the social costs. In fact, some farmers should be using more fertilizer if they are to maximize their profits.

Let me illustrate the conflict between farmers and society in the use of fertilizer by using a typical production function and price lines (Fig. 3). The fertilizer input, for a given level of other inputs, is shown on the x axis and the output of the crop is shown on the y axis. The slope of the price line CD is a function of the price of fertilizer divided by the price of the crop. Its tangency with the production function at point I indicates that the optimum quantity of fertilizer from the farmers' point of view is OB since they do not include the social costs in their computations. This may be substantially in excess of the socially optimum level OA. By including social costs, as well as the costs of fertilizer, in computing the slope of our price line EF, the price line becomes steeper and thus our socially optimum level of fertilizer usage is reached at a considerably lower usage. Only if there are no externalities or social costs would the optimum levels of fertilizer use for these two groups coincide. Incidentally, the socially optimum level of OA would suggest the *restriction* of fertilizer rather than a complete *ban*.

Any reduction in the use of fertilizer would result in the substitution of other inputs such as land, labor, and machinery[3]. Implicitly we can assume that costs of production would increase or farmers would have made these substitutions before. As consumers, it is important that we recognize that prices would

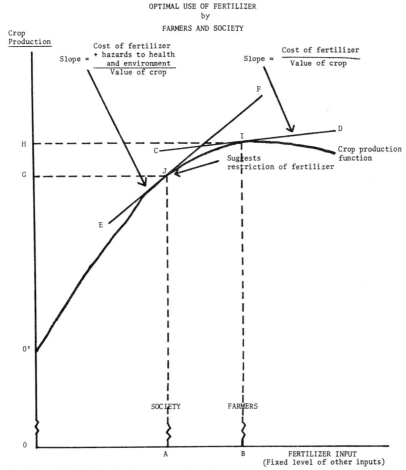

OPTIMAL USE OF FERTILIZER
by
FARMERS AND SOCIETY

Fig. 3 Optimal use of fertilizer by farmers and society

rise if adjustments in production practices result in decreased total output because of higher production costs or inability to maintain production with some restriction of resources.

Alternatives for Reducing Pollution

There are many who would choose to do nothing to reduce pollution from fertilizer because they are not affected adversely by its use. They would feel disadvantaged by restrictions on fertilizer use if it reduced supplies and raised prices of food and fiber.

There are, however, some positive alternatives which can be considered to reduce *environmental deterioration*. They range from various ways of reducing the quantities of fertilizer moving into the environment to removal of plant nutrients from bodies of water by *membrane techniques* and biological treatment. My discussion of alternatives focuses primarily on ways of reducing quantities of fertilizer used for crop production.

Implicit in what follows is the assumption that pollution is directly related to the quantities used. Alternatives for reducing pollution from fertilizer are of two kinds:

1) technological changes in the production and application of fertilizer, and 2) direct limitations or restrictions on use.

In the first general category, we could include:

1) Development of fertilizer having slower release of nutrients to enable more complete utilization of nutrients by plants.

2) Improvement in cultural practices and timing of applications to prevent leaching and erosion.

3) Improvement of methods of application such as placement of fertilizer closer to the plant root zone to improve utilization.

4) Attainment of lower levels of pollution through smaller but more frequent applications of fertilizer.

The second category of ways of reducing the potential pollution from fertilizer includes direct limitations or restrictions.* When considering limitations on the use of fertilizer, it is important to recognize that farmers will be affected differently depending on their normal use. For example, in 1970, Corn Belt farmers applied an average of 110 pounds of nitrogen per acre on about 31 million acres of corn land (Table 4). But on 34 percent of the corn acreage, they applied less than 101 pounds of nitrogen; on 24 percent of the acreage, they applied more than 150 pounds.

To illustrate further the effects of fertilizer restriction, let us consider corn production in the Corn Belt in 1970 (Table 5). Because of the corn blight in 1970, we use the 1969 average yield of 94 bushels per acre. Also, we use a constructed corn

Table 4 Corn for grain: Pounds of nitrogen per acre, percentage of acres harvested, yield per acre, and costs by quantities of nitrogen per acre, Corn Belt, 1970[1]

Quantity nitrogen per acre	Average nitrogen per acre[2]	Acres harvested	Yield per acre[3]	Production costs per bushel[4]
Pounds	Pounds	Per cent	Bushels	Dollars
1970 use	110	100	94	1.00
200 or more	225	4	111	.88
151–200	171	20	108	.89
101–150	121	42	99	.95
51–100	73	14	87	1.04
0– 50	25	20	70	1.23

[1] In 1970, 30,934,000 acres of corn were harvested for grain in the Corn Belt which includes Ohio, Indiana, Illinois, Iowa, and Missouri.

[2] Pounds per acreage adjusted from mid-point of range to obtain an average of 110 pounds per acre, 1970 actual use.

[3] Yields per acre are based on 1969 experience because of the reduction of yields in 1970 by the corn blight.

[4] Production costs per bushel are based on fertilizer costs per pound of 4.4 cent for N, 8.1 cent for P_2O_5, and 4.1 cent for K_2O and all other costs held constant at $82.28 an acre for all levels of fertilizer use. Quantities of P_2O_5 and K_2O were consistent with the levels of nitrogen used.

response curve for *nitrogen* for all farmers which shows 59 bushels per acre with no nitrogen and levels off at 110 bushels with 225 pounds of nitrogen (Fig. 4). We assume for simplicity that additional land or other inputs will not be substituted to maintain production of *corn*, and that farmers not affected by maximum limits will maintain their 1970 application rates of nitrogen.

A limitation of 200 pounds of nitrogen per acre would affect only 4 percent of the acreage (Table 4). It would reduce the quantity of nitrogen by 1 percent, and have little effect on yields (Table 5). Cost of production would decrease 2 cents per bushel which suggests that these farmers have slightly exceeded the optimum use of fertilizer.** A 1 00-pound limit

---

* Meyer and Hargrove (Iowa) analyzed the effect of fertilizer restrictions on food costs, farm incomes, and crop yields (5).

** In reality, some farmers at all levels of use could have exceeded the optimum use of fertilizer. However, in the example, the same production function is assumed for all farmers.

Estimated Response of Corn Yields to Alternative Levels
of Nitrogen, Corn Belt, 1970

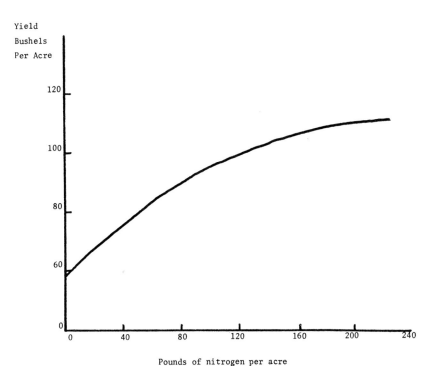

Fig. 4 *Estimated response of corn yields to alternative levels of nitrogen, Corn Belt, 1970*

Table 5   Corn for grain: Effects of alternative limitations of the use of 1) nitrogen on total use of nitrogen, production of corn, and costs, Corn Belt, 1970

Alter-natives	Maximum quantity of nitrogen per acre	Nitrogen used Total	Reduction from 1970 use	Corn production Yield per acre	Total	Reductions from 1970	Cost 4) Per bu	4) Per bu acre
	Pounds	1,000 tons	Per cent	Bushel	Mil. bu.	Per cent	Dollars	
	1970 use	2) 1,701	0	3) 94	2,908	0	1.00	93.66
1	200	1,686	1	94	2,907	5)	0.98	92.28
2	150	1,582	7	92	2,845	2	1.00	91.96
3	100	1,240	27	89	2,746	6	1.02	90.88
4	50	678	60	78	2,419	17	1.13	88.52
5	0	0	100	59	1,832	37	1.39	82.28

1) Reductions in the use of P_2O_5 and K_2O were consistent with that of N.
2) Based on an average of 110 pounds of nitrogen applied to 30,934,000 acres of corn in 1970.
3) The 1969 average yield was used because of the severe reduction in the 1970 yield from corn blight.
4) Production costs per bushel are based on fertilizer costs per pound of 4.4 cent for N, 8.1 cent for P_2O_5 and 4.1 cent for K_2O and all other costs held constant at $82.28 an acre for all levels fertilizer use. Quantities of P_2O_5 and K_2O were consistent with the levels of nitrogen used.
5) Less than 0.1 per cent.

would affect 66 percent of the corn acreage and reduce fertilizer use by 27 percent. Corn production would decline only 6 percent and production costs would increase 2 cents per bushel.

A complete ban on all nitrogen without additional land brought into production would reduce corn production more than a third. It would cause major adjustments in the location and methods of corn production, increase production costs, and increase food prices. For example, under our present conditions of plentiful food supply and inelastic demand, one might expect a 10-percent decrease in the supply would bring a 25-percent increase in price [8]. The increase in U.S. corn prices since January 1970 because of the decline in supply caused by the *corn blight* indicates that this estimate of elasticity of demand is reasonable [7].

If pollution from fertilizer is directly related to the quantities used, then pollution costs (social costs) from additional quantities of fertilizer may be high relative to returns to individual producers. Restrictions on the use of fertilizers by farmers who are approaching the economic maximum might accomplish the greatest reduction in the use of fertilizer with minimum costs to both producers and society. In other words, optimum use of fertilizer both from the standpoint of the producer and society might be found if restrictions could be applied for all those farmers who are operating in the range above some specific rate of application, 150 pounds of nitrogen for instance, where their returns for the extra fertilizer are negligible. Farmers who normally apply less than the limit would tend to increase their use of fertilizer. This would lessen the reduction in the use of fertilizer but the overall productivity and utilization by plants should be improved.

References

[1] Control of Agriculture – Related Pollution. A report to the President submitted by the Secretary of Agriculture and the Director of the Office of Science and Technology. Washington, D.C., 1969

[2] Environmental Quality. The First Annual Report of the Council on Environmental Quality transmitted to Congress. Washington, D.C., 1970

[3] Ibach, D. B.: Fertilizer Use in the United States, Its Economic Position and Outlook. USDA Agr. Econ. Rpt. No. 92, 1966

[4] Kneese, A. V.: The Economics of Regional Water Quality Management. The John Hopkins Press, Baltimore, 1964

[5] Mayer. L. V., S. H. Hargrove: Food Costs, Farm Incomes, and Crop Yields with Restrictions on Fertilizer Use. Iowa State University, Center for Agricultural and Economic Development, CAED 38, 1971

[6] Wadleigh, C. H.: Wastes in Relation to Agriculture and Forestry. USDA Misc. Pub. 1065, 1968

[7] Feed Situation. ERS, USDA. Washington, D.C., 1971

[8] Heady, E. O.: A Primer on Food, *Agriculture and Public Policy*. Random House, New York, N.Y., 1967

Economic Importance of Chemical Crop Protection in Relation to its Ecological Impact

W. Salzer

Vorstandsmitglied der Farbenfabriken Bayer, 56 Wuppertal-Elberfeld

When one embarks on a subject like this, consideration must first be given to the background aspects of the ecological importance of insect pests and diseases of crops. In every epoch of human history, right down to the present day, insect pests and plant diseases have had a revolutionary influence on the ecology and economy of countries and provinces. Records show that through the ages, from antiquity to the present era, insect pests and plant diseases have all too often been responsible for famines, scarcity and economic ruin. They have also been very largely responsible for mass emigrations as well as for changes in consumer habits. Present-day national and regional eating and drinking habits are often closely related to the fact that a nation had to reconcile itself to a completely changed supply situation brought about by catastrophic harvest failures. An example of the social effect of a plant disease is illustrated by the fact that Britain today drinks far more tea than coffee, a situation which has existed only since the 1870's. At that time, coffee plantations in the crown colony of Ceylon were laid waste by the coffee rust disease *(Hemileia vastatrix)* so that planters in that country were compelled to change over to the cultivation of tea, and ever since tea has dominated the British market[1]. The shift of predominance in cocoa production from Central and South America, the home of the cocoa tree, to Africa is also due to the devastating outbreaks of plant diseases in America. To quote a very recent example, skeletonizing of forests by insect pests could result in a revolutionary ecological change of large areas for long periods to come if an equal substitute for DDT is not found quickly for the control of these pests.

This is a highly interesting topic which has often been discussed but I must now leave it and turn to the actual subject of my paper, the economy of crop protection measures. I propose to examine the actual situation and the potential situation, in other words what has been achieved and what could be achieved when crop protection measures are expediently applied in accordance with the present state of technical knowledge and technical possibilities, giving consideration to direct costs and those brought about by the ecological risk. Apart from the catastrophies which I have already mentioned and on which we have reports dating from biblical times down to the present day, hardly any estimates of relative crop losses from insect pests, plant diseases and weeds were published until recently. It was the awareness of the world problem of food shortage in the so-called underdeveloped parts of the world, on the one hand, and the compulsion to adopt more rational methods of agricultural production in the highly industrialized countries, on the other hand, that first led to the systematic collation of data on these questions, from which estimates and calculations could be drawn up for individual countries and individual crops. Work in this direction has been in progress longest and on the most intensive lines in the United States, being confined at first to individual studies. But since the late 1920's, there has been a regular publication of data by the U.S.D.A.[2, 3, 4].

Table 1

| Year of USDA publication | Losses in U.S.A. from | | Total loss |
| | Insect pests | Plant diseases | |
	(in million)	(in million)	(in million)
1937	$ 491	$ 298	$ 789
1954	$ 1,066	$ 2,914	$ 3,980
1965	$ 3,685	$ 3,251	$ 6,936

Table 1 shows the development of the annual crop losses, expressed in U.S. dollars, from 1937 to 1965. The losses from weeds are not taken into account in these figures which likewise provide no indication of the extent to which crop protection was practiced in the period involved. The increase in total losses is largely due to the enormous increase in yields per unit area, for which numerous factors like fertilization, irrigation, soil cultivation, improved varieties, etc., are responsible, i. e. also when the percentage losses remain constant the absolute figures greatly increase. The price development of agricultural products is of course also reflected in these figures.

Table 2

| Crop | Yield per unit area (in 100 kg/hectare) | | | | Per cent increase since 1939 |
	1939	1954	1964	1968	
Wheat	10.1	12.8	17.7	19.2	90
Corn	20.2	28.2	39.3	49.3	144
Rice	23.9	31.6	45.9	49.6	107
Cotton	2.7	4.5	5.8	5.8	115

Table 2 gives the yields per unit area for four of the major crops grown in the United States[5].
A cautious evaluation of the estimates from all over the world shows that the annual world crop production loss is of the order of U.S. $ 70,000 to 90,000 million. In the Federal Republic of Germany where agriculture is more notable for its degree of intensity rather than for the absolute order of magnitude, the estimated loss of crop production is more than DM 3,000 million equivalent to just under 20% of the total production. This example demonstrates that despite a good standard of crop protection measures, which is certainly the case in the German Federal Republic, complete prevention of crop losses is seemingly unattainable or, as we shall have to show, might even be uneconomical.
The breakdown of crop production losses in the different regions of the world is presented in Table 3 below:

Table 3

Region	Total loss in %
Europe	25
Oceania	28
North and Central America	29
U.S.S.R and P. R. of China	30
South America	33
Africa	42
Asia	43

It is evident from this table that losses are greatest in regions with inadequate supply of food[6].
The losses from the different causal factors can also be estimated approximately, it being found that insect pests account for 14% of the loss of potential world crop production, *plant diseases* for 12% and weeds for approximately 9%.
Table 4 gives the estimated losses for the different commodities in million tons, in U.S. $ thousand million and as percentages of the potential production.
I shall now turn away from this global review of crop losses which covered the 60 most important crops since a survey of all these crops can, of course, be no

Table 4

Commodity	Loss in million tons	in $ 1000 million	Per-centage
Cereals	506	34	35
Potatoes	129	5	32
Sugar beets	636	6	45
Sugar cane	78	6	28
Vegetables	56	6	29
Fruit crops, incl. Citrus Fruits and Grapes	6	4	37
Coffee, Cocoa, Tea, Tobacco, Hops	42	5	33
Oil crops Fiber crops and natural Rubber	8	4	32

more than of a superficial nature, and now examine more closely the economic importance of chemical crop protection measures in rice farming which I consider to be a particularly suitable example for the following reasons:

1) Rice is one of the world's most important food crops.

2) For rice, very reliable data are available concerning the influence of crop protection measures on yield.

3) Rice growing very clearly demonstrates that food production cannot be practiced successfully without environmental change.

Table 5   Hectarages and yields of rice and wheat

World rice hectarage	132 million hectares
World rice production	284 million tons
World wheat hectarage	227 million hectares
World wheat production	332 million tons

Table 5 gives the hectarage and production figures for rice in 1968 in comparison with wheat, the second most important cereal.

The loss of the potential world rice production from insect pests, diseases and weeds is currently estimated at 46% as opposed to only 24% for wheat. This means that the annual quantitative loss

of rice amounts to 200 million tons. For a world population of approximately 3,500 million, this figure is tantamount to a quantity of 160 grams of rice per head and per day; in other words by eliminating rice production losses alone the food problem could be solved in a large part of the world. Of course this is a utopian calculation because, as I have already said, 100% protection is not possible. But what can be achieved in rice production will be shown by the following figures.

Table 6   Breakdown of rice hectarage and production at different yield per unit area levels.

5.0 tons/hectare:	2.6% of the hectarage and 6.8% of production
4.0–4.9 tons/hectare:	2.2% of the hectarage and 4.9% of production
3.0–3.9 tons/hectare:	1.3% of the hectarage and 2.5% of production
2.0–2.9 tons/hectare:	24.6% of the hectarage and 36.2% of production
1.0–1.9 tons/hectare:	65.1% of the hectarage and 47.4% of production
1.0 ton /hectare:	1.0% of the hectarage and 0.4% of production

Table 6 gives the average yield per unit area for different regions and the percentage share of total production. Countries with an average yield per unit area of over 4 tons per hectare account for approximately 12% of the total *rice production* realised on 4.8% of the total hectarage. Almost half of the total world production is obtained, however, in countries where the yield per unit area is below 2 tons per hectare.

Table 7

Country	Yield per unit area in tons/hectare	Per cent share of world rice hectarage	world rice production
India	1.6	27.0	20.9 = 1 : 0.8
Japan	5.7	2.5	6.6 = 1 : 2.6
Australia	7.2	0.02	0.08 = 1 : 4.0

Table 7 compares the yields per unit area and the percent share of world rice hectarage and world rice production for 3 countries, namely India, Japan, and Australia. These differences in yields per unit area are of course only partly due to the different levels of success achieved with crop protection measures.

Table 8 shows, however, that successful *crop protection* can be a factor of very considerable importance, this being evident on comparing the first and the final column of Table 8. It is also evident from this table that the most spectacular successes in rice production in the last 17 years were achieved in Japan[7].

Fig. 1 gives a plot of the average yields per unit area obtained in Japan from 1926 to 1968. Even if we leave out of consideration the enormous drop in production in 1945, which was due to the war, it is nevertheless noticeable that the annual yields extremely varied until 1953. Apart from inclement weather, particularly frost, these variations were chiefly caused by insect pests and plant disease attack. After 1953, rice growers in Japan began to make systematic and successful use of insecticides and fungicides, followed later by herbicides, which led to outstanding increases in production, with the result that Japan, a densely populated country with a large proportion of unfertile mountainous land, changed from a rice-importing country to

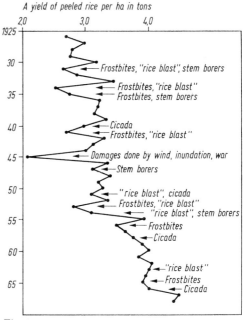

Fig. 1

one with a surplus rice production. Japan is, therefore, an excellent example of what can be achieved by carrying out expedient programs of crop protection in combination with other cultural practices. This example also clearly underlines the enormous importance of applying chemicals for the protection of rice crops, and furthermore indicates that it would be quite possible for the theoretically calculated production loss of 200 million tons per year to be reduced to a level of 60 million tons. In other words, there is no reason why the present 46 % loss of world

Table 8

Country	Actual yield per unit area in 100 kg/hectare	Per cent losses of production			Potential yield per unit area, by preventing losses in 100 kg/hectare
		from insect pests	from diseases	from weeds	
India	16.3	36	10	11	37.9
South Korea	33.3	15	8	10	49.7
Japan	51.5	1.6	3.4	8.6	59.5

rice production could not be reduced to a level of 14 % which is the current loss figure in Japan. The successes achieved in Japan also show that the widely discussed possibility of changing the eating habits of some Asian nations from rice to wheat which is less problematic to grow, need not be considered. It is not by chance that *rice* became the staple food in these regions. The extensive rice-growing areas of the Far East largely originated as the result of erosions (Mekong delta, Ganges delta, etc.). Of all the different kinds of cereals, rice was best able to develop under the conditions of the heavy monsoon rainfall in these regions. Rice growing also makes for water storage and protection against further erosion.

A well-calculated example of what can be achieved in crop production is provided by the long-term figures given in Table 9 below for potato production in the United States[8].

MacNew reports that the great gains made during the 1941/45 period came from putting into practice the knowledge gained concerning use of fertilizers, improved varieties and irrigation, as well as more extensive use of pesticides. However, the potato production gains of 1945 to 1950 were primarily a triumph of two new pesticides, namely DDT for the con-

trol of potato leafhopper and Nabam for the control of late blight. This increase in the yield per hectare also permitted the total hectarage to be reduced and to release some of it for other purposes, and to concentrate potato culture in the better-adapted areas, positive effects which must be credited to successful crop protection measures.

This purely quantitative review of crop losses is, however, not sufficient to define the economic value of crop protection measures. The quality of the produced commodities must also be taken into account in the calculation. In the case of apples, scab infection and codling moth infestation, to give but two examples, make the produce almost unsalable or at least cause a considerable loss of returns. The average results of a four-year study on three apple varieties at our experimental station in Bergisches Land showed that more than 50 % of the apples in the untreated plots were scab-infected so that only about 30 % of the harvested produce qualified for classification in the top commercial grade A. In the plots sprayed for scab, less than 1 % of the apples were infected so that nearly 80 % of the harvested crop was of grade A. Similar figures are reported from Illinois in the United States, where statistical records of codling moth infestation of apples have been kept since 1885. These records show that nearly 70 % of unsprayed fruit was damaged by codling moth in 1885 and that the percentage infestation is still the same today in untreated orchards while the level of infestation in sprayed orchards is now only about 2 %. Similar figures are on record for other crops. Mention should also be made of the fact that contamination of foods with fragments and metabolic products of insect pests, microorganisms or weeds may likewise present health hazards to consumers. *Ergotism, aflatoxin,*

Table 9

Period	Area harvested in hectares	Yield per hectare in 00 kg
1901–1905	1,262,000	61.7
1911–1915	1,407,000	67.6
1921–1925	1,360,000	71.6
1931–1935	1,422,000	72.3
1941–1945	1,141,000	94.7
1945	1,093,000	104.2
1946	1,052,000	125.2
1947	851,000	124.5
1948	854,000	144.8
1950	684,000	170.3

and the *quinones* of the cuticle of the granary weevil are widely cited terms into which I need not enter in detail.

What Does Chemical Crop Protection Cost?

Firstly, let us give consideration to the size of the pesticide market, which perhaps is often overrated. The total turnover on pesticidal active ingredients ex factory is not at all easy to determine because many companies, particularly the big ones, also sell products of other firms so that a simple addition of sales would give excessively high figures. Taking this fact into consideration, it is estimated that the total turnover ex-factory is in the region of U.S. $ 2,000 million for the whole world with the United States market accounting for about 40—50%, the European market for 20%, and the Far East market for about the same percentage. These are estimates which reflect only the approximate order of magnitude.

We already spoke about the correlations between the possible quality of harvest produce and crop protection measures: Wine production in the Mediterranean departements of France provides a very good example for demonstrating the dependence of economically expedient crop protection measures upon the quality of the harvest produce. These nine departments have a wine grape hectarage of 604,000 hectares, of which

465,000 hectares produce "vins courants"

81,000 hectares produce "vins d'appellations simples"

58,000 hectares produce "vins d'appellations contrôlées".

Therefore, less than 10% of the total hectarage produces high-quality wines which bring returns of 8,000 to 10,000 NF per hectare and more. Here, a full spray program is carried out at a cost of about 500 NF per hectare for the products. The vins d'appellations simples realize 4,500 to 5,000 NF per hectare at a cost of about 250 NF per hectare while the vins courants realize between 3,000 and 4,000 NF per hectare with hardly any expenditure for crop protection measures, i. e. less than 50 NF/hectare.

Besides questions of quantitative and qualitative yield, there is yet another aspect which has much bearing on crop exports, namely the quarantine regulations of the importing countries. A good example of this is provided by infestation of Spanish oranges by the Mediterranean fruit fly *Ceratitis capitata*. According to D. J. Pastor Soler of the Spanish Ministry of Agriculture, an average of 1—2%, sometimes even as much as 2.5% of the harvested crop was found to be infested in the period 1960—1965, which, for a production of about 2 million tons of citrus fruits, is equivalent to a loss of 20,000 to 50,000 tons worth 142 to 355 million pesetas.

On account of the outstanding importance of the export, the actual financial loss must, however, be rated considerably higher. In many countries of Europe, *Ceratitis* is a "quarantine insect", i. e. the imports are subject to official inspection. If infested fruits are found in a consignment, not only are the infested fruits themselves confiscated to prevent introduction of the pest but the entire shipment is rejected. Consequently, the financial loss is many times higher than the loss of the actually infested fruits because by the time a consignment has reached the frontier of the country of destination the price of the fruit has already been increased by the transportation costs incurred as well as by markup. According to G. Leib of the Federal German Ministry of Food, Agriculture and Forestry, 30 shipments of citrus fruits from Spain, weighing a total of 268,636 kg, were rejected in 1963 due to *Ceratitis capitata* infestation at the frontiers and ports of the Federal Republic of Germany. In 1964, 38 shipments weighing a total of 710,584 kg were rejected for the same reason.

In 1965—66, a new type of control measure was introduced for the first time, in which very low dosages of an organophosphate active ingredient were applied in combination with a special bait material. The outstanding advantage of this treatment is that the fruit flies are lured to the applied spray droplets. In 1966, this method of control was made compulsory by the Spanish authorities. The overall result of the control measures is reflected in the big decrease in the rejected tonnage of Spanish citrus fruit imports by the inspection authorities of the German Federal Republic, as shown below in Fig. 2 [9]:

The pesticide costs per hectare for the control of Mediterranean fruit fly in Spanish citrus groves are extremely low. The highest costs for pesticides are incurred in intensive deciduous fruit growing, amounting to between DM 800 and 1,000 per hectare.

A further example of the economic importance of crop protection is provided by cereal farming. Every loss from causal factors is a direct deduction from the net return which is low in terms of percentage, and can prove decisive as far as profitability is concerned, in other words whether production is at all worthwhile. At a 100 hectare farm in Westphalia, records and accounts were kept of all crop protection measures from 1965 to 1968, and in addition every treatment included parallel untreated plots which were later separately harvested. The returns were determined in each case (Table 10) [10].

With this expenditure of about DM 5,000, in other words an average of DM 50 per hectare, the farm income was raised by DM 18,100. If we add to this figure a

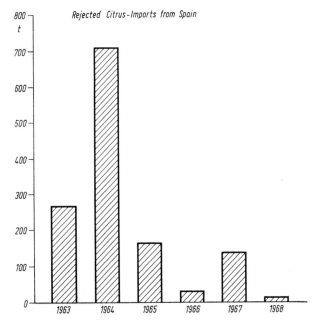

Fig. 2

Table 10   The costs for 1968 were as follows

Herbicides, including application	DM 4,431.—
Seed treatment	DM 268.—
Mildew control	DM 439.—
Total cost	DM 5,138.—

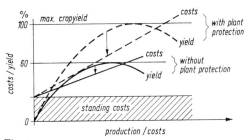

Fig. 3

sum of DM 616 saved on grain drying as a result of the weed control, then the increase in return amounts to DM 18,716. After subtracting the costs of DM 5,138, the additional net profit realized by carrying out crop protection measures is DM 13,538 equivalent to approximately DM 136 per hectare. It seems that a sum of the order of DM 120 to 200 can, under normal conditions, be regarded as that amount which can be credited as increased profit to the use of crop protection measures in cereal farming in Central Europe. All recent publications give similar results. The example shows to what great extent emphasis is placed on economic considerations in intensive cereal farming. It is also economic considerations which forbid growers from chasing after the final remaining loss percentages. In West European wheat growing, we today reckon with an average loss of 3% from insect pests, 2% from diseases and still as much as 9% from weed infestations. Perhaps this loss from weeds will drop to a still lower level as the result of improved *herbicide* applications so that in due course it might be expected that with a loss quota of 7 to 9% the limit of economic weed control will be reached thus creating an optimal condition at least for the time being.

We have thus arrived at a point at which, according to Weltzien of Bonn, it is not a matter of chasing after the final remaining loss percentages but rather of obtaining an economically optimal result, in other words of bringing expenditure and returns into the right relation to each other. What exactly is meant by this is

made clear by the following Fig. 3 taken from a publication of Weltzien[11]: I believe that this train of thought also accommodates ecological aspects. For if one were not to insist on obtaining levels of control very closely approaching 100%, it is obvious that the frequency and dosage rates of pesticide applications could be reduced and integrated methods would have greater prospects. It would perhaps be a good idea for the regulatory authorities also to apply this concept more sometimes. We have in mind, for example, the nematode problem in potato growing.

In weed control, the achievement of a *tabula rasa* need not be the sole aim, either, for economic and other reasons. By other reasons, we mean, for example, the functions which a controlled weed competition can have as a means of protection against siltation and erosion of soil. Controlled suppression of damage and *not* eradication seems to be the more recommendable principle for both economic and ecological reasons.

Having spoken at length about the increased crop yields and profits that can be realized by the expedient use of chemical pesticides, let me now say a few words about the cost aspect of the ecological impact on the environment.

As far as the costs concern those incurred by the manufacturer, they must be included in the product costs I have already mentioned. In the manufacture of chemical pesticides, a strain may be imposed on the environment through the emission of gaseous substances and the disposal of waste water. For some years past, we have been in the middle of a phase of technological change directed towards the realization of a new production technique with the objective of avoiding harmful or objectionable emissions and reducing waste water disposal and eventually eliminating it. The problem is not to avoid harmful emissions but to minimize *objectionable*, in other words bad-smelling emissions, which for all concerned is an exhausting struggle. The outsider can hardly imagine what effort it costs to achieve a success. Several years ago we had to set ourselves the task of building formulating plants which a pesticide manufacturer needs all over the world, that operate *free* of waste water. We succeeded in doing this for the first time in Asia Minor where we had to build a plant in an area without a drain outfall. The experience we gained will be used step for step in other countries. In the chemical factories themselves, waste water treatment is today fundamentally the final stage of the manufacturing process. In fact, in the newer factories waste water treatment is already being practiced on an adequate scale. In older plants, it is more difficult to carry out due to lack of space and often can be accomplished only by basically changing the process which, of course, takes time.

When pesticides are used in agriculture, it is feared that they may be a potential economic risk by causing damage to game, fish, as well as to groundwater and surface water. The statistics show that in areas where systematic use is made of chemical pesticides they have no influence on game populations. Studies on several groups of pesticidal compounds have revealed that there is likewise no need to fear that the applied pesticides will sink through the soil down to the groundwater because the top soil layers retain the products until they have broken down. If pesticides are at all carried from the site of application to other locations this usually happens in the form of dust, aerosols or else they are transported in surface water. However, only the persistent compounds are capable of remaining longer in surface water. Usually, however, the concentrations involved will not be of biocidal levels which necessitate taking cost-incurring measures although there are widely cited examples of where local accumulations caused damage to fish, but these were due to often unsolved cases of accidents and not to use of the products in agriculture. The charge that pesticides were the cause of the great fish kill in the River Rhine has proved to be unfounded. Nevertheless, one should always remain fully alert to the risks involved in transportation and to the possibilities of local damage from pesticides, which is usually of an accidental nature.

The costs of crop protection must, of course, also include state expenditure on protection of the public and on advice extended to the agricultural community on the use of pesticides. We have attempted to make a general estimate of the figures involved. In West Germany, the Federal Government has to spend DM 3—4 million per year on the Biologische Bundesanstalt. But this sum does not include the costs for the regional crop protection offices maintained by the State governments, nor the costs incurred by the food control offices and the Bundesgesundheitsamt. We assume that the total expenditures of state departments in West Germany at present amount to a sum about 4 to 5 times higher than the figure just mentioned. The situation will be similar in comparable countries. In

Japan the costs, we have found, are lower, while in the United States they are considerably higher both relatively as well as in terms of absolute figures. However, these costs, in their order of magnitude, are certainly of a lower level in all countries than the avoidable crop losses.

Have I forgotten any essential factors associated with crop protection measures and which are of economic importance? If so, perhaps they can be brought up in the following discussion so that we can obtain as complete a picture as possible.

Before closing, I should like to say a few words about the research costs that are necessary for discovering and developing new pesticides. The demand for still better, still safer and less toxic active ingredients is well known. Work in this direction is in progress at many research centers throughout the world. But how great are the prospects of success in relation to the high standards that have to be satisfied by these new products? H. Hoffmann, the successor of G. Schrader and head of a crop protection research laboratory at Elberfeld, has made some calculations of these prospects for insecticides, especially organophosphates with which we have had extremely extensive experience. According to Hoffmann, the prospects of finding a marketable product among newly synthesized insecticidal compounds are 1 in 10,000. If screening is concentrated only on compounds which have an acute mammalian toxicity of over 25 mg/kg, the prospects drop to 1 in 20,000.

If this toxicity level is 100 mg/kg, statistics show that the prospects undergo a further drop to 1 in 30,000, at 250 mg/kg to 1 in 50,000, and should a toxicity level of over 1000 mg/kg be required they are as low as 1 in 250,000. These ratios hold only for insecticides and especially for

those with *cholinesterase* mechanisms which exist in insects just as in mammals. When you consider these figures and, at the same time, bear in mind how public opinion reacts to successful research in these fields, you will perhaps show understanding for the news that several chemical companies, particularly in the United States, have given up their research activities in the crop protection field. It is worthwhile thinking over how this trend which we consider to be disadvantageous, could be countered.

References

[1] Ordish, G.: Untaken harvest. Constable et Comp. Ltd., London 1952

[2] U.S. Dep. of Agriculture: Losses in agriculture: A preliminary appraisal for review. Agric. Res. Serv. USDA (ARS–20–1) 1954

[3] U.S. Dep. of Agriculture: What we spend on crop pests. Agric. Res. 3: 14, 1955

[4] U.S. Dep. of Agriculture: Losses in Agriculture. Agric. Res. Serv., Agric. Handbook No. 291, Washington 1965

[5] FAO Production Yearbook 1968, Rome 1970

[6] Cramer, H. H.: Pflanzenschutz und Welternte, Pflanzenschutz-Nachr. Bayer 20: 1967/1

[7] Jung, H. F., H. Scheinpflug: Reisanbau und seine Pflanzenschutzprobleme in Japan, Pflanzen-schutz-Nachr. Bayer 23: 243–271, 1970/4

[8] McNew, G.: Pest control in relation to human society. In: New developments and problems in the use of pesticides. Nat. Academy of Sciences, Nat. Research Council publication 1082, Washington 1963

[9] Koppelberg, B., H. H. Cramer: Die Bekämpfung der Mittelmeerfruchtfliege, Ceratitis capitata Wied., in Spanien. Pflanzenschutz-Nachr. Bayer 22: 164–174, 1969/1

[10] Rieth, G., G. Schulte: Mehrjährige Untersuchungen über die Wirtschaftlichkeit von Pflanzen-schutzmaßnahmen in 2 landwirtschaftlichen Betrieben in Nord-Westdeutschland, Zeitschrift für Pflanzenkrankheiten und Pflanzenschutz 77: 1970

[11] Weltzien, H. C.: Neue Wege im Pflanzenschutz? Landwirtschaft – Angewandte Wissenschaft, Vorträge der 20. Hochschultagung der Landwirtschaftlichen Fakultät der Universität Bonn am 4. und 5. Oktober 1966 in Bonn

Economic Aspects of Pollution Abatement in Industrial Chemicals Production

L. J. Revallier

Director Chemical Research and Development DSM, Geleen, The Netherlands

Summary. Notwithstanding the efforts made over recent years, the chemical industry has not yet succeeded in complying with all the demands imposed by environmental control. The first thing to be done by the plants now in operation is to make up for the backlog existing on the point of *pollution abatement*.

In many cases, however, it is more difficult, and hence more expensive, to correct a current situation than to make appropriate provisions during design or construction of new plants.

Considered from the purely technological angle, the majority of the problems bound up with the abatement of the air and water pollution caused by a chemical industry are, in fact, not insoluble.

However, the question facing us is "would the gain be commensurate to the capital outlay?" The gain — higher purity of air and water — is difficult to assess. Yet, a pronouncement on this issue is definitely called for if we are ever to establish the urgency and priority of the various technical provisions that are needed. This is all the more cogent since the combat against pollution is in some cases a matter of choosing between alternatives.

As to the costs we may ask ourselves: should they be borne entirely by the polluter, or by the national economy as a whole? A country where high requirements are made on environmental hygiene, and the industry has to pay all the expenses needed for satisfying them, will find itself in a more unfavorable competitive position than a country where less strict standards are applied.

In this lecture an impression is given of the costs for eliminating organic substances from industrial waste water by means of a biological purifying plant.

A hazard involved in the production of fertilizers is that NH_4^+ and NO_3^- ions are liable to find their way into the waste water and entail hypertrophication. It is now possible to design an installation in which carbon and nitrogen compounds are eliminated at the same time. The provisional conclusion is that nitrate ions can be eliminated at relatively low cost but that the elimination of Kjehldahl nitrogen is very expensive.

Certain variations of processes aimed at biological elimination of organic material seem to be attractive, such as the recovery of feed yeast from waste water containing organic acids.

To form an idea of the cost involved in combating air pollution the costs of desulphuration of tail gas from a 450 t/a sulphuric acid plant is given. Catalytic reduction of the nitrous oxides emitted by a large nitric acid plant would cost several million dollars. However, the purification process is not sufficiently dependable yet. Still, the primary throw-out from new nitric acid facilities can already be curbed through an improvement of the absorption process.

Finally, the production of acrylonitrile and of Ziegler polyethylene will be presented as examples of new processes the development and design of which has gone hand in hand with the search for provisions capable of combating deterioration of the environment.

Zusammenfassung. Trotz der Bemühungen der letzten Jahre ist es der chemischen Industrie noch nicht gelungen, sämtlichen Forderungen auf dem Gebiet des Umweltschutzes Folge zu leisten. Erste Aufgabe der bestehenden Unternehmen ist, den Rückstand bei der Bekämpfung der Luft- und Wasserverschmutzung wettzumachen. In vielen Fällen aber ist es schwieriger und daher auch aufwendiger, eine bestehende Lage zu berichtigen, als geeignete Vorkehrungen bei dem Entwurf oder der Konstruktion neuer Anlagen zu treffen.

Aus rein technologischer Sicht gesehen, ist die Mehrzahl der Probleme, verbunden mit der Bekämpfung der Luft- und Wasserverschmutzung durch die chemische Industrie, im Grunde genommen nicht unlöslich.

Die Frage jedoch, mit der wir uns auseinanderzusetzen haben, ist: Wird der Effekt in angemessenem Verhältnis stehen zu den Investitionskosten? Der Effekt — nämlich reinere Luft und reineres Wasser — läßt sich immer (noch?) schwer bewerten. Eine Aussprache hierüber ist jedoch zur Bestimmung der Dringlichkeit und Priorität der verschiedenen technischen Vorkehrungen dringend notwendig. Dies trifft um so mehr zu, als bei der Bekämpfung der Verschmutzung manchmal zwischen Alternativen zu wählen ist.

Zu den Kosten können wir uns fragen: Sind diese völlig vom Verschmutzer oder von der Wirtschaft in ihrer Totalität zu tragen? Ein Land, das auf dem Gebiet des Umweltschutzes hohe Anforderungen stellt und in dem die Industrie alle damit verbundenen Kosten zu tragen hat, wird sich in einer ungünstigeren Konkurrenzlage befinden, als ein Land, das auf diesem Gebiet weniger strenge Normen kennt.

In vorliegendem Bericht wird ein Eindruck vermittelt von den Kosten zur Beseitigung organischer Substanzen aus industriellem Abwasser durch biologische Reinigung.

Mit der Düngemittelproduktion ist die Gefahr verbunden, daß NH_4^+- und NO_3^--Ionen in das Abwasser gelangen, wodurch Hypertrophie eintritt. Es ist jetzt möglich, eine biologische Reinigungsanlage zu entwerfen, in der organische Stoffe und Stickstoffverbindungen gleichzeitig beseitigt werden können. Die vorläufige Schlußfolgerung ist, daß Nitrationen bei verhältnismäßig geringem Kostenaufwand entfernt werden können, daß aber die Beseitigung von Kjehldahl-Stickstoff mit hohen Kosten verbunden ist.

Bestimmte Abarten von Prozessen, welche eine biologische Beseitigung organischer Substanzen bezwecken, sind anscheinend attraktiv, z. B. die Gewinnung von Futterhefe aus organische Säuren enthaltendem Abwasser.

Damit man sich ein Bild von den mit der Bekämpfung der Luftverschmutzung verbundenen Kosten machen kann, folgen im Bericht die Kosten einer Entschwefelung von Restgas aus einer 450-jato-Schwefelsäureanlage. Über die Salpetersäurefabrikation wird berichtet, daß die katalytische Reduktion der Stickoxyde im Abgas mehrere Millionen Dollar kosten würde. Ein solcher Reinigungsvorgang ist aber noch nicht zuverlässig genug.

Einem starken Ausstoß von Abgasen aus neuen Salpetersäureanlagen kann bereits durch eine Verbesserung des Absorptionsprozesses Einhalt geboten werden. Zum Schluß werden die Produktionen von Acrylnitril und Zieglerpolyäthylen als Beispiele solcher neuen Verfahren genannt, bei denen bereits im Entwicklungs- und Entwurfsstadium Vorkehrungen, auf dem Gebiet des Umweltschutzes, getroffen worden sind.

Introductory Remarks

For nearly all industries it is true that not only their wastes but, also their main products enter the environment at some moment. The way in which the various industrial chemicals influence nature differs from one product to another. For example, agricultural chemicals like fertilizers and pesticides have a direct impact on the environmental conditions. Other chemicals, such as solvents that evaporate during use, also have a direct influence. With bulk products, such as polymers and man-made fibers, environmental degradation does not occur during use, but only upon disposal of the used or worn-out articles made of them. The monomers employed as feedstocks and intermediates for these polymers and fibers, such as ethylene, propylene, butadiene, styrene, acrylonitrile, terephthalic acid, caprolactam, phenol, formaldehyde, urea etc., hardly get into nature as individual products; their deteriorating action is limited to the stage in which they are synthetized and processed into other commodities.

In the present paper I want to consider with you the environmental pollution caused by chemical processes, together with the economic aspects of its abatement, not the problem of the disposal of used products, solid wastes, refuse etc.

The importance of certain industrial chemicals is evident from Fig. 1, which gives an idea of the world production of polymers in the second half of this century. At the end of the seventies the output of man-made fibers will exceed that of wool

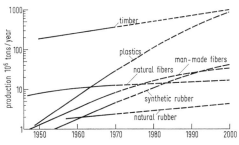

Fig. 1 World production of polymers

and cotton. Synthetic rubber ran abreast with natural rubber as far back as the early sixties.

Impressive is the current and future volume of production for all sorts of plastics. By 1980, the present figure may well have gone up to 100 million tons a year, which is equivalent to one-fifth of that for a natural product like wood. The output of synthetic polymers, which now stands at roughly 10 kg a year per capita of world population, would then amount to some 30 kg/year.

It is almost inconceivable that this trend will be called to a halt by a production stoppage under the threat of the inherent environmental degradation.

If such a stoppage were to come about without any alternative being open, it would mean the abrupt termination of the commonly desired increase in prosperity. It is very doubtful, however, whether any such alternatives can be found and, if so, whether they will present an improvement on the point of environmental control. For example, for products like plastics that come in a nearly endless variety of types and modifications covering a very wide range of application, it is virtually impossible to think of any natural analogs. Efforts on these lines might seem more to the point with regard to man-made fibers and rubbers, although even here success will at best be partial since their uses are much more varied than those of cotton, wool and natural rubber. To this must be added that the cultivation of cotton and rubber would take away large areas of agricultural land from their present destination, i.e. maintaining and improving the world food supply.

Last but not least, it is open to question whether production of *'natural' products, would contribute less to environmental pollution* than production of *synthetics*. The fact that natural wastes can in many cases be turned into compost is not of any

great significance. In the processing stage it is not even uncommon for natural products to pollute the environment even more than their synthetic analogs.

I will try to make this clear by means of the following example.

The Netherlands are using some 60,000 tons/year of polyethylene for packaging purposes. The pollution concommittant with the production of this quantity corresponds to approximately 2500 population equivalents. If this polyethylene were to be replaced by packaging paper, roughly 200,000 tons/year of paper would be needed, the production of which would increase pollution by a factor of 150, so to 400,000 population equivalents.

A similar situation would arise in the case of rubber. Production of 1 ton of EPT rubber gives a run-off of 6 kg of COD*, the corresponding figure for natural rubber being approximately 240 kg, so 40 times as high.

It is surprising that the BOD** content of the effluent flows discharged in the processing of "natural fibers", such as cotton, wool and natural silk, is comparable with that for such synthetics as polyacrylonitrile and polyester fiber, viz. at 200—300 kg/ton, versus 100—500 kg/ton.

This leads to the conclusion that the production of all these materials is extremely unlikely ever to be stopped, and that the major economic problem on this point boils down to the question: "How to make a clean Chemical Industry and how to pay for it?"

Looking at the chemical industry on the whole, we note that the efforts made in recent years failed to satisfy all the demands which public opinion — whether or not inspired by statements from hygienists and ecologists — is making with regard to environmental control. Especially in the plants now in operation there is

* COD = Chemical Oxygen Demand.
** BOD = Biological Oxygen Demand.

quite some leeway to be made up. In many cases it is much more difficult (and, hence, more expensive) to correct an existing situation than to provide appropriate facilities during the construction of new installations, where the design can be tailored to the stricter demands that are now considered essential.

Looked at from the purely technological angle, the majority of the problems bound up with the abatement of air and water pollution are not insoluble for the chemical industry. The problem, however, is "What degree of purification is required? Should nuisance or environmental degradation, be tolerated in order that products which are desirable by themselves can be manufactured?" Finding an answer to this question could be facilitated by assessing the value of commodities that are now growing scarce, such as pure water, pure air and unspoiled nature. This approach is followed by Hueting [1] and others. This is a most laudable initiative, since quantified assessments on this issue are indispensable if we are ever to be capable of rationally establishing the priority of the various technical provisions that have to be made.

Cooling of an industrial plant can be effected in several ways:

— by means of flowing or stagnant water,

— by means of cooling towers,

— by means of air

Ways and means by which the respective nuisance hazards involved (i. e. thermal pollution, mist formation and noise production) could be evaluated separately would be most helpful.

As another example, let us consider the case where the production of chemicals (e.g. acrylonitrile and caprolactam) is attended by undesired side reactions and formation of by-products. Biological oxidation of the by-products, resulting in a residual degree of water pollution (e. g. by nitrates) could be a remedy here. Incineration, which is an other way of tackling the problem, entails a residual degree of atmospheric pollution (e.g. by nitrogen oxides).

Depending on the concentration of the effluent, either the one or the other approach will be the more economical for the manufacturer. However, the ultimate choice to be made will have to depend also on the relative value of non-contaminated water and non-contaminated air.

In addition to these, some further problems of choice may come up.

In some cases waste digestion may yield a valuable by-product. An example in point is feed-yeast grown on a medium of organic-chemical effluent. From the manufacturer's point of view, the preparation of such by-products becomes attractive already if the relative expenditure is lower than the cost of the alternative purification, or than the levies possibly to be attached to pollution. However, the manufacturer concerned must be prepared to place such a rather unusual commodity on his products list, and be sure of finding a market for it.

It does happen that the standards of environmental hygiene have a stimulating effect on the search for alternative procedures; this brings us to the use of sulphur. Sulphur is a very important material in chemicals production. The annual output of elementary sulphur or pyrites in the USA now stands at 16 million tons, but 12 million tons are discharged into the atmosphere as a constituent of flue and vent gases [2]. From the mere point of view of resource management, so regardless of the environmental aspects, this is an evident example of wasteful exploitation and dissipation. To this it must be added that sulphur (or, rather, sulphuric acid) although a very important auxiliary in the chemical industry, does not form an essential component of many products, and therefore buyers of such products are not prepared to pay a price for it.

This seems to justify the expectation that in the years to come the chemical industry will make ever greater efforts to keep the sulphur in the process cycle, which involves the neccesity of developing a number of novel processes.

Industry has declared itself agreeable to the demand that the polluter is to pay the cost of the necessary purification measures. It cannot, however, avoid building these expenses into the sales prices of its products. Provided all suppliers of a given product were to satisfy the same requirements this would not cause any great difficulties. However, this is far from being the case, even on a national scale, with the result that e.g. an industry operating in the interior finds itself in an unfavorable competitive position towards one established on the sea-shore.

Recognition of these circumstances, along with some financial compensation or cost-sharing arrangements, may be well demanded of the national government. The matter, of course, also has its international aspects.

It is evident that quite a lot of research and development work remains to be done on the perfection of pollution abatement by the chemical industry. Also on this point active assistance by the government cannot be dispensed with. The manufacturer who will be the first having to comply with stricter purification requirements and, hence, will be obliged to strike out along new routes, is bound to meet with numerous practical difficulties and should be eligible to receive some sort of pioneering subsidy.

Environmental Disturbances: Physical and Chemical Causes

Environmental disturbances differ widely in nature. Some sort of classification can be arrived at by making a distinction between physical and chemical pollutants. Normally, chemical pollutants are considered the more objectionable of the two. At a first impression, however, it seems that the nuisance, or damage, due to physical causes is much more difficult to suppress than the harmful effects of chemical pollution. In principle, the latter can often be tackled in many more ways than, say, excessive production of heat or noise. An example of acoustic pollution is the noise nuisance concommittant with the starting-up of production units of ever larger capacities, with turbo-compressors, flares and steam-release as the major offenders.

Although nuisance due to physical causes does not fall outside the scope of my subject, I would yet like to restrict myself to the discussion of pollutants of chemical origin.

Chemical Pollution of Water and Air

In dealing with this subject I will set out mainly from the experience gained in a chemical complex of DSM over the last 5—10 years. From the viewpoint of environmental hygiene, this complex is rather unfavorably situated, i.e. some 200 km from the coast and 5 km from the Maas, a pluvial river which, in summertime, carries only very little water and into which the effluents from the region are discharged. The surrounding area is densely populated, the number of inhabitants being some 600,000. As there are no other major industries in the neighborhood, the polluter can easily be identified, which might be looked upon as a drawback. An advantage, however, is that the effect of any measures taken can be readily observed and evaluated. This has enabled us to assemble a number of data that might be useful for other industries too.

Water Pollution (e. g. by Carbon and Nitrogen Compounds)

Most organic substances, like alcohols and acids, but also poisons like nitriles and phenols, can, in the same way as domestic effluent be broken down biologically by means of atmospheric oxygen and bacteria, the latter acting as digestive agents. The resulting end products are carbon dioxide and water. All effluents in a given area that contain such waste materials can be handled by a central, biological purifying plant.

Smaller industries will normally find it more economic to be connected with a public treatment plant for domestic waste water[3]. Relatively large industries, however, cannot depend on such a system: capacity expansion troubles in the production processes and drainage of abnormally large or excessively concentrated effluents may give rise to serious problems. These considerations induced DSM in the early sixties to build its own installation for *biological purification* of waste water; originally, only the effluents from the coking plants were treated in it[4].

In the preliminary research stage, experiments were undertaken on a 1:60 scale along two alternative routes: the activated-sludge method with a short retention time (3—5 h) and the Pasveer method (oxidation pond), which uses a long retention time (2—3 days). The latter clearly proved to be preferable, notably for the following reasons:

1) lower susceptibility to operational troubles and surge-loading,

2) relatively lower investment costs.

It is precisely in a chemical industry where surge loads induced by all sorts of causes are liable to occur; leakage overflowing of tanks etc. may involve the sudden drainage of large volumes of waste materials into the sewer systems and, from there, into the purifying plant, with the result that the bacterial population may be completely or partially destroyed. The survey in Table 1 gives you an idea of the diversity of the contaminants that may get into the effluent of a chemical industry.

Table 1 Contaminants in effluent of various production plants

coke plant	phenol, cyanide, rhodanide, ammonium
cyclohexanone plant	sodium salts of lower mono- and dicarboxylic acids
naphtha crackers	sulfides
high density polyethylene plant	catalyst residues, lower alcohols
low density polyethylene plant	catalyst residues
caprolactam plant	by-products of caprolactam production
acrylonitrile plant	cyanide, cyanic hydrines, nitriles, amm. sulphate

The installation — of 30,000 m^3 volume — has been operating since 1964. The aerating potential, which has already been increased beyond the original figure of 250,000 population equivalents, will, in the course of this year, be raised further to 500,000 population equivalents.

Figure 2 shows a survey of the plants which, over the seven years the unit has been operating, have drained waste water into it. In December 1968 DSM took its last coke oven battery out of operation. Since then, the pond has been handling chemical effluents only.

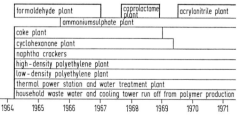

Fig. 2 Effluent streams purified in the Pasveer pond

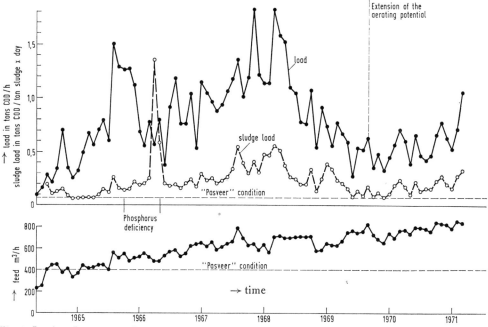

Fig. 3 Load on Pasveer pond

Figure 3 gives you an idea of how the plant has been loaded over the years. As you see, the "Pasveer conditions" have been frequently deviated from to a considerable degree. The *purification efficiency* varies from 75—85% for COD, the corresponding figure for BOD under normal conditions being upwards of 98%. A special problem to be dealt with in such an industrial purifying plant is the limited diversiveness of the feed. Whereas domestic run-off contains all the elements needed for making the bacteria flourish, industrial effluent may be deficient in one or several of them. It was soon recognized at DSM that shortage of phosphor tends to render bacterial sludge less active and too slimy. At the moment, inorganic phosphate must be supplied daily in order to ensure good performance.

Every time new products are placed on the list of commodities, it must be examined whether their wastes can be biologically broken down and, if so, whether they will not interfere with the biological digestion of the other waste materials.

This calls for extensive and prolonged experimental work on a laboratory scale. A difficulty encountered in these studies is how to get hold of sufficient, representative waste when the new product is still under development. In most cases the amount available is inadequate to permit experiments to be conducted on an appropriate scale.

Nevertheless, we have succeeded in building up know-how that enables us in most cases to predict what will be the impact of a given effluent on the performance of the oxidation pond.

It even seems worthwhile usefully to employ the current know-how also outside the company and, occasionally, to undertake contracts for research in this particular field. This takes us to the issue of the cost of the oxidation pond. Research and development in behalf of the constructional work has cost us approximately $ 150,000, the total investments amounted to $ 700,000—800,000 (1964). The expenses of research during operation have run to another $ 150,000.

Table 2 gives the annual costs incurred since 1965, together with a breakdown over the principal headings. As you see, the expenditure per ton of COD oxidized has gone up considerably since 1969. This is because in that year the loading of the pond was reduced and the aerating potential was increased. These measures had to be taken in order to provide sufficient surplus potential for coping with peak loads. Before 1969, when no surplus potential was available, the plant frequently got overloaded. The resulting stench formation gave rise to complaints from the people in the neighborhood.

Table 2   Costs of the Pasveer ditch

year	costs per year	costs per ton COD oxidized
1965	$ 165,000	$ 42
1966	$ 180,000	$ 47
1967	$ 265,000*	$ 56
1968	$ 210,000	$ 39
1969	$ 220,000	$ 111
1970	$ 330,000*	$ 167

* Including additional costs for maintenance.

Break-down of the costs over the principal headings

electric power	20%
phosphoric acid, chemicals	5%
operation and super vision	5%
maintenance	25%
depreciation	30%
miscellaneous	15%

This is a clear example of a case where a financial sacrifice was made to prevent inconvenience.

A new, important problem of water pollution is caused by the simple nitrogen compounds NH_4^+ and NO_3^-, particularly because they entail the hazard of *hypertrophication* of surface waters. It is understandable that the national authorities are trying to reduce the discharge of industrial effluents into public waters by the provision of drainage limits, and also by the imposition of levies. In the Netherlands such a levy is already payable for the drainage of all oxygen-consuming substances, viz. COD and Kjeldahl nitrogen. For waters supervised by the Government, the current levy (1970) stands at $ 8/ton for COD and at $ 40/ton for Kjeldahl nitrogen. In a number of years it will be raised to a multiple of these amounts. Drainage of NO_3^- nitrogen will not be charged, but drastically limited.

It is particularly the producers of nitrogen fertilizers that are hit by these measures, because the ingress of NO_3^- and NH_4^+ into the effluents can hardly be prevented in the present installations.

With regard to this problem of waste disposal we are not out of the wood yet. A normal biological treatment plant is not capable of eliminating the nitrate and ammonium ions in an effective manner. Still, biological processes by means of which the necessary conversion can be brought about have long been known. Hence, in the present state of the art, the purification might — in principle — technically be realized in the way set forth below.

As we know from what happens in the river, ammonia can be oxidized to nitrate by the action of certain bacteria (nitrification). In an oxygen-poor medium, this nitrate ion can then be converted to nitrogen (N_2) (denitrification). It is now possible to design an installation in which carbon and nitrogen compounds are eliminated at the same time. A diagram of such a plant is represented in Fig. 4.

From experiments conducted in the laboratory and on a semi-technical scale it has appeared that the underlying principle can be industrially realized. Tentative calculations have shown, however, that

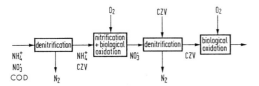

Nitrification : $NH_4^+ + 2\,O_2 \longrightarrow NO_3^- + H_2O + 2\,H^+$
Denitrification: $2\,NO_3^- + 2\,H^+ \longrightarrow N_2\uparrow + H_2O + 5\,[O]$

Fig. 4 Biological purification of waste water with removal of ammonium and nitrate

for an industry like DSM a technical installation of this type would cost more than 5 million dollars, which means that further economic optimization will be necessary. This will involve rather complicated calculation work because there exist a number of alternative approaches. However, once the necessary technical data have become available and these are weighed against the above-mentioned drainage levies, the problem may well be soluble.

Starting from a few simplified assumptions, we can draw some provisional conclusions:

1) elimination of Kjeldahl nitrogen is very expensive, amounting to, say, $ 500 per ton of N. An amount of this size would suffice for taking such drastic measures in the plant as would reduce the amount of effluent to a minimum;

2) nitrate N can be eliminated at relatively low cost because part of the COD is removed at the same time. An appropriate quantity of nitrate N may even be desirable because removal of COD by means of nitrate present in the effluent is cheaper than removal with atmospheric oxygen.

I gave you this example to make it clear that, just as in chemicals production, economic optimization is necessary and possible also with regard to processes aimed at the prevention of pollution. Normally, the chemical industry will, by virtue of its experience, be capable of finding a solution on its own account, but, as in most other cases of process selection, much research work will be necessary as well as evaluation of the final product: pure water.

Feed Yeast

It will have become clear from the foregoing how important it is to have the disposal of alternative possibilities for solving the pollution problems in the chemical industry. It is for this reason that I wish to refer here to an interesting procedure for the recovery of feed yeast starting from an effluent with a high percentage of organic chemical pollutants. In this process the mixture of organic acids formed as a by-product in the oxidation of cyclohexane to cyclohexanone is converted by means of fermentation. This oxidation of cyclohexane is an important step in the preparation of caprolactam, an intermediate for nylon-6. The by-products are organic mono- and dicarboxylic acids, which can be either converted in the oxidation pond or be subjected to a fermentation treatment for recovery of high-quality feed yeast.

The economics of the process depend on several factors, the influence of which appears from Table 3.

Table 3

costs per ton yeast	production capacity (yeast)		
	2,400 t/yr	6,000 t/yr	20,000 t/yr
Fixed costs (operation, maintenance, depreciation, interest)	$ 220	$ 170	$ 120
Variable costs (chemicals, utilities)	− 80	− 80	− 80
Oxanone effluent	X	X	X
Total production costs	$ 300 (+X)	$ 250 (+X)	$ 200 (+X)

From these figures it can be concluded that

1) only if alternative processes for treating the oxanone effluent are not available or too expensive, the choice will be in favor of yeast production,

2) the fixed costs decrease with expansion of the production capacity; this might induce different companies to combine their effluent flows.

At the current low yeast price of approximately $ 200/ton (EEC price for 1970), which even shows a downward trend, the process is not practicable for DSM.

Atmospheric Pollution (e. g. by Sulphur and Nitrogen Oxides)

In the introduction I already pointed out that the amount of sulphur wasted into the atmosphere in some troublesome form is nearly as large as the amount extracted from the earth for useful industrial application[2]. It is not surprising, therefore, that the interest is focussed on the pollution of air by sulphur dioxide. By far the major source of SO_2 is the flue gas produced by combustion of sulphur-containing fuels. Coal can hardly be desulphurated, while desulphuration of viscous cheap fuel oil is still difficult and costly. Desulphuration of flue gases is not simple either and, in addition, relatively expensive, because the gases contain no more than 0.2 % by volume of SO_2 besides a high percentage of CO_2 and water vapour. However, since this problem is not peculiar to the chemical industry, we shall not discuss this further.

A pollutant that does come within the scope of our subject is the final gas formed in the production of sulphuric acid, one of our major bulk chemicals. Here, the situation is much more favorable than it is for flue gases. For, this vent gas can easily be stripped of SO_2 and SO_3 by means of an ammonium solution. Combustion of the resulting salt solution

yields water, nitrogen, and concentrated SO_2 that can be recycled in the process[5]. What is even more attractive in some cases is utilizing the ammonium sulphite as such, e.g. for preparing hydroxylamine, one of the intermediates for caprolactam. Table 4 shows a cost-price estimate for desulphurating the end gas of a sulphuric acid plant having a capacity of 450 tons/day. Such a desulphuration, of course, does not make up for the feedstock expenses, let alone earn back the investment cost of the installation. Nevertheless, it is to be regarded as a very effective measure against environmental degradation (95 % efficiency).

Table 4 Two-stage scrubbing of tail gas from a 450 t/d sulphuric acid plant

Investments	$ 500,000
Annual Costs	
– Ammonia water	– 30,000
– Desilicified water	– 3,000
– Electric power	– 30,000
Total	$ 63,000
Annual Result	
– Sulphate solution	$ 1,000
– Sulphite solution	– 35,000
	$ 36,000

Attempts at suppressing the emission of nitrogen oxides from nitric acid plants of the current design have been much less successful so far.

Reduction of these compounds, among which the dioxide is a nuisance if only because of its brown color, can, in principle, be effected with natural gas over a noble metal catalyst. The expenses are high, however, and there is great uncertainty as to the question whether the catalyst will remain sufficiently active in the long run. Anyway, in a number of cases this approach has turned out a down-right failure. In the new nitric acid plants, the problem has fortunately become much less serious, because, with the high absorption efficiencies now attainable, the emission rate can be cut down to one quarter of that for the older facilities.

Abatement of Environmental Pollution from New Plants

Reduction of water or air pollution caused by an existing industrial plant is normally anything but simple. As a rule the process regime, the lay-out of the installation and the infrastructure no longer allow drastic alterations to be made. During the development of new chemical procedures, as well as during the design of new plants, on the other hand, one has the opportunity to tailor the installation more or less closely to the demands imposed by environmental control.

I shall elucidate this point with reference to two recent examples: the production of acrylonitrile and high-density polyethylene. A brief survey of the measures taken and their consequences is illustrated in Table 5. The principal difference between the two cases is that with acrylonitrile the remedy was effected in the effluent treatment, and with high-density polyethylene in the process itself.

Acrylonitrile

In the Sohio process, acrylonitrile is prepared in a single stage from propylene, ammonia and air, by means of a fluidized catalyst. By-products are hydrogen cyanide, acetonitrile, several other nitriles and tarry products.

A highly simplified flow sheet is shown in Fig. 5.

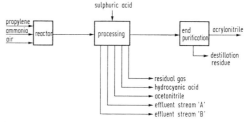

Fig. 5 Production acrylonitrile by the Sohio-process

The gas stream leaving the reactor is scrubbed with sulphuric acid to remove excess of ammonia. This yields ammonium sulphate, 99% of which goes into effluent B. The residual gas — mainly nitrogen — is directly blown off. In the standard Sohio process all other waste streams are pyrolyzed in one large furnace by means of gas-fired pilot burners. In our situation this practice was most objectionable, because the ammonium sulphate is thus converted to SO_2 and, probably, some SO_3. The production of SO_2 would be 1 ton per hour; in addition, remainders of catalyst are liable to get into the atmosphere.

We solved the problem as follows. Acetonitrile and distillation residue are burnt and hydrogen cyanide is processed to other compounds. The dilute effluent stream A, which contains only some 10 g/l of COD and a little ammonium sulphate, and which would need much fuel for evaporation of the water, is now dis-

Table 5   Abatement of environmental pollution from new instillations

product	operation involved	measure	result
acrylonitrile	effluent treatment	separation of organic and inorganic wastes	– sale of inorganic products – incineration and biological breakdown of org. products
high-density polyethylene	production process	sharp reduction of catalyst concentration	– no metal ions in effluent – 94% less COD in effluent

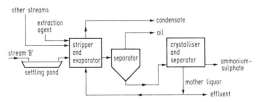

Fig. 6 Processing of waste water containing $(NH_4)_2SO_4$ by means of extraction and crystallization

Table 6 Annual costs of the ammonium sulphate treatment in the acrylonitrile plant

Interest and depreciation	$ 250,000
Operation and maintenance	– 100,000
Laboratory	– 30,000
Extraction agent	– 100,000
Steam, electric power, cooling water	– 300,000
Biological oxidation	– 50,000
Combustion	– 20,000
Other costs	– 100,000
Total	$ 950,000

charged into the biological treatment plant to be thoroughly broken down. Stream B, made up of an organic part and an inorganic part, is carried through a number of extraction and crystallization stages to be split up into (see Fig. 6):

— ammonium sulphate, which is sold,

— an oil, which is burnt,

— small quantities of condensate and liquid effluent, which go to the biological treatment plant.

The cost, shown in Table 6, is compensated to roughly 30 % by the sale of ammonium sulphate, which leaves a loss of some $ 600,000 a year. This is about equal to the cost of the natural gas which in the original Sohio process would be needed as a fuel for evaporating all of the water. The advantage of the new version, however, is that the emission of SO_2 has been reduced to nil, and that the ammonia becomes available as fertilizer.

High-Density-Polyethylene

High-density polyethylene is made in addition by a process developed by Professor Ziegler, which uses aluminum alkyls and titanium chlorides as catalysts.

Since the invention of the process, a great many variants have been worked out. A highly simplified diagram of the process, which may be considered as rather conventional already, is illustrated in Fig. 7.

Since the catalyst must not only be deactivated but also be washed out of the product, the waste water carried to the biological oxidation plant contains metal ions and oxidizable substances (COD). Some time ago we embarked on a new and thorough study of this polyethylene process. One of the results achieved by it is a novel, highly active catalyst, which can be used in much lower concentrations than those used up until now: an economic advantage. The operation of washing the catalyst out of the product can now be dispensed with altogether. As a result, hardly any metal ions will get into the effluent water. Another advantage is that only a very minor quantity of organic effluent is released during purification of the dispersing agent, i. e. at most 1.5 kg per ton of polyethylene, which means a reduction to 6 % compared with the original process.

Fig. 7 Production of high-density polyethylene by normal Ziegler processes

This is another example of an essential improvement which can be used when a new more economic plant has to be built. Here industrial interests run parallel to improvement of the environmental quality.

Conclusion

I trust to have succeeded in making it clear that the chemical industry has already acquired a fair amount of theoretical and practical knowledge in the field of environmental control, and also that a great deal of this know-how has already been practically applied, not infrequently at the cost of high capital outlay. Very often success is evident; there are, of course, occasionally disappointments, however.

Looking into the future, we may wonder whether it will be possible for the chemical industry to control or, rather, significantly to reduce the environmental pollution it has caused. Many people are highly pessimistic on this point, mainly because this branch of industry is still vigorously expanding, and it is felt that the degree of pollution will increase more or less in proportion with the growth of production. Yet, there is ground for optimism, precisely because of this growth. For, expansion goes hand in hand with a relatively fast obsolence of installations and processes, a circumstance which favours the introduction of important economic improvements, some of which are bound to have a beneficial effect also on the environmental conditions.

If the growth of chemical production is to continue at the rate of 10% a year, the production volume will in ten years from now have gone up to 2.5 times the current figure. At least 60% of this output will then come from installations not yet in existence at the moment. If we succeed in reducing the pollution from the old plants still in operation by that time to 50%, and in making the new plants three times cleaner than the present ones, the environmental degradation will not exceed beyond the present level. In view of the experience we have gained over the past few years, this is definitely within the range of possibilities. We must of course not shut our eyes to the fact that there is no royal road to the improvements we are aiming at; the only way of achieving them is through continuous, purposeful and intensive efforts in which chemical research and economic evaluation will have to play an important role. In the plants now in operation, it must be endeavored to fit the improvements into the processes as far as the, usually limited, possibilities allow. Much can be gained here also through well-conceived disciplines and instruction of the operating staff. For example, wastes that can be collected should not, for mere convenience sake, be flushed into the sewer.

As to new plants now being developed or designed, the following objectives can be formulated:

— high yields and high selectivities,
— great attention for waste products (are they toxic, degradable, or combustible; do they release toxic or corrosive vapours during combustion?),
— work out processes in waterfree medium,
— maximum recycling of process flows,
— high concentration of effluent flows,
— dependable equipment and accessories,
— separate sewer systems (in many cases three may be needed, i.e. for process effluent recycle process water and non-polluted water, e.g. rain water),
— elimination of mixing condensors, to avoid unnecessary contamination,
— adequate buffer capacity for coping with operational troubles.

Returning to our point of departure, we may conclude by saying that the branch of the chemical industry under discussion will, also during its further expansion, be capable of keeping environmental pollution within reasonable limits.

There are possibilities at our disposal to convert pollutants with a high degree of efficiency, or to suppress their emission. New installations can be made cleaner than they are today. Industries can profit from exchange of experience and know-how, while much is to be gained through more intensive research. Nor can industry do without assistance and stimulating measures from the government.

References

[1] Hueting, R.: Economische Statistische Berichte 55: No. 2730, 80, 1970, 55: No. 2740, 344, 1970
[2] Teller, A. J. Air Pol. Control. Ass. 19: No. 11, 839, 1969
[3] Plas, J. C.: H_2O, 4: No. 2, 26, 1971
[4] Adema, D.: The largest oxidation ditch in the world of the treatment of industrial wastes. Proc. 22nd Industrial Waste Conference, Eng. Bull., Purdue University 52: No. 3, 717, 1968
[5] Chemiebau Zieren, NOA 6.504.692 and 6.507.567

Environment Program of the Federal Republic of Germany

H. P. Mollenhauer

Ministerialrat, D 53 Bonn – Bad Godesberg, FRG

The Federal Cabinet has formed a special *Cabinet Committee for Environment*, which is being assisted by an Interministerial Steering Committee for the Elaboration of the Environment Program. The very close interministerial collaboration of the responsible sections of various Ministries is typical of this organizational form.

The work is mainly focussed on the Ministry of the Interior, while other ministries collaborate according to their respective responsibilities.

I would like to restrict myself to one thought, which has been entitled the *"Immediate Action Program"* (Sofortprogramm) of the Federal Government. It refers to the occurrence of chemicals in the environment.

The numerous publications concerning the environment show, that the study of the chemical contamination of the environment is based upon the knowledge gained in the control of pesticides.

Pesticides belong to the large group of substances which have been developed by scientists and technologists for the purpose of combatting noxious living creatures such as insects, weeds, or fungi (biocides). Although they are used intentionally, it is unavoidable that — unintentionally — some residues remain in the treated cultures and foods. There are, however, residues or rather quantities of substances, with similar *toxicological significance for man*, which enter into the so-called food chains and finally foods by various ways from the environment.

From the view point of the consumer, it does not matter by which way or whether intentionally or unintentionally his health is endangered through the intake of chemicals and other substances.

All of the possibilities mentioned must be taken into account and evaluated toxicologically in their completeness, as a base for effective counter measures.

Therefore, the Federal Government in its "Immediate Action Program" has developed a new concept, which embraces all substances which enter into food chains and man's food from the various media such as water, soil, air.

A comprehensive approach is aimed at involving administration and control in their responsibilities, and involving sciences and research, especially toxicology, who must change over from the study of individual simple substances to the complexity of the "overall toxic situation" (toxische Gesamtsituation).

With this aim in view the term "environmental chemical" has been coined, which is gradually also being accepted in international circles (OECD).

Environmental chemicals (Umweltchemikalien) are substances which occur in the environment by reason of human activity and which may occur in quantities capable of harming living beings, including animals, plants, microorganisms, and man.

They include *chemical elements* and compounds of organic or inorganic nature, and of synthetic or natural origin. *Human activity* may be affected directly or indirectly, intentionally or unintentionally. The *harmful effects* of the substances may be of an acute or chronic nature; they may also occur by way of accumulation, or chemical changes, or synergism.

Biocides are normally understood to be

substances used in combatting noxious animals and plants including microorganisms such as fungi, weeds, insects. Typical examples are pesticides, fungicides, herbicides.

These definitions make no reference to environmental or other media where they may occur; they are included in the comprehensive toxicological evaluation, as soon as they occur in the environment including foods. This evaluation must also cover preventive measures, that is to say, not only the actual occurrence of chemicals in the environment is to be considered, but also their manufacture, application, and disappearance.

In this connection it should be understood that environmental chemicals do not only stem from the application of biocides, but that they are other important sources of chemical pollution emitting substances such as mercury-, lead-, or sulphurcompounds, which must be just as carefully controlled as pesticides, for instance.

These are examples for the badly needed comprehensive consideration of substances from various sources.

It would, for instance, not be sufficient to set a permissible maximum limit for a certain substance in foods without taking into consideration the intake of the same substance from other sources in the environment.

In order to broaden the protection of the consumer, existing regulations are at present being amended with the aim of safeguarding man against the effects of *environmental chemicals*.

Concerning the use of *pesticides in agriculture*, effective regulations already exist concerning sale and application and also residues in foods. As much as public opinion is opposed to any residues of pesticides in foods, a complete ban would not be feasible at present.

Within the foreseeable future, man will not be able to do without biocides, however, their use must be reduced to a practicable minimum. This requires, for instance, further development of and support for methods of *biological plant protection* in conjunction with an unavoidable measure of pesticides (integrated plant protection).

The use of chemical pesticides will be more closely watched: It is not sufficient, for instance, to aim at destroying a certain kind of insects with the aid of chemicals, but we must also consider what effects such action may have on the biological environment of the particular species of insects, and what other consequences the remaining residues of the chemical may have for the environment at large.

The use of pesticides must also be kept in a sensible proportion to the economical damage which is to be prevented: i.e. pesticides should only be applied when really necessary.

The environment program of the Federal Government aims generally at controlling the distribution of chemicals in the environment more closely.

As a closing remark, let me quote from a paper "The Global Village Pump" which Fred Wheeler has published in New Scientist, which reads about as follows: The phenomena caused by pollution of the environment have, up to now, mainly been looked upon as a performer in a circus with three arenas, less as the subject of one discipline of science.

Protection of Environmental Quality in Israel — Organizational Aspects

Ch. Resnick

Ministry of Agriculture, Jaffa, Israel

Although very small, the State of Israel contains a number of diverse geographical and climatic regions. In the northern Galilee, rainfall is quite heavy, about 700 mm per annum (versus 600 mm in London, for example). Summers, however, are warm and dry. This produces a typical Mediterranean-type climate. From north to south and from west to east, the country becomes more arid. *Rainfall* is very low in the southern half of the country which belongs, geographically, to the Sahara desert.

The population density is high within the coastal zone (from Acre in the North, to Ashkelon in the South). Most of the industry and much of the country's agriculture is centered in this area.

This combination of climate, relatively dense population in a small part of the country, rapid industrialization and intensive agriculture has already produced many of the symptoms of environmental degradation common to older and more developed countries. Israel's more pressing environmental problem is the use and reuse of its available *fresh water*. Over 90 % of the estimated water resources of the country are already being exploited while the reclaim and disposal of used water is at very early stage of development.

Israel has a relatively large and active scientific community. In recent years there has been a considerable increase in the interest and motivation of Israeli scientists towards environmental research. This has been brought about both by the increasing world-wide awareness of the urgency of preserving the biosphere and also by the appearance of local environmental problems.

Israel National Committee on Biosphere and Environment

The main coordinating authority for all aspects of the environment is the "Israel National Committee on Biosphere and Environment". This committee has recently been set up in accordance with a governmental directive. Its chairman, directorate and members are appointed by the National Council for Research and Development and the Israel Academy of Sciences and Humanities. The committee incorporates the "Israel Committee for the International Biological Program and the Biosphere" of the Academy.

The duties of the committee are:

1) To study the problem of the quality of the environment and the process of interaction between the *living environment* and man and to define the problems requiring action.

2) To recommend courses of action for dealing with these basic problems, including legislation and the training of academic and professional personnel.

3) To recommend guidelines for *research*.

4) To gather, edit, and publish information on the environment.

5) To represent Israel in the relevant *international organization such as IBP, SCOPE, and MAB*.

The membership of the committee includes scientists acquainted with the various aspects of the environment, representatives of the appropriate Ministries including the Ministries of Health, Justice, Agriculture, Development, Communications, Trade and Industry, Education, Housing and the Interior, and representatives of various public and scientific bodies, such as the Israel Ecology Association, the Nature Reserves Authority, the Society for the Preservation of Nature, the National Parks Authority, and the Petroleum Institute.

The committee operates through sub-committees in the following spheres: water quality; air-pollution; solid wastes; sea-water pollution (in cooperation with the Oceanographic and Limnological Research Company); soil-pollution; nature reserves; the influence of the environment on man; urbanization; legislation; education; and a sub-committee which reviews biological research in desert areas. It is intended in the future to set up additional sub-committees dealing with the pollution by noise, pollution by pesticides, sociological aspects of the environment and system analysis of eco-systems.

The committee is preparing reports on problems of environmental quality in Israel and also published an information bulletin called "The Biosphere" (in Hebrew).

It is hoped that with the cooperation of government, public organization and the scientific community, Israel will be able to prevent further deterioration of the environment for the benefit of her people.

Canadian Government Programs on Environmental Quality

H. Hurtig

Research Coordinator (Environmental Quality) Canada, Department of Agriculture

A new development in Canada is the establishment of a Department (Ministry) of the Environment. The enabling legislation is before parliament now. In the meantime the components of existing agencies going into the new department have been assembled temporarily in the *Department of Fisheries and Forestry*. They include:

Fisheries Service, Fisheries Research Board, Canadian Forestry Service, Canadian Forestry Institutes, Canadian Wildlife Service, Public Health Engineering Division, Air Pollution Control Division, Environmental Quality Directorate, Marine Sciences Branch, Canadian Hydrographic Service, Inland Waters Branch, Policy Research and Coordination Branch and Water Planning and Operations Branch.

Some restructuring on an organizational basis will, no doubt, take place.

Other ministries retain some major responsibilities for specialized statutory and research functions concerned, e. g. human health aspects of environmental quality (Department of National Health and Welfare), regulation of sale and research on pesticides and fertilizers (Department of Agriculture), transport, urban development, northern environment, etc.

In the Department of Agriculture new research programs are being developed on the environmental quality aspects of: animal and plant waste management; pesticide residue management; soil aspects of plant nutrients, heavy metals and trace elements; and hazard to agriculture of pollutants of non-agricultural origin.

Additional support for university research on environmental quality is being developed by individual departments through specialized committees, e. g. *National Committee on Water Quality Research,* in which there are representatives from other operating ministries with an interest in the subject.

A major recent development was the establishment of the National Research Council Associate Committee on Scientific Criteria for Environmental Quality. This committee has a scientific secretariat of ten professionals which will eventually grow to twenty. The secretariat services subcommittees on:

air, water, pesticides and organohalogens, metals, biology, physical and energy phenomena, and relevance.

The secretariat is also engaged with the *National Science Library* in building up a computerized information system on environmental quality. The subcommittees will address themselves to evaluation of scientific information on "cause-effect" relationship leading up to developing scientific criteria for environmental quality and defining needs for research to provide criteria. This information will be assembled into monographs and made available to operating agencies with statutory authority for developing and enforcing environmental standards. The target is to produce a national *"environmental code".*

Since Canada, like many other nations, is faced with requirements to participate in a growing number of international meetings of governments on the subject of environmental quality, an "Interdepartmental Committee on International Environmental Activities" has been established to provide coordination. Subcommittees have been established on the UN 1972 Stockholm Conference, UNESCO/Man and the Biosphere, NATO/CCRM Program, OECD/Environment Committee. Also, it is expected that very soon a National Committee for ICSU/SCOPE will be established.

Most Canadian provinces are developing programs on environmental quality and some have already established new ministries of that specialized nature.

At present, approximately 40 special regional interagency studies are in progress on a river-basin, or specialized area basis to develop environmental quality standards for regional multiple use of the local natural resources. It is expected that this will grow to 300 such studies if qualified people are available for such a program.

Canada has particular concern about northern environmental quality and determining just how fragile is the northern ecology and how long does it take to recover from various types of injury such as *oil spills,* travel over tundra, exploration, pipelines, mining, etc.

A long standing program carried out jointly with the USA comes under the "International Joint Commission Canada and United States". A special study of water pollution and remedial measures for Lake Erie, Lake Ontario, and the International Section of the St. Lawrence River was initiated in 1964. A three volume report was released in 1970 and a comprehensive summary · published in 1971 (Information Canada, Catalogue No. E95-1970, Ottawa, Canada, $ 1.75). A similar "Joint Air Pollution Study of St. Clair-Detroit Rivers Areas" for this International Joint Commission is now in progress.

Pollution Control in England*

Air Pollution

From Domestic Fires

Under the *Clean Air Acts* local authorities can declare Smoke Control areas, in which the burning of smoky fuel is prohibited. Over 4.75 million premises are now covered by smoke control orders. Local Authorities can also control other kinds of emissions for health and amenity reasons.

From Factories

Air pollution from the most dangerous industrial processes is controlled by the Alkali Inspectorate who require the use of "the best practicable means" of abating emissions. Emissions from about 100 works and processes are controlled in major industries, including iron and steel production, electrically supply, petroleum refining, and petro-chemical production, production and processing of ammonia, ceramic works, and cement works. An extension to the list of processes is in the pipeline. Regulations made in February 1971 impose limitations on emissions of grit and dust from furnaces.

From Motor Vehicles

Avoidable smoke from motor vehicles has been forbidden for many years (about 2000 prosecutions annually). Heavy vehicles are tested annually for this.

Regulations have just been made requiring new cars sold after 1 January 1972 to be fitted with a "crank case breather" which will reduce the emission of hydrocarbons (which smell) by 25—30%. Regulations setting a limit on emissions from diesel engines have also been introduced.

The Government are considering further measures to reduce vehicle emissions, possibly in concert with other European countries. *Photochemical smog* of the kind best known in Los Angeles is not a problem encountered in our climate.

In the last ten years the average levels of smoke and sulphur dioxide at ground level have fallen by about 60% and 30% respectively. There has been no London smog for seven years. Winter sunshine in central London has increased by 50% and is now virtually the same as that in the surrounding countryside. One result of these improvements is that our buildings can be kept clean.

Noise

From Vehicles

Vehicles have long been required to be fitted with efficient silencers (about 12,000 prosecutions annually). From April 1970, new vehicles have also been required to meet precise noise maxima. From 1973, higher noise maxima for vehicles will be required.

From Aircraft

Regulations made in 1970 will require future types of subsonic jet aircraft to conform to noise maxima of about half the levels from aircraft now in service. This is in accordance with an international agreement and possible further measures are being studied internationally.

The Department of Trade and Industry (DTI) regulates aircraft routing to

---

* This paper sets out the main framework of how pollution is controlled in England but is not comprehensive. Arrangements in Scotland and Wales are broadly similar but differ in organizational details.

reduce disturbance and provides grants for sound-proofing houses near airports. The decision to site the third London Airport on the coast will ensure that it does not give rise to serious *noise problems.*

From Industry

The Secretary of State for Employment has appointed a special sub-committee of the Industrial Health Advisory Committee to advise on this problem and on action to prevent damage to workers' hearing.

The Noise Advisory Council

This organization has issued a first report identifying a number of areas which it is now studying including aircraft noise, industrial noise and the operation of the law on noise, including the Noise Abatement Act.

Fresh Water

Emissions to rivers from industry and from local authority sewage works must be authorized by the River Authority which also has a general responsibility for the state of rivers in its area. In 1959 only 6% of our 20,000 miles of river were grossly polluted. A new survey which will help to identify priorities more precisely, is nearly completed and the first indications are that conditions have, on the whole, improved since the 1959 survey. One very encouraging change since that time is that fish have returned to the Thames at London. The Working Parties on Sewage Disposal and on the Disposal of Toxic Wastes have made recommendations on water pollution problems; the Central Advisory Water Committee has considered the organization of the water industry.

The Government will be announcing its conclusions on these reports soon. In the meantime the major task of increasing sewage treatment capacity to meet the increasing load, is well in hand — capital expenditure has been rising steadily and is expected to exceed £ 700 million over the next five years, an increase in real terms of 50% over the previous five years.

With some exceptions, the problem of detergent foam has already been overcome by the development of *"soft" detergents* for domestic use. The problems associated with phosphate containing detergents in the United States have not arisen acutely in Britain.

Seas and Coasts
Oil Pollution

Oil pollution of the sea is already covered by several international agreements. Amendments to the main agreement which sets out to prevent polluting discharges were signed in 1969; their implementation will make enforcement easier. Two new agreements were also signed in 1969, one making tanker operators liable for *oil pollution* damage and requiring appropriate insurance and the other establishing the rights of Governments to intervene if their coasts are threatened by oil pollution.

The Oil in Navigable Waters Act 1971 enacts the ammendments to the main agreement for the UK. It also improves the law in other respects, and increases the maximum summary penalty for illegal discharges of oil to a record £ 50,000. The Merchant Shipping (Oil Pollution) Bill will enable the UK to ratify the civil liability agreement.

Local authorities are responsible for clearing any oil which comes ashore and are helped by a Government grant. Action is also taken by the Department of Trade and Industry while the oil is out at sea, and the Government's Warren Spring Laboratory has developed the most modern methods of dealing with it. In the "Pacific Glory" incident, DTI action prevented any significant quantities of escaped oil coming ashore.

Other Substances

New discharges by industry from land to the sea are controlled under the planning system, and discharges by local authorities under the loan sanction system. *Dumping* within the 3 mile limit is regulated by the Sea Fisheries Committees, and outside the limit by the Ministry of Agriculture, Fisheries, and Food (MAFF) under a voluntary scheme. The Working Parties (see above) have made some recommendations on these matters.

> The effectiveness of the planning system was demonstrated by the recent decision to impose strict conditions on discharges from the Cleveland Potash Mine. The Working Party recommendations are being carefully considered and the Royal Commission on Environmental Pollution is making a full study of coastal pollution problems. Marine pollution can really only be dealt with on an international basis, and the Government have invited a group of experts to meet in London in June 1971 to discuss the action the United Nations Conference on the Human Environment in 1972 can take on marine pollution. We are working with our European neighbors on the problems of the North Sea.

Pesticides

The sale and use of pesticides is regulated by a voluntary control scheme run by MAFF. Under this scheme a large number of the uses of the most *persistent pesticides* have been phased out and residues of pesticides in wild life have shown a welcome decline. Bird populations are recovering.

> Under this scheme *DDT* will no longer be supplied for use in homes and gardens after September 1971.

Radioactivity

Stringent controls on the disposal of radioactive wastes are maintained by the DOE and MAFF. All discharges are within safety levels agreed internationally. Within nuclear power stations, the handling of radioactive waste is controlled by DTI.

Refuse Disposal

A working party on refuse disposal, which reported in April 1971 made recommendations for the better organization and control of these activities. The Government have already accepted the major, longer-term recommendations in principle and for the immediate future local authorities and others engaged in solid waste disposal have been asked to review their methods to ensure that tipping is being carried out to the highest possible standards.

Organization

Most pollution problems, including air pollution, freshwater pollution, refuse disposal, radioactive wastes, and oil on beaches are, at central government level, the responsibility of the new Department of the Environment (DOE). The control of oil pollution at sea and aircraft noise is the responsibility of the Department of Trade and Industry (DTI); the Ministry of Agriculture, Fisheries, and Food

(MAFF) is responsible for the control of agricultural chemicals like pesticides and for checking that food is not contaminated. MAFF also seek to protect our fisheries both in freshwater and at sea.

Research is undertaken by various *Government Laboratories* including the Water Pollution Research Laboratory (DOE); the Warren Spring Laboratory (DTI) (which is concerned with air pollution, oil at sea and waste disposal); the MAFF Fisheries Laboratories (which in addition to examining the effects of pollution on fish have a wide experience of marine pollution problems generally); the MAFF Infestation Control Laboratory and the Monks Wood Laboratory (Nature Conservancy) (both concerned with pesticides and toxic chemicals) etc.

The Secretary of State for the Environment has a general co-ordinating role, in which he is assisted by a Central Unit on Environmental Pollution in his Department. The Government is advised by a number of specialists bodies, such as the Noise Advisory Council and the Clean Air Council, and by a standing Royal Commission on Environmental Pollution. In their first report published in February 1971, the Commission suggested the choice of priorities for future Government action and identified some areas for their own further study. Their general conclusion was that the existing arrangements should be maintained and considerably improved but that the present situation does not call for panic measures.

Measures for Environmental Pollution Control (Japan)

K. Yanagisawa

Ministry of Health and Welfare, Tokyo, Japan

Japan's measures for environmental pollution control are based on the *Basic Law for Environmental Pollution Control* which was enacted in August 1967.

However, pollution has become more and more complex in recent years, and in the extraordinary session of the Diet held late in 1970, 14 laws and amendments were passed, including amendments to the Basic Law for Environmental Pollution Control, in order to strengthen the legal basis for pollution control.

The existing Basic Law defines environmental pollution as air pollution, water pollution, soil pollution, noise, vibration, ground subsidence, offensive odors and the damage to human health and living environment caused thereby.

The gist of the current measures for environmental pollution control is given below.

Methodology of Control Measures

Establishment of Environmental Quality Standards

Environmental quality standards are administrative guidelines with regard to the levels of air pollution, water pollution, noise, and soil pollution, the maintenance of which is desirable for the protection of human health and the conservation of the living environment.

At present, environmental quality standards are set for sulphur oxides and carbon monoxide which affect man's health, as well as for water pollution. Work is now in progress on the setting of standards for particulates in the air, noise and so on.

Formulation of Environmental Pollution Control Programs

The Environmental Pollution Control Programs are to be formulated by the national and local governments so that they may implement their various measures in a coordinated way and according to an overall plan to achieve the common goal of preventing environmental pollution.

In the formulation of the Environmental Pollution Control Programs, the Prime Minister is to designate the areas to which the programs are to apply, to make clear the fundamental policy with regard to such programs, and to instruct the Prefectural Governors to formulate such programs; the Governors are then to formulate their programs and submit them to the Prime Minister for his approval.

Measures to Encourage the Installation of Anti-Pollution Facilities

Long-term low-interest loans are made available by the Environmental Pollution Control Service Corporation, the Japan Development Bank, the Small Business Finance Corporation and so on. In addition, favorable tax treatment is given for smoke and soot treatment facilities, waste water treatment facilities, fuel oil desulphurization facilities, and the like.

Promotion of Research and Development

Research and development is being carried on at various local government institutes, universities, and other laboratories, with national research institutes taking the lead.

Individual Control Measures

Air Pollution

The Air Pollution Control Law (enacted in 1968 and amended in 1970) provides the basis for regulating the discharge of smoke and soot (sulphur oxides, soot and toxic substances such as cadmium, lead and hydrogen fluoride) and the discharge of particulates (other than those resulting from a conbusion process) produced as by-products of industrial operations and for imposing maximum permissible limits for *carbon monoxide, hydrocarbons, lead,* etc. in automobile exhaust.

With regard to *sulphur oxides,* emission standards are to be determined for each area in accordance with the level of pollution which has already been reached, while uniform national standards are to be set for the discharge of soot and toxic substances.

In the case of automobile exhaust, maximum permissible levels of carbon monoxide and other pollutants are to be determined when safety standards are established in accordance with the Vehicles for Road Transportation Law.

Water Pollution

In December 1970, the Government enacted the Water Pollution Control Law (to come into force from June 1971) which provides, among others, for 1) the establishment of quality standards for effluent for all public waters, 2) the strengthening of enforcement measures by providing for penalties for violations of the standards, 3) the improvement of the inspection system for public waters.

Marine Pollution

A new Marine Pollution Prevention Law was established in December 1970 incorporating the 1969 Amendments to the International Convention for the Prevention of Pollution the Sea by Oil, 1954. This has made possible the consolidation and improvement of the marine pollution control system by clearly providing for 1) tighter restriction on the discharge of *oil,* etc. from ships, 2) a ban in principle on the discharge of *wastes* into the sea, 3) measures relating to the reporting of marine pollution and the issuing of removal orders, and so forth.

Waste Disposal

The Wastes Disposal and Public Cleansing Law was passed in December 1970 in a total revision of the Public Cleansing Law. Under this new Law, wastes are divided into two categories, namely industrial wastes, stemming from industrial activities, and non-industrial wastes, including wastes from domestic sources. In the case of the former, the Law not only stipulates the responsibility of the enterprise for disposing of such wastes but also requires him to see to it that his products, containers, or the like can be easily disposed of when they become waste. In the case of the latter, while following the system of disposal in the Public Cleansing Law, each municipal government's area of disposal responsibility has been enlarged to encompass, as a rule, the entire area of the municipality, and residents are required to cooperate with the municipal government's disposal operations.

Noise and Ground Subsidence

Under the Noise Regulation Law which is the principal legislation for noise control, industrial noise which includes factory noise and construction noise is regulated as necessary. The Law also provides for the measurement and regulation of traffic noise levels, for instance, through the establishment of maximum permissible limits on *automobile noise.*

Special legislation has also been enacted to curb *aircraft noise*.

Measures to combat ground subsidence are being carried out in accordance with the Industrial Water Law, the Law Concerning Regulation of Pumping-up of Ground Water for Use in Building, and other Laws. As for the problem of offensive odors, comprehensive restrictions are currently being discussed and prepared.

Soil Pollution

"Soil pollution" was added to the Basic Law for Environment Pollution Control in December 1970, and to implement this Basic Law provisions, the Agricultural Land Soil Pollution Prevention, Law, Etc, was enacted. Under this Law, measures are being taken to prevent or discriminate the pollution of agricultural soil and to cleanse such soil of pollution caused by specific harmful substances, and to achieve the proper use of such polluted land, so that no agricultural or dairy products which might be harmful to human health will be produced and the normal growth of agricultural crops will not be hindered.

Agricultural Chemicals

The Agricultural Chemicals Regulation Law was amended in December 1970 to create stricter criteria to be applied in the registration of agricultural chemicals, to establish new provisions concerning registration revocation and sales bans, and to impose restrictions on the use of those chemicals which could contaminate agricultural products if improperly used. With regard to countermeasures for food contamination, on the basis of studies carried out since 1967, standards of safe use have been formulated and guidance has been given on safe usage in order to guard against the possible contamination of crops by agricultural chemicals.

WHO's Food Safety Programs and the Problem of Mercury as a Food Contaminant

Frank C. Lu

Chief, Food Additives Unit, World Health Organization, Geneva, Switzerland

Summary. The WHO activities in the field of food safety are outlined. These include the convening of Expert Committees on Food Additives and Pesticide Residues, the support of the Joint FAO/WHO Food Standards Program and the soliciting of information needed for evaluation of the food additives, contaminants, and pesticide residues.

The problem of mercury as a food contaminant and the way in which it is dealt with at an international level are briefly discussed. The main aspects of the problem are the following: the severity and irreversibility of the symptoms and signs in a large number of cases of methylmercury poisoning, the greater susceptibility of fetuses compared to adults, and the possibility of clinical manifestations not yet recognized as a feature of methylmercury poisoning. The problem is further complicated by the fact that an excessively prudent course, taken to ensure the safety of fish for the consumer, might adversely affect the health of populations dependent on fish as a main source of protein and of those dependent on the fishery industry as a main source of income.

Recently the World Health Assembly, the governing body of the World Health Organization, considered the growing accumulation of harmful agents in the environment, including food, and recognised the need to protect human health against such agents. It adopted a resolution urging the Director-General to intensify and expand the activities of WHO concerning food contaminants and related matters. These activities are outlined in the following sections. As an illustration, the problem of mercury as a food contaminant and the way in which WHO dealt with it are briefly described.

The Food Safety Programs of WHO

According to its constitution, the objective of WHO is "the attainment by all peoples of the highest possible level of health". In order to achieve its objective, the Organization has a number of functions. One of the functions is to develop, establish and promote international standards with respect to food. WHO has consequently initiated certain activities, including the elaboration of food standards, to ensure the safety of food.

Expert Committees

In 1953, the World Health Assembly expressed the view that "the increasing use of various chemical substances in the food industry has in the last few decades presented a new public health problem and might be usefully investigated". After further investigations, a Joint FAO*/

* Food and Agriculture Organization of the United Nations.

WHO Conference on Food Additives[1] was held in 1955. Following a recommendation of the Conference, a series of annual meetings of the Joint FAO/WHO Expert Committee on Food Additives were convened, the first being held in 1956. Over the years, the Expert Committee has provided toxicological evaluations on a large number of intentional food additives. Where appropriate, an acceptable daily intake was established. It has also dealt with some food contaminants, notably certain trace elements and residues of antibiotics in food. The results of the evaluation are summarized in the reports of the Committee[2-11].

Since 1961 similar meetings have been held on pesticide residues, with the object of protecting the health of the consumer and at the same time allowing the proper use of pesticides in accordance with good agricultural practices. Some 100 pesticides and their residues have been dealt with by the Joint FAO/WHO Meetings on Pesticide Residues. Toxicological evaluation and recommendation of tolerances and/or practical residue limits have been made on most of them. The evaluations and recommendations are listed in an annex to the report of the 1971 Meeting[12].

The Joint FAO/WHO Food Standards Program

In 1962, a Joint FAO/WHO Conference on Food Standards was held. In the following year, the two organizations established the Joint FAO/WHO Food Standards Program, with the Joint FAO/WHO Codex Alimentarius Commission as its principal organ. The purpose of the Program is to protect the health of the

Table 1   Members of the Codex Alimentarius Commission

Europe		Latin America (cont.)	South West Pacific
1. Austria	27. United Kingdom	49. Uruguay	74. Australia
2. Belgium	28. Yugoslavia	50. Venezuela	75. New Zealand
3. Bulgaria			76. Fiji
4. Cyprus	North America	Africa	
5. Czechoslovakia	29. Canada	51. Algeria	Asia
6. Denmark	30. USA	52. Burundi	77. Ceylon
7. Finland		53. Cameroon	78. China
8. France	Latin America	54. Central African Rep.	79. India
9. Germany, Fed. Rep.	31. Argentina	55. Congo, People's Rep.	80. Indonesia
10. Greece	32. Barbados	56. Egypt, Arab. Rep. of	81. Iran
11. Hungary	33. Bolivia	57. Ethiopia	82. Iraq
12. Iceland	34. Brazil	58. Gambia	83. Japan
13. Ireland	35. Chile	59. Ghana	84. Jordan
14. Israel	36. Colombia	60. Ivory Coast	85. Korea, Rep. of
15. Italy	37. Costa Rica	61. Kenya	86. Kuwait
16. Luxembourg	38. Cuba	62. Madagascar	87. Lebanon
17. Malta	39. Dominican Republic	63. Malawi	88. Malaysia
18. Netherlands	40. Ecuador	64. Mauritius	89. Pakistan
19. Norway	41. Guatemala	65. Morocco	90. Philippines
20. Poland	42. Guyana	66. Nigeria	91. Qatar
21. Portugal	43. Jamaica	67. Senegal	92. Saudi Arabia
22. Romania	44. Mexico	68. Sudan	93. Singapore
23. Spain	45. Nicaragua	69. Togo	94. Syrian Arab. Rep.
24. Sweden	46. Paraguay	70. Tunisia	95. Thailand
25. Switzerland	47. Peru	71. Uganda	96. Yemen, People's
26. Turkey	48. Trinidad & Tobago	72. Zaire, Rep. of	Dem. Rep. of
		73. Zambia	

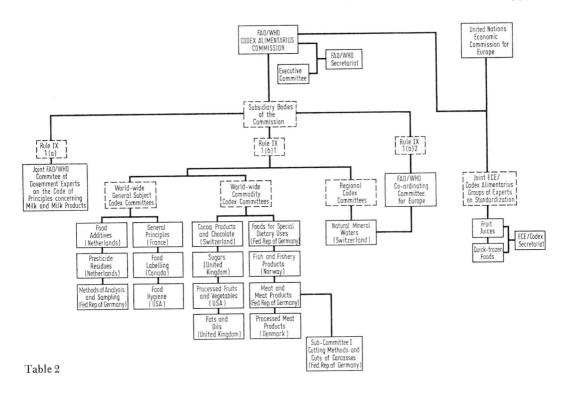

Table 2

consumers and to ensure fair practices in the food trade. Membership of the Commission is open to the Member States and Associate Members of WHO and/or FAO and comprises all eligible nations which have notified the Director-General of either WHO or FAO of their desire to be considered members of the Commission. By June 1972, 96 countries had become members (Table 1). In addition, a few other countries, although not members of the Commission, are participating in its work.

The Commission is assisted in its work on the elaboration of food standards by a number of committees. Some of these deal with particular food commodities, others with general subjects applying to foodstuffs in general (Table 2). Of particular interest to the problem of food contamination are the Codex Committee on Food Additives, which reviews provisions not only for intentional additives but also contaminants, and the Codex Committee on Pesticide Residues which recommends, on a world-wide basis, tolerances for particular pesticides in specific foods. A very close working relationship exists between these two Codex Committees and the corresponding Expert Committees.

Assessment of Consumer Risks

It is generally recognized that all chemicals are toxic to animals and man if large enough amounts are administered. Consequently, in evaluating the food additives and pesticide residues, the Expert Committees on Food Additives and Pesticide Residues have, where appropriate, allocated "acceptable daily intakes" (ADI) for man. The ADI of a chemical is defined as the daily intake which, during an entire life-time, appears to be without appreciable risk on the basis of all known facts at the time of evaluation. It is expressed in milligrams of the chemical per kilogram of body-weight.

To assess the consumer risk of a chemical, ideally one should compare, over a sufficiently long period, the actual intake of the chemical with the acceptable intake. It would be a formidable task to do so. The Expert Committee on Food Additives, at its sixth session in 1961[2], suggested a procedure by which the intake of a food additive might be estimated. The validity of the procedure was reaffirmed by the Committee at its tenth session in 1966[6]. Using this procedure, WHO has made estimates of the intakes of a number of food additives. Similar estimates have also been made on the potential intakes of pesticide residues. In a great majority of substances the estimated intakes provide assurance regarding health of the consumer; in the other cases, further investigations are indicated.

Miscellaneous Activities

In order to solicit comprehensive and up-to-date information on food additives and related subjects, WHO maintains contact with experts in many parts of the world, either informally or by appointing them as members of the WHO Expert Advisory Panel on Food Additives. It has designated a number of WHO Collaborating Laboratories in the field of toxicology and chemical analysis, and supported research to generate data that are vitally needed.

The Problem of Mercury as a Food Contaminant

As outlined above, food contaminants are also dealt with within the framework of the WHO food safety program. The various facets of the problem of food contaminants as illustrated by the example of mercury are described in the following sections.

Sources of Mercury and its Levels in Food

Mercury is present in air, water, and soil, usually at extremely low levels. However, activities of man in certain industries may raise the levels of mercury, especially in their vicinity. It has been estimated that mercury is used in 80 types of industry in at least 3000 different ways[13]. There are, in addition, agricultural and medical uses of mercury, although contributing to a much smaller extent.

As a result of its ubiquitous presence in the environment, small amounts of mercury are found in all foods. In general, the average level of mercury in food is around 20 ppb but varies considerably depending on the type of food and the level in the environment. Certain predatory fish and fish caught in contaminated waters may contain higher levels (0.5—5 ppm or even higher). Furthermore, gross contamination of food can occur as a result of misuse of seeds treated with fungicides containing mercury. The precise chemical nature of the mercury in most foods, other than fish is not known, since the level is usually very low. However, it is believed that, except under special circumstances, most of it is in the form of inorganic mercury. Other forms of mercury, such as aryl, alkyl, and alkoxyalkyl mercury compounds, may be found in foods when they are directly or indirectly exposed to these agricultural chemicals, especially when mishandling of treated seeds is involved.

On the other hand, a significant portion of the mercury in fish is in the form of methylmercury, and where the mercury level is high virtually all of it is in this form. Metallic mercury is not a problem as a food contaminant, though it may present an occupational hazard to workers in certain industries.

Toxicity of Mercury and Its Compounds

Much information has been accumulated on the toxicity of various forms of mercury. This has been reviewed and summarized in some recent publications [13, 14]. Consequently, only the highlights are outlined in the following paragraphs.

Metallic mercury is virtually not absorbed from the gastrointestinal tract. Inorganic mercury compounds are absorbed but are relatively rapidly excreted and do not cross the blood-brain barrier as readily as methylmercury compounds. Thus, they present much fewer health hazards than the latter.

Methylmercury is readily absorbed in the gastrointestinal tract. It is relatively evenly distributed in the tissues, including the central nervous system. Much higher levels are found in the kidneys, liver and hair. It is excreted slowly, with a half life of about 70 days; at a constant daily intake, equilibrium is reached only after six months of exposure.

Hunter et al. (1950) [15] observed four cases of poisoning in workers using methylmercury compounds in the manufacture of fungicidal dusts. The symptoms and signs were severe generalized ataxia, dysarthria and gross constriction of the visual fields. Similar clinical manifestations have been reported following the use of methylmercury thioacetamide dermally in the treatment of fungal diseases [16].

Two major outbreaks of methylmercury poisoning have occurred in recent years in Japan. During the period from 1953 to 1960, a total ol 111 cases of a special syndrome appeared in the Minamata Bay area of Japan. The voluminous literature arising from this incident is largely cited in the Report of the Study Group of Minamata Disease (1968) [17]. Up to the time of the writing of the Report, 29 of 59 adults, 10 of 30 children and 2 of 22 congenital cases had died. In 1964 and 1965 a similar outbreak occurred in Niigata, where 26 diagnosed cases were reported in 1966 [18]. Now the total number of cases in the Niigata region exceeds 100, though many of them have only mild symptoms and signs [19]. In both outbreaks, industrial pollution of water and subsequent contamination of fish and shellfish by methylmercury was shown to be the cause of poisoning.

There are a number of reports of poisoning following accidental ingestion of various types of organic mercury compounds, including methylmercury. Such intoxications have resulted both from the ingestion of treated seed grains or indirectly, after consuming meat from animals which had been fed grain treated with alkylmercury compounds [14].

The earliest clinical manifestation was in general sensory disturbances of the extremities, the lips and the tongue. This was followed by tremor, ataxia, concentric constriction of the visual fields and impaired hearing. Often there was a long latent period between exposure and appearance of symptoms and signs, which were irreversible, although the motor functions did improve somewhat after medical rehabilitation, especially in children. On the other hand, the symptoms and signs worsened in some patients, especially adults, even after cessation of consumption of fish [17].

The high incidence of congenital cases of methylmercury poisoning in the Minamata area suggests that the fetus may be at greater risk than the adult. This view is further supported by the fact that the mothers of affected offspring showed little or no clinical manifestation of Minamata disease [17].

Chromosome analysis of cultured lymphocytes from normal controls, and individuals with high mercury levels in the

Table 3   Correlation of Levels of Mercury in Blood and Hair

No. of subjects	Blood level (x) Hg in µg/g	Hair level (y) Hg in µg/g	Linear regression*	Reference
12	0.004–0.65	1–180	$y = 280x - 1.3$	Birke[23] (Sweden)
51	0.004–0.11	1– 30	$y = 230x + 0.6$	Tejning[22] (Sweden)
45	0.002–0.80	20–325	$y = 260x + 0$	Tsubaki[19] (Japan)

* A different relationship between the levels in hair and blood was found in Finland[24] ($y = 140x + 1.5$)

erythrocytes indicated that increased chromosome breakages could be correlated with increased mercury levels[20].

From pathological examinations of the fatal cases, methylmercury appeared to affect almost all parts of the central nervous system. The greatest effects, however, were observed in the cerebellar cortex and the cerebral hemispheres, especially the gyrus precentralis and area striata. These effects included various degrees of degenerative changes up to complete disappearance of neurones, glial proliferation, hydrocephalus, perivascular edema and demyelination[17]. More recently it has been found that the peripheral nerves from patients with Minamata disease showed demyelination and even destruction of nerve fibers[21].

Indicators of Exposure to Mercury

The extent of exposure to mercury can be estimated by determining the levels of mercury in the blood or hair. Urinary levels are now generally recognized as unreliable indicators. Methylmercury in blood is mainly located in the erythrocytes, whereas other forms of mercury are relatively equally distributed in the blood cells and plasma. Consequently, the mercury level in erythrocytes provides an accurate indication of the magnitude of exposure to methylmercury. On the other hand, after a single exposure, the blood level reaches a maximum in 3—6 hours. It then declines gradually, but somewhat more quickly than that of the whole body.

The magnitude of the error associated with the fluctuation of blood levels in determining the extent of exposure, especially in those individuals who only occasionally eat fish containing relatively high levels of mercury, has not been precisely assessed.

As an indicator of exposure to mercury, the hair level has a number of advantages. It is about 230—280 times higher than that in the blood, thus much easier to determine with precision (Table 3). It reflects the average exposure over a period of time. In many cases, the analysis can be made in a number of sections of the hair, thereby providing in indication of changes in the extent of exposure to mercury. Furthermore, it can be collected much more easily than blood.

Levels of mercury in hair and whole blood of patients with methylmercury poisoning in Niigata were determined in a retrospective study extending over several months after the onset of clinical manifestations[25]. The rate of decrease in mercury levels in these patients studied corresponded to a biological half-life of about 60 days. Extrapolation of the mercury levels in these and other Niigata patients to the time of onset of the symptoms indicated that poisoning generally occurred at mercury levels in hair ranging between 200 and 1000 µg/g, but in one case the hair level was as low as 50 µg/g. The extrapolated blood levels were 0.2—2.0 µg/g.

Relationship between Levels in Hair and Blood and Intake of Mercury

Many studies have been conducted on the levels of mercury in hair and on the frequency of intake of fish in contaminated control regions. Figure 1 provides an interesting illustration of the correlation between the mercury level in hair and the frequency of eating contaminated fish, and the much lower levels found in individuals in the control areas where the fish was "uncontamined". It further illustrates the variation of individual sensitivity to the toxic effects of mercury; although most of the patients had much higher levels of mercury in the hair than the apparently normal subjects, there was considerable overlapping of the levels in the two groups.

Estimation of the intake has been made on the basis of the consumption of fish and the levels of mercury in the fish. The relationships between the estimated intakes and levels in hair and blood can be expressed by the linear regression shown in Table 4.

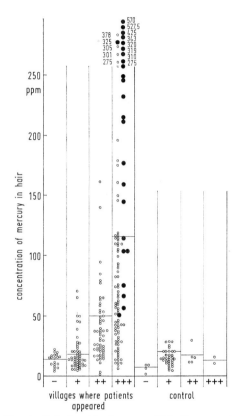

Fig. 1   Relationship between Fish Consumption and Levels of Mercury in Hair*

○   individuals with no apparent manifestations of intoxication
●   patients of methylmercury poisoning

Amount of fish eaten:

−   none
+   once a week
++   several times per week or less frequently but in large amount
+++   every day or very large amounts occasionally

*   Tsubaki (1972) personal communication.

Evaluation for "Acceptable" Intakes

The 1967 Joint FAO/WHO Meeting on Pesticide Residues [27] reviewed the animal toxicity data of various organo-mercurial compounds used in agriculture and considered that there was insufficient information to establish acceptable intakes for the residues of these compounds in food. It stressed that any use of mercury compounds which increases its levels in food should be strongly discouraged. No recommendations for tolerances were made. However, guidelines of 0.02 to 0.05 ppm were suggested for food crops.

The Joint FAO/WHO Expert Committee on Food Additives considered the matter of mercury as a food contaminant at its meetings in 1966 [6] and 1970 [10]. No acceptable intake was allocated because of lack of information. However, in 1972, the Committee [28] established a "provisional tolerable weekly intake" of 0.3 mg/kg total mercury, of which not more than 0.2 mg/kg should be present as methylmercury (expressed as mercury), equivalent to 0.005 and 0.0033 mg/kg bodyweight respectively. This figure was arrived at using the lowest mercury level of 50 µg/g in hair observed at the onset of

Table 4   Relationship of Levels in Hair and Blood and Intake of Mercury

No. of subjects*	Hg intake (x) mg/day	Hg in hair (y₁) μg/g	Hg in blood (y₂) μg/g	Reference
735	0–0.8	$y_1 = 150x + 1.66$		Kojima & Araki[26]
6 + 26**	0–0.8		$y_2 = 1.4x + 0.003$	Birke et al.[23]
139 + 26**	0–0.4		$y_2 = 0.6x + 0.011$	Tejning et al.[22]

* In individuals without clinical manifestation of methylmercury poisoning, but when the Niigata cases are included, the relation becomes $y_1 = 166x - 1.23$.

** Twenty-six cases from Tejning's subjects who did not consume fish.

poisoning in Niigata, the estimating the intake of mercury corresponding to this level in hair using the first equation in Table 4, and finally applying a suitable margin of safety. The Committee relied mainly on the information obtained in man in its assessment of the health hazards of mercury. This is consistent with its standing policy of giving preference to human data whenever they are available. In this particular case, although there are deficiencies in the data in several respects information is available on populations with an extremely wide range of exposure to methylmercury, including even cases of mild up to fatal poisoning.

The figure is expressed on a weekly basis because of the cumulative nature of this contaminant and because of its uneven distribution in diets of each individual over a period of time. It is provisional because of the uncertainties about the extent of variation of individual susceptibility to mercury, about the accuracy of the estimated intakes, and about the possibility of the existence of other manifestations not at present recognized as a feature of methylmercury poisoning. Despite these considerations, the Committee did not feel justified at present to use a larger margin of safety, for fear of depriving unnecessarily a large population of an important source of protein. Severe restriction of consumption of fish might adversely affect the health of the popu-lations dependent on fish as a main source of protein and of those dependent on the fishery industry as a main source of income.

Control Measures and Food Standards

As mentioned above, several outbreaks of poisoning have been reported as a result of unwitting diversion to the human food chain of seeds treated with mercurial fungicides. The toxicological aspect of the problem is relatively straightforward but its control has proved very difficult especially in certain parts of the world.

The problem of methylmercury in fish is a much more complex problem. In certain bodies of water, removal or drastic reduction of the polluting source has been effective in reducing the levels of mercury in fish, in other areas, it has not been effective. Furthermore, any control measures are unlikely to have an effect on the levels of mercury in fish caught in the open seas, such as tuna and swordfish.

In order to prevent the consumer from an excessive intake of methylmercury from fish, other measures are required. These consist mainly of permitting the sale of fish containing mercury up to a certain level, but also possibly a recommendation on the frequency of the consumption of fish. Many countries have established permissible levels, mainly as "administrative guidelines". These vary

Table 5 Summary of Some Current Permissible Levels of Mercury in Fish

Country	Maximum Permissible Mercury Level in ppm	Remarks
Federal Republic of Germany	0.1	Maximum permissible in all food including fish products
	0.5	Specific tolerance for dogfish, tuna, and swordfish
Belgium	0.5	
Canada	0.5	Administrative guideline
New Zealand	0.5	
Spain	0.5	
USA	0.5	Administrative guideline
France	0.7	Provisional
Italy	0.7	Only for imports
Denmark	1.0	
Japan	1.0	Administrative guideline
Sweden	1.0	Based on one meal of fish per week

from 0.1 ppm to 1.0 ppm, although 0.5 ppm is the level in half of those countries where a level has been set.

Because of the differences of the national permissible levels and the uncertainty about the magnitude of the health risks involved, WHO had been asked to take up the matter by the Codex Alimentarius Commission and its Committee on Food Additives, as well as the Executive Board of WHO. Now that the Expert Committee on Food Additives has agreed on an intake of mercury which can be considered as "tolerable" from a health point of view, albeit on a provisional basis, it is hoped that the Codex Committee on Fish and Fishery Products would make a comprehensive review of the data on the levels of mercury in various species of fish harvested in different parts of the world and then make recommendations on the levels of mercury that might be considered as permissible. Such recommendations would be transmitted to the Codex Committee on Food Additives for consideration in the light of the intake that is considered tolerable by the Expert

Committee on Food Additives. If endorsed by the Codex Committee on Food Additives, these recommended levels of mercury in fish would be further considered by the Codex Alimentarius Commission and Member States culminating in the possible adoption of international standards. The relationship between the Expert Committee on Food Additives, the Codex Committee on Food Additives and a Codex committee dealing with a food commodity, in this case fish and fishery products, is described in some detail in a recent paper by the author[29].

References

[1] First Joint FAO/WHO Conference on Food Additives. FAO Nutrition Meetings Report Series, 1956, No. 11; Wld Hlth Org. techn. Rep. Ser., 107, 1956
[2] Evaluation of the Toxicity of a Number of Antimicrobials and Antioxidants: Sixth Report. FAO Nutrition Meeting Report Series, 1962, No. 31; Wld Hlth Org. techn. Rep. Ser., 228, 1962
[3] Specifications for the Identity and Purity of Food Additives and their Toxicological Evaluation: Emulsifiers, Stabilizers, Bleaching and

Maturing Agents: Seventh Report. FAO Nutrition Meetings Report Series, No. 35, 1964; Wld Hlth Org. techn. Rep. Ser., No. 281, 1964

4 Specifications for the Identity and Purity of Food Additives and their Toxicological Evaluation: Food Colours and Some Antimicrobials and Antioxidants: Eighth Report. FAO Nutrition Meetings Report Series, No. 38, 1965; Wld Hlth Org. techn. Rep. Ser., No. 309, 1965

5 Specifications for the Identity and Purity of Food Additives and Their Toxicological Evaluation: Some Antimicrobials, Antioxidants, Emulsifiers, Stabilizers, Flour-treatment Agents, Acids and Bases: Ninth Report. FAO Nutrition Meetings Report Series, No. 40, 1966; Wld Hlth Org. techn. Rep. Ser., No. 339, 1966

6 Specifications for the Identity and Purity of Food Additives and their Toxicological Evaluation: Some Emulsifiers and Stabilizers and Certain Other Substances: Tenth Report. FAO Nutrition Meetings Report Series, No. 43, 1967; Wld Hlth Org. techn. Rep. Ser., No. 373, 1967

7 Specifications for the Identity and Purity of Food Additives and their Toxicological Evaluation: Some Flavouring Substances and Non-Nutritive Sweetening Agents: Eleventh Report. FAO Nutrition Meetings Report Series, No. 44, 1968; Wld Hlth Org. techn. Rep. Ser., No. 383, 1968

8 Specifications for the Identity and Purity of Food Additives and their Toxicological Evaluation: Some Antibiotics: Twelfth Report. FAO Nutrition Meetings Report Series, No. 45, 1969; Wld Hlth Org. techn. Rep. Ser., No. 430, 1969

9 Specifications for the Identity and Purity of Food Additives and their Toxicological Evaluation: Some Food Colours, Emulsifiers, Stabilizers, Anticaking Agents, and Certain Other Substances: Thirteenth Report. FAO Nutrition Meetings Report Series, No. 46, 1970; Wld Hlth Org. techn. Rep. Ser., No. 445, 1970

10 Evaluation of Food Additives. Specifications for the Identity and Purity of Food Additives and their Toxicological Evaluation: Some Extraction Solvents and Certain Other Substances; and a Review of the Technological Efficacy of Some Antimicrobial Agents: Fourteenth Report. FAO Nutrition Meetings Report Series, No. 48, 1971; Wld Hlth Org. techn. Rep. Ser., No. 462, 1971

11 Evaluation of Food Additives. Some Enzymes, Modified Starches, and Certain Other Substances: Toxicological Evaluations and Specifications and a Review of the Technological Efficacy of Some Antioxidants: Fifteenth Report. FAO Nutrition Meetings Report Series, No. 50, 1972; Wld Hlth Org. techn. Rep. Ser., No. 488, 1972

12 FAO/WHO (1971) Pesticide Residues in Food; Report of the 1970 Joint Meeting of the FAO Working Party of Experts on Pesticide Residues and the WHO Expert Committee on Pesticide Residues. FAO Agricultural Studies, No. 87; Wld Hlth Org. techn. Rep. Ser., No. 474, 1971

13 National Institute of Public Health, Stockholm (Sweden). Methyl Mercury in Fish. Report from an Expert Group. Nord. Hyg. T. Suppl. 4, 1971

14 Lu, F. C., P. E. Berteau, D. J. Clegg: The Toxicity of Mercury in Man and Animals, in "Mercury Contamination in Man and His Environment", to be published by FAO/IAEA/ILO/WHO, 1972, IAEA Technical Report Series, No. 137

15 Hunter, D., R. R. Bomford, D. S. Russell: Poisoning by Methyl Mercury Compounds. Quart. J. Med. 9: 193, 1940

16 Okinaka, S., M. Yoshikawa, T. Mozai, Y. Mizuno, T. Terao, H. Watanabe, K. Ogihara, S. Hirai, Y. Yoshimo, T. Inose, S. Anzai, M. Tsuda: Encephalomyelopalthy due to an organic mercury compound. Neurol. 14: 69, 1964

17 Kutsuna, M. (Ed.): Minamata Disease, Study Group of Minamata Disease, Kumamoto University, Japan, 1968

18 Irukayama, K.: The Pollution of Minamata Bay and Minamata Disease, in "Advances in Water Pollution Research". Vol. 3, p. 153, 1966

19 Tsubaki, T.: Personal communication, 1972

20 Skerfving, S., A. Hansson, J. Lindsten: Chromosome breakage in human subjects exposed to methyl mercury through fish consumption. Arch. Environ. Hlth 21: 133, 1970

21 Etoh, K.: Pathological changes of peripheral nerves in human Minamata disease: An electronmicroscopic observation. Adv. neurol. Sci. 15: 606–618, 1971

22 Tjning, S.: Mercury contents in blood corpuscles, blood plasma and hair in persons who had for long periods a high consumption of freshwater-fish from Lake Väner. Report 67 08 31 from Department of Occupational Medicine, University Hospital, S-221 85 Lund, Stencils, 1967

23 Birke, G., A. G. Johnels, L.O. Plantin, B. Sjöstrand, T. Westermark: Mercury poisoning through eating fish? Läkartidningen 64: 3628–3637, 3654, 1967

24 Sumari, P., A.-L. Backman, P. Karli, A. Lathi: Health studies of Finnish Consumers of Fish. Nord. Hyg. T. 50: 97, 1969

25 Tsubaki, T., T. Sato, K. Kondo, K. Shirakawa, K. Kambayashi, K. Hiroda, K. Yamada, I. Morune, S. Ueki, K. Kawakami, K. Okada, S. Chujo, H. Kobayashi: Diagnosis of Mercury Poisoning

I. Circumstances connected with the Outbreak of the Illness, in "Report on the Cases of Mercury Poisoning in Niigata", Ministry of Health and Welfare, Tokyo, Stencils, 1967

[26] Kojima, K., T. Araki: Unpublished report, summarizing relevant Japanese data, 1972

[27] FAO/WHO: Pesticide Residues in Food; Joint Report of the FAO Working Party on Pesticide Residues and the WHO Expert Committee on Pesticide Residues. FAO Agriculture Studies,

No. 73; Wld Hlth Org. techn. Rep. Ser., No. 370, 1967

[28] Evaluation of Food Additives. Evaluation of Certain Food Additives and the Contaminants Mercury, Lead and Cadmium: Sixteenth Report (1972). To be published

[29] Lu, F. C.: Toxicological Evaluation of Food Additives and Pesticide Residues and the "Acceptable Daily Intakes" for Man; the Role of WHO in Conjunction with FAO. Residue Reviews, 46, 81–93

Retrospective and Prospective Aspects in the Establishment of Air Quality Guides

The International Academy of Environmental Safety organized a Symposium on "Scientific Basis for the Establishment of Air Quality Standards".
President of the Symposium has been Prof. Dr. R. Truhaut, vice president Prof. Dr. L. Golberg. The meeting was attended by experts from 11 countries. The full papers will be published in the forthcoming volume of this series.
The recommendations of this Symposium are given here because of the special topical interest.
Over the past twenty-five years, air quality guides have been in use for the control of occupational exposure to airborne chemicals. The scientific basis for some of these standards has on occasion been less than adequate, permitting only an 'educated guess' to be arrived at. Also, deep division of opinion has existed over the experimental procedures, epidemiologic data and other methodological approaches to be used, as well as on overall interpretation of results as a basis for developing occupational standards. Nevertheless, in practice, compliance with these standards has proved to be a successful means of protecting the health of the occupationally exposed segment of the population.
In recent years the need has been recognized to extend similar protection to the general population, particularly the urban sector, by setting up air quality standards for the community. Such action is one of many manifestations of our increasing consciousness of man's responsibility to improve the quality and safety of his environment. However desirable the limitation of pollutants in community air undoubtedly is, the fact cannot be over-

looked that the procedures used for the establishment of standards for industrial air are not readily applicable nor adaptable to community air, where the safety considerations are rather different. Such differences involve the philosophy and concept of safety evaluation, the population at risk and the conditions of their exposure. What is more, the pollutants in community air should be considered not only in relation to human health, but also as they affect human welfare, which encompasses the potential for adverse effects on man's entire environment.
Before air quality standards can be promulgated, the potential of each pollutant in both of these respects should be clearly delineated, always remembering that the pollutant is present in ambient air in association with a multitude of other substances, notably particulates, that are likely to exercise profound effects on the biological properties of the individual pollutant. The biologically permissible limits for a pollutant are a function of its concentration in ambient air, the duration of exposure and the biochemical parameters and other epiphenomena utilized as indices of its effect. Such limits have to be set realistically, which means defining a zone that safeguards human health and welfare fully, but at the same time does not incorporate a safety factor so unreasonably high as to jeopardize needlessly man's activities and technological progress.
Looking to the future, there is no justification for anticipating a rapid drift to ecological disaster, so long as our common responsibilities are taken seriously by all concerned. The increasing capacity of analytical, biochemical, pathological, epi-

demiological, and other approaches to illuminate the complex problems of air pollutants means that rapid progress is possible in the attempts being made to remedy our present lack of full understanding of these problems. In this respect, studies involving controlled exposure of human volunteers to ambient air levels of pollutants are a particularly useful source of information. With the achievement of better understanding will come clearer indications of possible solutions. These will take the form of air quality standards based on careful definition of those facts which can be firmly established as well as those possibilities which may reasonably be entertained on scientific grounds, always distinguishing such possibilities from the multitude of purely hypothetical speculations that have proved so tempting to standard-setters in the past. The air quality guides established require interpretation in specific instances by qualified scientists.

Subject Index

Information for Authors

Language of Publication

English.

Contents

Original scientific articles, reviews, presentations by governmental agencies, surveys and other communications with global environmental aspects of chemistry, toxicology, engineering, physics, etc.

The editors are particularly interested in receiving original scientific papers or reviews dealing with the evaluation of safety of chemicals, drugs, natural products and physical agents on plants, animals, and man. Particular emphasis will be given to manuscripts that help define the chemical hazards involving the ecology of the food supply. Negative or confirmatory data will be published when of sufficient interest. The editors desire papers from investigators in different scientific disciplines and from industrial and govermental groups who are interested in the protection and safety of the biosphere from harmful chemicals.

The contributions should not have been published before and the author must possess the publication rights.

Articles should be concise. Submit two complete copies including the original in a typewritten form, double-spaced, with one-inch margins on all sides. The title should be limited to 15 words or 80 characters. The abbreviated running title should contain no more than 40 characters. Use generic names of chemicals wherever possible. Proprietary names and trade marks should appear only to identify the source of the chemical and subsequently only the generic name should be used. All abbreviations should be unpunctuated. The name and mailing address of the author(s) must be clearly indicated.

The manuscripts have to be submitted ready for press. The authors receive only page proofs for information and no corrections deviating from the manuscript are permitted. Return of such information proofs to the competent editors must be made within two weeks. Corrections are the responsibility of the editors of Environmental Quality and Safety.

Standards and References

All abbrevations and journal names should follow the style of Chemical Abstracts; Vol. 55, 1961. Chemical names should be according to the IUPAC Nomenclature of Organic Chemistry, Butterworths, London. References should be listed by number in the text. Entries in the reference list should include authors' names, name of the title, the journal, the volume, and the actual pages the article occupies in the journal, and date.

DuBois, K. P., F. K. Kinoshita, J. P. Frawley: Quantitative measurement of inhibition of alieste: ases, acylamidase, and cholinesterase by EPN and Delnav. Exp. Molec. Path. 12: 173-284, 1968

Abstracts

As a rule a brief summary is to precede original articles and of reviews. It may be omitted, however, if the structure of the contribution makes it unnecessary.

Keywords

In the running text of the manuscript up to five keywords per page should be underlined. They will be printed in italic letters and serve as the basis for the subject index.

Tables and Figures

Tables and figures should be completely understandable even without reading the text. Every table should have a title directly above it. Every figure should have a legend. Figures and tables should be identified consecutively with Arabic numerals. All charts and graphs must be done with black ink on coordinate paper.

Photographs

Photographs are desirable wherever necessary to substantiate and illuminate the text. Black and white photographs and photomicrographs may be submitted as glossy prints. Do not clip or mark the photograph in any way. All drawings should be done on heavy, white drawing board with black India ink. Authors should write their names on the back of each glossy print.

If the number of black-and-white half tones is above average a surcharge for printing costs shall paid upon the editors' request. Color illustrations will be accepted only against surcharge.

The editors reserve the right to accept contributions and to make formal corrections or shortening, where necessary, after contact with the authors.

Manuscripts

Manuscripts from Europe, Great Britain, Ireland, Africa and the Far East should be submitted to Dr. W. Klein
Institut für ökologische Chemie
D 5205 St. Augustin 1
Postfach 1260
Manuscripts from the Americas and Australia should be submitted to
Dr. I. Rosenblum
Dept. of Pharmacology
Albany Medical College of Union University
ALBANY, N. Y. 12208
U. S. A.